MW00464069

Biostatistics

Biostatistics

A Bayesian Introduction

George G. Woodworth
University of Iowa
Department of Statistics and Actuarial Science
Iowa City, IA

A JOHN WILEY & SONS, INC., PUBLICATION

Library of Congress Cataloging-in-Publication Data:

Woodworth, George G.
    Biostatistics : a Bayesian introduction / George G. Woodworth.
      p. cm.
    Includes bibliographical references and index.
    ISBN 0-471-46842-8 (cloth : acid-free paper)
      1. Bayesian statistical decision theory. 2. Biometry. I. Title.

  QA279.5.W66 2005
  519.5'42—dc22   2004047408

Printed in the United States of America.

10 9 8 7 6 5 4 3 2 1

To

Carrol, Griffin, and Zoë

John and Rebecca

Richard Savage and Milton Sobel

Contents

Preface

Statistics and Scientific Reasoning

Bayesians take the position that statistical science ("statistics" for short) is the science that deals with the evidentiary value of observations. Statistics, in other words, is related to *epistemology*, the study of how it is we come to know things about the world.

Working scientists have puzzled philosophers since David Hume. The puzzle is the problem of induction: "How do human beings form opinions about unobserved matters of fact? ... Are beliefs formed in this way justified? Does someone who 'reasons' as we normally do really have reason to believe his conclusions about the parts of nature he has not observed?" (Princeton University, Philosophy 203(W), Syllabus, 1999)

We are all familiar with the experience of becoming more and more convinced, as evidence accumulates, that a thing is true; however, the paradigm of scientific reasoning in general and statistical reasoning in particular that dominated most of the twentieth century was incapable of explaining how this could happen.

The Failure of Falsification and Significance Testing

Twentieth century philosophy of science was dominated by the views of Karl Popper, who proposed that

[S]cientific theories ... are not inductively inferred from experience, nor is scientific experimentation carried out with a view to verifying or finally establishing the truth of theories; rather, *all knowledge is provisional, conjectural, hypothetical* — we can never finally prove our scientific theories, we can merely (provisionally) confirm or (conclusively) refute them; hence at any given time we have to choose between the potentially infinite number of theories which will explain the set of phenomena under investigation. Faced with this choice, we can only eliminate those theories which are demonstrably false, and rationally choose between the remaining, unfalsified theories. ("Karl Popper," © 2002 by Stephen Thornton, in *Stanford Encyclopedia of Philosophy*[1], quoted with permission.)

---

1. *Stanford Encyclopedia of Philosophy*, Edward N. Zalta, ed. The Metaphysics Research Lab Center for the Study of Language and Information, Ventura Hall, Stanford University, Stanford, CA 94305-4115, http://plato.stanford.edu/contents.html, 2002

Popper, along with Fisher, Neyman, and Pearson, the founders of twentieth century *frequentist* statistical methodology, held that a theory is scientific only if it makes predictions that could turn out to be incorrect. Such theories are said to be "falsifiable." Evidently, this inspired the idea that the way to do science is to seek evidence by which to "reject" a theory. Presumably after this rejection business had gone on long enough, the last theory standing must be the truth. Popper spoke only of absolute rejection — observing something that was impossible under a particular theory. However, many scientific theories (Mendel's laws for one) make only probabilistic predictions and therefore cannot be absolutely refuted by any amount of data. To cope with this, statisticians proposed "significance testing;" the idea is to compute the probability that the theory in question (Mendel's for example) would produce the observed data. A theory incapable of producing the observed data is logically false. By extension, so the frequentist statisticians believed, a theory under which the observed data are unlikely is likely to be false. This is clearly wrong, since the observed data might be even less probable under any other plausible theory.

Although Popper instructs us to "rationally [choose] between the... unfalsified theories," neither he nor anyone else was able to explain satisfactorily how to objectively weigh the strength of evidence for or against the each of the currently "unfalsified theories." Consequently, the notion that science proceeds by falsifying incorrect theories could not explain how humans practice scientific (or everyday) induction.

Motivated in part by their perception that falsification — whether absolute or statistical — cannot be the basis for scientific reasoning, some statisticians and philosophers, notably Harold Jeffreys, Bruno De Finetti, and Leonard Savage, began to establish mathematically rigorous foundations for Bayesian epistemology, a system of rationally updating beliefs in the face of evidence. The key features of this program are:

- All probability is degree of belief and not a property of objects.
- Under fairly weak definitions of rationality, degree of belief must obey the laws of probability.
- Scientific "models" predict data, either absolutely or probabilistically.
- Via Bayes' rule it is possible to calculate how data shifts the degrees of belief among competing models.

William Talbott said this about Bayesian epistemology:

The combination of its precise formal apparatus and its novel pragmatic self–defeat test for justification makes Bayesian epistemology one of the most important developments in epistemology in the 20th century, and one of the most promising avenues for further progress in epistemology in the 21st century. ("Bayesian Epistemology," © 2001 by William Talbott, in *Stanford Encyclopedia of Philosophy*[1], quoted with permission.)

It is important to understand that Bayesianism is a *normative* theory of knowledge, one that *prescribes*, rather than *describes*, how to use evidence rationally. Following the rules of probabilistic reasoning guarantees that data are used in a rational, internally consistent, way. In that sense probabilistic reasoning is an extension of logic. Bayesian probabilistic reasoning does not claim to describe how people use data in everyday life; in fact, there is ample experimental evidence that people often make errors in probabilistic reasoning. However, that does not invalidate it as a guide to rational use of data, any more than the fact that people are often illogical invalidates the rules of logic.

About This Book

Entry-level undergraduate statistics courses must take seriously the need to teach students the foundations of statistical science from the beginning. Foundations are ignored at our peril — one reason for the baffling persistence of demonstrably inconsistent conventional statistical practices is the failure to examine their foundations (or lack thereof). After a conventional (frequentist) statistics course or two, students come to view conventional statistical practices as somehow natural and self-evident when in fact they are neither. The only remedy is to provide entry-level textbooks and other teaching materials that give students the tools to think about the basis of statistical and scientific reasoning and that present modern Bayesian methods.

This textbook is intended for students in the biological sciences. It does not require calculus, although students need to be proficient in basic algebraic manipulations and to know how to communicate with a computer (pull-down menus, click–drag, etc.). At my university this group of students has had one or more basic science courses and one or more university-level mathematics courses and plenty of computer experience.

I believe that it is important to connect statistics and scientific method in a deep way for this group; therefore, this book presents the axiomatic basis of statistical reasoning via Dutch book thought experiments. This book covers the standard biostatistics topics: inference on rates; relative risks; odds ratios; means; linear, logistic, and proportional hazards regression; survival analysis; meta-analysis and hierarchical models. It shows how to compute, interpret, and report Bayesian statistical analyses in practice. A recurring theme is teaching students how to reinterpret conventional statistical reporting (confidence intervals, margins of error, one-sided p values) in approximate Bayesian terms where possible.

F tests and Chi-Squared tests, which produce only p values, are not included in this text, because their p values, while interpretable as the posterior probability of the highest density region tangent to the null hypothesis, do not answer meaningful inferential questions. Nonparametric, exact, and inefficient procedures are not included because they violate the likelihood principle; nonnormal data are handled by transformation rather than by nonparemetric procedures. Adjustments for multiple comparison are not covered here, because this textbook is mainly about the evidentiary value of observations and adjustments for multiple comparisons are in the domain of decision analysis and therefore depend on the loss function. The Waller-Duncan method is available in SAS software is a Bayesian solution to the multiple comparison problem.

This book provides explicit instruction on two statistical packages, the SAS[2] System® and WinBUGS. SAS software comprises an extremely versatile conventional statistics package along with data management facilities and is widely used in biomedical research; WinBUGS, it is probably fair to say, is the most commonly used program for routine Bayesian data analysis.

After completing this text, students should understand the principles of statistical reasoning, be able to read and interpret statistical reports in the biomedical literature that use the techniques covered in this book, and carry out simple statistical analyses using SAS and WinBUGS software.

Supplemental materials, data sets, computing exercises, and notes on new developments in Bayesian statistical practice will be posted on the textbook website. A link to the website can be found at http://www.stat.uiowa.edu/~gwoodwor/.

I am grateful to Jay Kadane for many helpful discussions of Bayesian inference and pedagogy, and to Kate Cowles for almost daily discussions of all aspects of Bayesian practice, for listening to my rants, and for a careful reading of the manuscript.

GEORGE G. WOODWORTH

Iowa City, Iowa
January 2004

---

2. SAS and all other SAS Institute Inc. product or service names are registered trademarks or trademarks of SAS Institute Inc. in the USA and other countries. ® indicates USA registration.

1

Introduction to Statistical Science

1.1 THE SCIENCE OF EVIDENCE

Statistical science is concerned with the evidentiary value of observations. Another way to put this is to say that statistics is concerned with how data change beliefs — the purpose of statistical science is to help people change their beliefs in a rational way when confronted with new information.

A familiar example is diagnostic testing. Nate got a tattoo while on spring break in Florida. He didn't give it another thought until he was refused as a blood donor two months later. The phlebotomist explained that he had to wait a year to make sure he didn't get hepatitis B from the tattoo. That got him worried, so he ordered a home test kit for hepatitis B virus (HBV) from a European website. The website said that the sensitivity of the test was 0.99 and the specificity was 0.995 — we'll learn what that means in a minute.

The website gave this advice: "If the [test] result is negative ...you may conclude that you are not infected." In case of a positive test result, it advised, "Do not interpret a positive reaction from the ... rapid hepatitis B screen test as absolute evidence of the presence of HBV. If the test result is positive, we advise you to consult a doctor. If necessary, change your lifestyle and habits to exclude the possibility of infecting other people, until you are certain you are not infected with hepatitis B." That seems odd — a negative test is convincing proof that Nate does not have hepatitis B virus, but a positive test is not convincing proof of infection. Probabilistic reasoning explains why.

Hepatitis B is rare among those who are not intravenous drug users — about 2 cases per 100,000 people. Studies suggest that getting a tattoo from an operator who follows accepted hygiene standards does not greatly increase the risk. Let's assume that Nate believed that his risk was about 3 in 100,000. Fig. 1.1 shows what would happen if 10 million young, healthy college students who got a tattoo from a professional used the rapid test kit to check for HBV. About 300 of the 10 million (i.e., 3 in 100,000) would actually have HBV, and the rest would not. The HBV test has 99% sensitivity, which means that it will catch 99% of the HBV cases (297 of the 300 cases) and miss the rest. The test has 99.5% specificity, which means that 99.5% of the noninfected people will test negative, but 0.5% of them will be false positives.

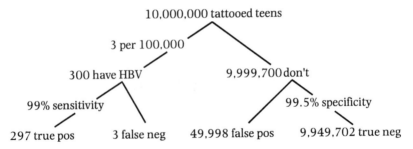

Figure 1.1   Tattooed Teens

Now let's look at Nate's situation. Suppose he tests negative. There are 9,949,705 people like him — negative. Of these only 3 have HBV, so there are only 3 chances in 9,949,705 (about 1 in 3.30 million) that a person who tests negative actually is infected. On the other hand, suppose Nate tests positive. There are 50,296 people like him — positive. Out of this group, only 297 really do have HBV (about 1 in 170). That means that even if Nate tests positive, there is still only about a 0.6% chance that he is actually infected.

Before Nate bought the test kit, it was reasonable for him to believe that his chances of having HBV were 3 in 100,000. If the test came back negative, he could believe that his chances of HBV were negligible. But if the test result came back positive, he would believe that his chances of HBV were 1 in 170. The odds are not overwhelmingly bad, but they certainly represent a level of risk he could not ignore, considering the serious consequences of infecting someone else.

1.2   DEGREES OF BELIEF

Belief is not an all or nothing proposition; there are degrees of belief ranging from things you believe "beyond a doubt," such as my belief that the Holocaust happened, to absolute disbelief, such as my disbelief that Elvis was kidnapped by aliens. In between these extremes are degrees of belief such as "more likely than not." For example, I think it is more likely than not that a woman will be elected president of the United States within the next 25 years.

In statistical science, degree of belief is expressed in terms of probability, for example a 30% chance of rain. But in everyday life, particularly in legal matters, words or phrases are used to express degrees of belief. For example, 16 states have passed laws permitting the civil commitment of sex offenders who have completed their sentences. The majority of these laws require the jury to decide if it is *likely* that the offender would commit another violent sexual offense if released. With lawyers using words and statisticians using numbers to express degree of belief, it might seem that statistical reasoning cannot be used to assess evidence in court; however, that is not the case. In fact, surveys indicate that many probabilistic words and phrases have reasonably stable numerical meanings. Kadane (1990), commenting on Mosteller and Youts's 1990 summary of studies of how people use probabilistic expressions, suggested 11 terms that span the interval between total disbelief and total belief in 10% steps. Table 1.1 lists these phrases and the corresponding numerical probabilities. Thus, for example, the word *likely* corresponds to about a 65% to 75% chance. So under sexual offender laws, juries must determine if there is about a 70% or greater chance that the offender will commit a new crime if released.

Table 1.1 offers chance phrases and frequency phrases. They both describe the probability that something is true; however, frequency phrases refer to things that recur with some regularity, like rain or homicides or automobile accidents, and emphasize the rate of occurrence. For example, it *almost never* rains in Death Valley but a penny lands heads *as often as not*. There are no phrases to precisely characterize extremely small or extremely large chances; in such cases we must resort to numbers such as "1 in 10 million" to describe degree of belief.

1.3 KNOWING IS BELIEVING

Scientists and jurors, among others, are in the business of seeking knowledge. Scientists are convinced that complete knowledge of nature is ultimately possible and that the pursuit of knowledge is an end in itself. Jurors are instructed that they can convict only if they know the defendant is guilty beyond a reasonable doubt.

What is knowledge? What does it mean to know something? Philosophers define knowledge as *justified true belief* or *reliably produced true belief* (Steup 2002). Justification (or reliable production) of belief is what statistical science is about.

Table 1.1 Quantification of probabilistic expressions.

Numeric Probability	Frequency	Chance
0 to 5%	Almost never[a]	—
5% to 15%	Seldom	*Low chance*
15% to 25%	Infrequent	*Unlikely*
25% to 35%	Sometimes	—
35% to 45%	*Less often than not*	Less than even chance
45% to 55%	*As often as not*	Even chance
55% to 65%	More often than not	*Better than even chance*
65% to 75%	Often	*Likely*
75% to 85%	*Very frequent*	High probability
85% to 95%	*Almost always*	Very high probability
95% to 100%	—	(Virtually) certain

a. Source: Mosteller and Youtz (1990) Table 2 and Kadane (1990) Table 2. Italicized terms are from Kadane's list; others are from Mosteller's list. Reprinted with permission from the Institute of Mathematical Statistics.

As a consequence, working statisticians have one foot in epistemology (the philosophy of knowledge) and one foot in science. It is their job to develop rules and methods for rationally updating beliefs in the face of data. As a practical matter, that means that we need to figure out how to quantify belief and how to quantify the way that data support some beliefs and not others. That's how this book is organized. First we'll quantify belief in terms of probability; then we'll learn how to use the laws of probability — Bayes' rule in particular — to use data to revise beliefs.

It is important to understand that we are presenting a *normative* theory of knowledge — we are presenting rules of probabilistic reasoning that will guarantee that data are used in a rational, internally consistent way. In that sense, probabilistic reasoning is an extension of logic. Logic is a set of rules that will produce true statements starting from statements assumed to be true and relationships known to be true. For example given that "all animals will die" and "all humans are animals," then it follows logically that "all humans will die." Logic is *prescriptive*, not descriptive — it prescribes a set of rules to follow to avoid logical errors but does not claim to describe how people actually reason. The fact that some people are sometimes illogical does not make the rules of logic incorrect.

The same can be said about probabilistic reasoning. It is prescriptive in that it provides rules for rationally updating beliefs in the face of data. It does not claim to describe how people use data in everyday life. In fact, it is well known that people do not always follow the rules of probabilistic reasoning. Here is an example. "As a college student, Lisa was in the debate club and was active in environmental and other liberal causes. Which of these statements is more probable about Lisa today: (a) Lisa is a lawyer, (b) Lisa is a lawyer specializing in environmental law." Most people think that (b) is more probable than (a); however, that belief violates the laws of probability because environmental lawyers are a subset of lawyers.

People who do not consistently follow the rules of probabilistic reasoning are said to be *incoherent* — their beliefs do not "cohere" (hang together). The fact that people are sometimes incoherent does not invalidate the laws of probability any more than the fact that people are sometimes illogical invalidates the laws of logic. The point is that if you wish to make rational use of data — neither to over react nor under react to data — then you must follows the laws of probabilistic reasoning (Savage 1972). That is the basis of statistical science.

What is the cost of being irrational or incoherent? Mostly it results in some degree of inefficiency, but the consequences can occasionally be worse. In everyday life incoherence can lead you to make decisions that will cost more than necessary in the long run, or that will place you at a higher risk of disease or injury than necessary, or that will lead you to support public policy that is less fair or less effective than it could be. But the consequences can be severe: incoherence can lead a juror to support an unjustified verdict. Incoherence can cause a scientist to reach one conclusion when another conclusion actually has more support from the data (D'Agostini 1998). In medical research, incoherence can expose patients to unnecessary risks (Kadane 1996).

1.4 OTHER VIEWS OF STATISTICAL SCIENCE

An online search for the phrase "what is statistics" produces over three thousand hits. An impressionistic survey of a few dozen of these produced the following two themes:

- Statistics is about gaining knowledge from data:

   "... learning from data..."
   "... drawing conclusions from data..."
   "... interpreting data..."
   "... assist other scientists and researchers to make informed decisions
        in the face of uncertainty..."
   "... extracting useful knowledge from both experiments and data..."
   "... turning raw data into arguments that can resolve
        profound questions..."
   "... the science of drawing conclusions from data with the aid
        of the mathematics of probability..."

- Statistics is about collecting and organizing data:

   "... collecting data..."
   "... organizing data... "
   "... extracting information from [masses of data] in compact form... "

Kendall (1950) described statistics

   ... as the matrix of quantitative knowledge of nearly every kind, as the
   principal instrument yet devised by man for bringing within his grasp the
   terrifying complexity of things and relations-between-things, and a powerful
   illuminant of the process of rational thought itself...

This book takes the position that statistical science is about rationally converting data into knowledge; thus it more or less agrees with Kendall and more or less agrees with the various ways of saying that statistics is about getting knowledge from data, but disagrees with the various ways of saying that statistics is about gathering and organizing data.

Gathering and organizing data can be useful and even fun, but it is not statistical science. It is certainly true that data are often collected and reported for no specific purpose; examples are sports statistics, census data (other than the population counts required by the constitution for the purpose of apportionment), causes of death as reported on death certificates, and levels of the Dow Jones, Standard and Poors, and NASDAQ averages. This sort of data can turn out to be extraordinarily useful, like Tycho Brahe's meticulous naked-eye observations of the position of Mars, or can turn out to be irritatingly ambiguous. For example, FBI uniform crime reports are compiled from data voluntarily supplied by local police departments. Crimes not reported to local police departments do not appear in the statistics. Thus, an increase in reported rapes may be due to more rapes being committed or it may be due to more rape victims being willing to make a complaint.

The problem with the idea of the scientist as a pure observer is that it does not explain how she decides what to observe. Imagine being let loose in a biology lab and told to "observe." What are you supposed to do? Look at something under a microscope? Watch the interactions between your fellow students? Look out the window and count the number of cars going by? Working scientists use theories to focus their attention — theories suggest useful observations. Einstein's theory of relativity implies that light is deflected by gravity, Newton's older theory implies that light travels in straight lines in a vacuum. Without Einstein's theory it is unlikely to have occurred to anyone to observe a sun-grazing star during an eclipse (Coles 2001).

As a corollary, it is unlikely that generic "stats" and data series will be precisely what is needed to resolve a future scientific or policy question. This is not meant to be an argument against gathering data for its own sake. The point is that it is not possible to guarantee in advance that any particular data set or descriptive statistic will be useful.

1.5 PREVIEW OF STATISTICAL REASONING

Individuals (scientists or consumers of science) entertain many theories about nature (plate tectonics, big bang, neo-Darwinism, string theory, tanning beds don't cause melanoma, etc.). People are rarely absolutely certain that a particular theory is true; instead, they have some degree of belief in that theory but are open-minded about modifying or even discarding a theory in favor of a new one (Newton vs. Einstein, for example).

Using thought experiments (Dutch book arguments), it can be proved that if a person wishes to be rational, his or her degrees of belief must obey the laws of probability. Thus it is correct to refer to degrees of belief as "subjective probabilities."

This insight can be exploited to give a method (Bayes' rule) for rationally revising beliefs in the face of new evidence.

Some traditional statistical tools, such as confidence intervals, are useful approximations to coherent statistical arguments; others, such as p values, are not.

1.6 EXERCISES

1.1 Suppose that Nate tested positive for hepatitis B. The test catches 99% of hepatitis cases, so why is Nate not 99% certain that he has hepatitis?

1.2 Jane thinks that the sentence "A woman will be elected president in 2024," has a better than even chance of being true. Quantify her degree of belief.

1.3 Explain how logic is prescriptive and not descriptive. Why is this a logical error: "If it walks like a duck and quacks like a duck, then it is a duck?"

1.4 Precipitation is any form of water from the sky (snow, rain, hail, sleet, etc.). Why is this a probability error, "It's winter, so the probability of snow tomorrow is greater than the probability of precipitation tomorrow."

1.5 Ed thinks there is a 90% chance that angels exist and Ellen thinks there is only a 1% chance. Why is this not a probability error?

1.7   REFERENCES

Coles, Peter, "Einstein, Eddington and the 1919 Eclipse," http://arxiv.org/abs/astro-ph/?0102462, 2001.

D'Agostini, G., "Bayesian Reasoning Versus Conventional Statistics in High Energy Physics," *Maximum Entropy and Bayesian Methods:Garching, Germany 1998*, ed. Linden von der. et al., Kluwer Academic Publishers, Dordrecht, 1999.

Kadane, Joseph B. "Comment: Codifying Chance", *Statistical Science*, Vol. 5, No. 1 (Feb. 1990), pp. 18–20.

Kadane, Joseph B. *Bayesian Methods and Ethics in a Clinical Trial Design*. New York: John Wiley & Sons, 1996

Kendall, M. G., "The Statistical Approach," *Econometrica*, Vol. 17, pp. 127-145, 1950.

Mosteller, F. and Youtz, C., "Quantifying Probabilistic Expressions", *Statistical Science*, Vol. 5, No. 1, (Feb. 1990), pp. 2–12.

Savage, Leonard, *The Foundations of Statistics*, 2nd ed., New York: Dover Publications, 1972.

Steup, Matthias, "The Analysis of Knowledge," in *Stanford Encyclopedia of Philosophy*, Edward N. Zalta, ed. The Metaphysics Research Lab Center for the Study of Language and Information, Ventura Hall, Stanford University, Stanford, CA 94305-4115, http://plato.stanford.edu/contents.html, 2002

2

Probability

2.1 CASINO PROBABILITIES

We're going to start with casino examples, not because they are important, but because everyone agrees on the probabilities, which eliminates a complication that we'll deal with in the next chapter. Roulette is a popular game of chance, and we'll use it to explore the ideas of probability. The Wizard of Odds (http://www.wizardofodds.com/games/roulette.html) describes roulette this way,

> A U.S. roulette wheel consists of 38 numbered slots: numbers 1 to 36, a zero, and a double zero. The betting layout consists of every individual number as well as a host of "outside" or combinations of numbers. After the players make their bets the dealer spins the wheel and a ball and after several seconds the ball will land in one of the numbered slots. ("Roulette," © 1997-2003, Michael Shackleford, reproduced with permission.)

Bets available under United States roulette rules are shown in Fig. 2.1 along with the usual house odds on various bets. House odds are always expressed as the ratio of the house's stake to the player's stake in the bet. For example the house odds on a four-number bet are 8 to 1, which means the gambler puts up $1, the house puts up $8, and the winner (gambler or house) takes, $9. The probability of winning this bet is usually calculated as 4 in 38, or 10.53%, because 4 of the 38 slots are winners for the gambler. This would be written as $P(\text{Win}) = 0.1053$. But is that really the probability? It is based on the assumption that the 38 slots are equally likely, and, after all, a roulette wheel is a mechanical device; it is *possible* that it might not be perfectly balanced and that some numbers might come up more often than others. The point is that the probability is, in part, based on a subjective judgement that the wheel is fair and balanced. Most people seem to believe that each slot on the wheel is as likely as any other, and the Wizard of Odds thinks that belief is justified,

> There are books out there that promise big riches by tracking the outcomes on a roulette wheel over thousands of spins and then examining the distribution to see if the wheel is biased toward one side. I don't deny that this has been done in the past but I doubt it is a practical way to beat the casinos today. There are products the casinos can use today that track every spin and

on a periodic basis run the distribution through a statistical test to check for bias. Although I am not an expert on the subject I believe wheel tracking to be a waste of time today. ("Roulette," © 1997-2003, Michael Shackleford, reproduced with permission.)

The Wizard believes that for practical purposes, the ball is equally likely to land in any slot; in other words he advocates the belief that each slot in the wheel has 1 chance in 38 (probability $1/38 = 0.026315$). This is a belief that most people share — roulette wheels at big casinos are probably so close to perfect mechanical balance that it is pointless to think otherwise. In any case, we are going to compute probabilities based on that belief.

Similar beliefs are the basis of computing probabilities for games of chance that involve some sort of mechanical randomization — spinning a roulette wheel, shuffling cards, rolling dice. For example, the probability that a five-card poker hand is a straight flush (five cards of the same suit in numeric sequence, ace high or low) is reported to be $36/649,740$, or about 0.00006. The calculation assumes that any combination of five cards is as likely to be dealt as any other.

So, for the purpose of computing casino probabilities we'll assume that every slot on a roulette wheel is equally probable, every side of a die is equally probable, and every possible five-card poker hand is equally probable.

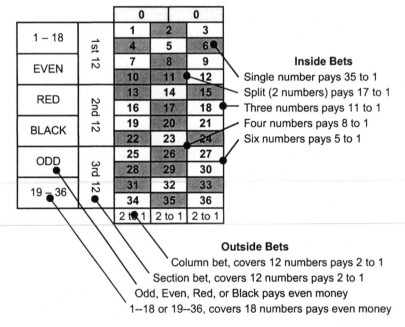

Figure 2.1   U.S. roulette betting layout (shaded = black, dotted = green, white = red).

2.2 PROBABILITY

Probability refers to sentences, like "The ball will land in a red slot" or "The ball will land in slot 27." If you believe that the slots are equally probable, then you should believe that the probability of a sentence is the proportion of slots that make the sentence true, or, as it is sometimes put, probability is the number of *favorable alternatives* divided by the *total* number of alternatives. In roulette for example, the ball could land in slot 1, or slot 2,...., or slot 36, or slot 0, or slot 00, a total of 38 possible alternatives. A bet on "odd" has 18 favorable (winning) alternatives: slot 1, slot 3, etc., so $P(\text{Odd}) = 18/38 = 0.4737$. A bet on Odd is a bet that the sentence "The ball will land in an odd slot" is true. Probabilities *always* refer to sentences; however, it is tedious to write out full sentences, and we'll generally use symbols instead. For example the letter R might stand for the sentence, "The ball will land in a red slot."

2.2.1 Combining Sentences with And, Or, and Not

Sentences can be logically combined to form new sentences. For example, suppose that symbol R stands for "The ball will land in a red slot," and symbol E stands for "The ball will land in an even slot." The symbol $R \cap E$, pronounced "red *and* even" for short, stands for the rather awkward sentence, "The ball will land in a red slot *and* the ball will land in an even slot." A more natural way to say the same thing is "The ball will land in an even-numbered red slot."

Some people seem to be confused about the precise meaning of "a bet on Red and Even." It does not mean putting $1 on Red and $1 on Even, what it *does* mean is placing a bet that will win only if both conditions are true; the only way to bet on Red and Even is to put $1 on each of the even-numbered red slots (slots 12, 14, 16, 18, 30, 32, 34, and 36 on the roulette table shown in Fig. 2.1 on page 10).

The probability that the sentence $R \cap E$ is true is the number of favorable alternatives (slots 12, 14, 16, 18, 30, 32, 34, and 36) divided by the total number of alternatives, in other words $P(R \cap E) = 8/38 = 0.215$, or 21.05%.

The symbol $R \cup E$ is pronounced "red *or* even," and stands for the compound sentence "The ball will land in a red slot *or* the ball will land in an even slot." This *or* is *inclusive*, meaning Red or Even *or both*. There are 28 slots in Fig. 2.1 on page 10 that are red or even[3] or both (slots 1, 2, 3, 4, 5, 6, 7, 8, 9, 10, 12, 14, 16, 18, 19, 20, 21, 22, 23, 24, 25, 26, 27, 28, 30, 32, 34, and 36); consequently, the probability of $R \cup E$ is $P(R \cup E) = 28/38 = 0.7368$, or 73.68%.

The symbol $\sim R$, pronounced "not R," and the symbol R^c, pronounced "R complement," both stand for the sentence "It is *not* true that the ball will land in a red slot." A more natural way to say that is "The ball will *not* land in a red slot."

---

3. In roulette, 0 and 00 are neither even nor odd.

The symbol Ω (omega) stands for any sentence that is always true (for example, "Elvis is either alive or dead.") and its complement, the symbol ∅ (null), represents any sentence that is always false (for example, "I have two brothers and I have no brothers").

2.2.2 Conditional Probability

This book will make the case that probability describes a person's belief about what is likely to be true. Since probability describes belief, it ought to respond to information. Conditional probability is probability altered by new information.

Imagine this: You managed to get a bet down on Even, but while the ball was in motion a tall person elbowed you out of the way. The ball has landed in a slot, but you can't see the roulette wheel and it's too noisy to hear the croupier. You ask the tall person in front of you, "Where did it land?" He says, "On red." That is not the information you were looking for, but nevertheless it is information. Let's see how it has changed your probability of winning.

You still don't know where the ball landed, but you can eliminate the black slots and the green slots, because you know the ball is in a red slot. That leaves 18 equally probable alternatives (slots 1, 3, 5, 7, 9, 12, 14, 16, 18, 19, 21, 23, 25, 27, 30, 32, 34, and 36). What are the chances that your bet on Even is a winner? The reasoning is simple; you know there are 18 alternatives and 8 of them are winners (the winners are the even red slots: 12, 14, 16, 18, 30, 32, 34, and 36). So, you have 8 chances out of 18 of winning; in other words, the *conditional probability* of Even given Red is 8/18 =0.444 or 44.4%.

The symbol for conditional probability of E given R is P(E|R). The word "given" means "assumed to be true as a basis of calculation, or reasoning." Thus the statement P(E|R) = 0.444 means, assuming that you have been told that R is true, there is 44.4% probability that E is also true. The formula for conditional probability is,

$$P(E|R) = \frac{P(E \cap R)}{P(R)} \tag{2.1}$$

Equation (2.2) shows the essential idea how to prove Equation (2.1). The the fraction 8/18 is the conditional probability we derived two paragraphs above by counting alternatives. The number 38 is the total number of slots on the wheel, so for example, 18/38 is the number of red slots divided by the total number of slots, which is the definition of P(R). Similarly 8/38 is the number of even red slots divided by the total, which is the definition of P(E ∩ R).

We've demonstrated that receiving the information "the ball is in a red compartment" actually changed the gambler's belief about his chances of winning his bet on Even. Before the spin, he thought he had 18 chances in 38; thus *prior* to the spin the probability was P(E) = 18/38 = 0.474, or 47.4%. After

$$P(E|R) = \frac{8}{18} = \frac{8/38}{18/38} = \frac{P(E \cap R)}{P(R)} \qquad (2.2)$$

he was told that the ball was in a red compartment, he concluded that he had 8 chances in 18 of winning, so $P(E|R) = 8/18 = 0.444$, or 44.4%. The conclusion is that information changes probability. In this situation learning that the ball is in a red compartment didn't help the gambler, because bets are not permitted after the wheel is in motion; however, conditional probability *is* useful in another casino game, blackjack. In blackjack cards are dealt one at a time and bets are allowed after each card is exposed. Therefore there is a flow of new information to the gambler. In 1962 the mathematician Edward O. Thorpe developed his famous card-counting strategy that enables a blackjack player to use that information to identify situations in which the conditional probability of winning is large enough to make betting profitable to the gambler.

2.3 LAWS OF PROBABILITY

The laws of probability are rules for computing probabilities of composite sentences created by combining individual sentences using *and*, *or*, and *not*. These laws are the product rule, the addition rule, the complementation rule, and the sure thing rule.

2.3.1 The Product Rule

This rule is used to calculate the probability of a compound sentence consisting of two sentences connected by the word "and." The rule states that the probability that sentences R and E are both true is,

$$P(E \cap R) = P(E|R) \cdot P(R) \qquad (2.3)$$

The product rule is a rearrangement of the definition of conditional probability in Equation (2.1). It is obtained by multiplying both sides of Equation (2.1) by $P(R)$.

For example, two cards are dealt out of a standard 52-card deck. What is the probability that both cards are kings? Let K1 stand for the sentence "the first card is a king," and K2 stand for the sentence "the second card is a king." The first card is dealt out of a deck of 52 cards, 4 of them kings, so the probability that the first card is a king is $P(K1) = 4/52$. Given that the first card is a king, the deck now contains 51 cards, 3 of them kings. Therefore, the conditional probability of a second king is $P(K2 \mid K1) = 3/51$. So, according to Equation (2.3), the probability of two kings is

$$P(K1 \cap K2) = P(K2 \mid K1) \cdot P(K1) = \frac{3}{51} \cdot \frac{4}{52} \cong 0.0045$$

2.3.2 The Addition Rule

This rule is used to calculate the probability of a compound sentence consisting of two sentences connected by the word "or." The rule states that the probability that either R is true or E is true or both are true is,

$$P(R \cup E) = P(R) + P(E) - P(R \cap E) \tag{2.4}$$

Fig. 2.2 suggests how to prove the addition rule. The figure arranges the 38 slots in into five sets: Odd Red, Even Red, Odd Black, Even Black, and Green. The sentence R is true if the ball lands in an Even Red or an Odd Red slot as depicted by the oval surrounded by a dashed line; consequently $P(R) = (8 + 10)/38$. The sentence E is true if the ball lands in an Even Red or an Even Black slot, as depicted by the oval surrounded by a solid line; consequently, $P(E) = (8 + 10)/38$. The sentence $R \cup E$ is true if the ball lands in an Even Red, an Odd Red, or an Even Black slot; consequently $P(R \cup E) = (8 + 10 + 10)/38$.

The sum $P(R) + P(E)$ equals $(8 + 8 + 10 + 10)/38$ and is larger than $P(R \cup E)$ because it counts the 8 Even Red slots twice. To get the correct count it is necessary to subtract the probability of the Even Red slots; that is, $P(R \cap E)$ must be subtracted from $P(R) + P(E)$ to get the correct value of $P(R \cup E)$.

	Even	Odd	Neither
Red	12, 14, 16, 18, 30, 32, 34, 36	1, 3, 5, 7, 9, 19, 21, 23, 25, 27	
Black	2, 4, 6, 8, 10, 20, 22, 24, 26, 28	11, 13, 15, 17, 29, 31, 33, 35	
Green			0, 00

Figure 2.2   Gray area: $R \cup E$, solid oval: E, dashed oval: R. Here $R \cap E$ is in both ovals and is therefore counted twice in $P(R) + P(E)$.

2.3.3   The Complementation Rule

This rule is used to compute the probability that a sentence is false; that is, the probability that it is *not* true. The probability that R is not true is

$$P(\sim R) = 1 - P(R) \qquad (2.5)$$

Here's the idea of how to prove the complementation rule: There are 18 red slots and $38 - 18$ slots that are not red; therefore the probability of not Red is $(38 - 18)/38 = 1 - 18/38 = 1 - P(R)$.

2.3.4   The Sure Thing Rule

This rule states that if a sentence is always true then it has probability 1, meaning that it is 100% certain. In logic, a sentence that is always true is called a *tautology*. In probability an always true sentence is called a *sure thing* and is represented by the symbol Ω (omega). The probability of a sure thing is,

$$P(\Omega) = 1 \qquad (2.6)$$

This seems self-evident: if something is always true then there must be 100% probability that it is true. Here is the idea of a proof. The sentence S: "The ball will land in one of the 38 slots," is always true, so $S = \Omega$. There are 38 ways for S to be true, so the probability of S is $38/38 = 1$.

2.4   STATISTICAL INDEPENDENCE

The sentences R and E are said to be statistically *independent* if knowing that R is true does not change the probability that E is true. Thus, the sentences R and E are independent if and only if

$$P(E|R) = P(E). \qquad (2.7)$$

It is important to understand that Equation (2.7) is the definition of independence, *not* a law of probability. $P(E|R)$ is not always equal to $P(E)$; however, when they are equal (and that must be verified), E and R are statistically independent. When Equation (2.7) is not true, E and R are said to be *dependent*, and in that case learning that R is true can improve or worsen the chances of winning a bet on E. For example, for the roulette wheel depicted in Fig. 2.1 on page 10 Even and Red are *dependent* because Equation (2.7) is not true; $P(E|R) = 0.444$ is not equal to $P(E) = 0.474$.

2.5   FAIR BETS

A bet is *fair* if neither party to the bet has an advantage, or *edge*. For example, suppose Bill and Jane bet on a coin toss with Jane calling "heads." If Bill bets \$1 but insists that Jane must bet \$2, then the wager is obviously unfair to Jane because she stands to lose twice as much as Bill. As we'll discover, a bet on a sentence S is fair if the amounts of money, or *stakes*, bet by each player are directly proportional to the probabilities of winning. In other words, for the bet to be fair to both parties it is necessary that

$$\frac{\text{Jane's stake}}{\text{Bill's stake}} = \frac{\text{Jane's probability of winning}}{\text{Bill's probability of winning}}. \tag{2.8}$$

Now consider a roulette bet on Red. Instead of Bill and Jane, the participants are a gambler and the house. The probabilities of winning are $18/38$ for the gambler and $20/38$ for the house and the amounts that they bet are called *stakes*. According to Equation (2.8) the fair ratio of stakes is 10 to 9,

$$\begin{aligned}
\frac{\text{house's stake}}{\text{gambler's stake}} &= \frac{\text{house's probability}}{\text{gambler's probability}} \\
&= \frac{20/38}{18/38} \\
&= 10/9 \tag{2.9}
\end{aligned}$$

So, to be fair, the house should stake \$1 and the gambler only \$0.90.

Here is an intuitive argument that those stakes are fair: On average the house will win 20 of every 38 bets and the gambler will win the other 18. The winner receives the total stake of \$1.90, so the balance sheet for 38 bets is

	Player	House
Invests	38·\$0.90 = \$34.20	38·\$1.00 = \$38.00
Wins	18·\$1.90 = \$34.20	20·\$1.90 = \$38.00
Nets	\$0	\$0

On average, each side expects to win back exactly what it invested. In the short run one side or the other could be ahead at any given moment, but the expectation is that neither side has an advantage, a point to which we will return in Section 2.8.

Of course, casinos do not offer fair bets; for example, according to Fig. 2.1 on page 10, the casino offers 1 to 1 odds for a bet on Red which means that the house's stake is $1 and the gambler's stake is $1. The balance sheet for this unfair bet is

	Player	House
Invests	38·$1.00 = $38.00	38·$1.00 = $38.00
Wins	18·$2.00 = $36.00	20·$2.00 = $40.00
Nets	−$2.00	$2.00

On average the house expects to earn a $2 profit on each $38 it invests at a cost of $2 from the player. This is the *house edge*, and it amounts to an expected gross profit margin of 2/38, or about 5.3%.

2.6   FAIR PRICES

It is going to be useful in the next chapter to work with *standard bets* that is, bets with a total stake of $1.00 (gambler's stake plus house's stake). For example, we calculated above that the fair stakes for a roulette bet on Red are $1 for the house and $0.90 for the gambler, for a total stake of $1.90. To convert that into a standard bet, divide each stake by the total stake, so that the standard house stake is $1.00/1.90 = $0.53 and the standard gambler's stake is $0.90/1.90 = $0.47. The gambler's stake in a standard bet is called the *fair price* of the bet. The fair price of a standard bet is particularly easy to calculate: It is simply the probability that the gambler will win the bet

$$\text{fair price of a standard bet} \ = \ \$P\big(\text{Win}\big) \qquad (2.10)$$

In the next chapter, we consider "horse race" probabilities — probabilities that refer to unique, never to be repeated combinations of circumstances; for example, the probability of the sentence "The next presidential election will be won by a Republican." We can't compute these probabilities by counting the number of ways the sentence can be true, because it is hard to imagine how to make a list of the alternative ways it could be true, and in any case there is no reason to believe that the alternatives are equally probable, as was the case with slots in a roulette wheel. Instead, the probability of the sentence will be *defined* as the fair price of a standard bet that the sentence is true. For example, if I think there is 60% probability that the next presidential election will be won by a Republican, that means that I think that $0.60 is a fair price to pay for a standard bet on that sentence. The trick is to prove that fair prices behave like probabilities — a task we turn to in the next chapter.

2.7   HOUSE ODDS, FAIR ODDS, AND STATISTICAL ODDS

House odds are odds offered to gamblers by a casino or bookmaker. House odds specify the *stakes* in a gamble; thus house odds of 8 to 1 mean that the gambler puts up $1, the house puts up $8, and the winner takes $9. As we demonstrated above, house odds are never fair to the gambler. According to Equation (2.9) on page 16, if the gambler's probability of winning is P, then the fair house odds would be in the proportion $1 - P$ to P. For example, the probability that Red wins in roulette is $P = 18/38 = 0.4737$. Consequently fair house odds would be $20/38$ to $18/38$ or any convenient multiple, such as 20 to 18 or 10 to 9.

Statistical odds refer to chances, not stakes. Statistical odds of 2 to 1, for example, mean that the gambler has 2 chances of winning and the house has 1 chance of winning, and thus the gambler has 2 chances out of 3 of winning, or probability 0.667 of winning. Statistical odds are quoted as *gambler's chances to house's chances*, the opposite of the way house odds are quoted. Thus the statistical odds on Red would be stated as 18 to 20 or 9 to 10; other examples are shown in Table 2.1. Statistical odds are frequently expressed as a ratio: for example 9/10 or 0.9. Fair house odds, statistical odds, and probabilities are interconvertable using the formulas in Table 2.2.

Example 1 If the gambler's probability of winning is $P = 0.4$, then the statistical odds on winning are $0.4/(1 - 0.4) = 0.667$, and the fair house odds are 6 to 4, or 3 to 2. Example 2: if the statistical odds are $7/2$ (7 to 2), then the gambler's probability of winning is $P = (7/2)/(1 + 7/2) = 7/(2 + 7) = 0.778$, and the fair house odds are 2 to 7. Example 3: If the fair house odds are 20 to 18, then the statistical odds are $18/20 = 0.9$, and the gambler's probability of winning is $P = 0.9/1.9 = 0.474$.

Table 2.1   Odds, fair odds, probabilities, and fair prices.

Type of odds	Bet on red	Column bet[a]
House odds	$1 to $1	$2 to $1
Statistical odds	18 to 20, or 18/20	12 to 26, or 12/26
Probability[b]	$P = 18/38$, or 0.474	$P = 12/38$, or 0.316
Fair house odds	$20 to $18, or $10 to $9	$26 to $12, or $13 to $6
Fair price	$P = 0.474	$P = 0.316

a. Bet on all 12 numbers in a column (see Fig. 2.1)
b. Probability that the gambler wins the bet.

Table 2.2   Conversions among fair house odds, probability, and statistical odds

If you know this:	then the statistical odds[a] are	the probability[b] is	and the fair house odds[c] are
Fair odds: B to A	$\text{odds} = \dfrac{A}{B}$	$P = \dfrac{A}{A+B}$	B to A
Probability: P	$\dfrac{P}{1-P}$	P	$1-P$ to P
Statistical odds	odds	$P = \dfrac{\text{odds}}{1+\text{odds}}$	1 to odds

a. Odds on the gambler winning the bet: (chances of winning) / (chances of losing).
b. Gambler's probability of winning the bet.
c. Proportionality of house's stake to gambler's stake.

2.8   EXPECTED VALUE

The fair price of a standard bet that pays $1 if sentence S is true is $\$1 \times P(S)$. However, not all bets are so simple. For example a no-fault automobile insurance policy with $500 deductible is a bet that can pay many different amounts, depending upon the amount of damage to the vehicle. Thus for example, the truth of the sentence "A minor accident resulted in $1200 worth of repairs" produces a $700 payout.

The fair price of a bet with many different possible payouts is called the bet's *expected value*. Expected value is the weighted average of the payouts weighted by their probabilities. For example, Jane is playing roulette at the roulette table in Fig. 2.1 on page 10. She bets $1 on Red (which pays $2) and another $1 on 21 (which pays $36). Her *portfolio* of bets has three possible payoffs: If the ball lands in compartment 21, both bets pay off (because 21 is a red) and Jill receives a total of $38. If the ball lands in any other red compartment, she receives $2; otherwise, she receives nothing. Table 2.3 gives the details of Jill's portfolio.

After the spin, the value of Jill's portfolio will be either $38 or $2 or $0, depending on where the ball landed. The question is what is the fair price of the portfolio before the spin. To answer this we need to break down Jill's portfolio into things that we know how to price, that is, standard bets on single sentences. For example, we know that the fair price of a standard bet on "The ball will land in slot 21" is $\$1 \times P(21) = \$1/38$, or about 2.6 cents. But Jill actually stands to win $38 if the ball lands in slot 21, which means, in effect, that she owns 38 copies of the standard bet. Obviously, if the fair price of one copy is about 2.6 cents, then

Table 2.3   Expected value (fair price) of Jill's Portfolio

Sentence S	Payoff Value V	Probability P	Product $P \times V$
S_1: 21	$38	1/38 = 0.0260	$1.000
S_2: Red, not 21	$2	17/38 = 0.4474	$0.895
S_3: Black or Green	$0	20/38 = 0.5263	$0.000
		Expected value = sum of products:	$1.895

the fair price of 38 copies must be 38 times the price of a single copy, which works out to $38 \times \$1/38 = \$1$.

Jill also owns 2 copies of a standard bet on "The ball will land in a red slot but not slot 21," and the fair price of a standard bet on that sentence is its probability, $17/38$, so Jill's two copies of the standard bet have a total value of $2 \times \$17/38$. The other possibility is "Black or Green", but she owns no copies of that contract. Table 2.3 shows how to calculate the fair price, or expected value, of Jill's portfolio.

In general, the expected value of any portfolio is the weighted average value of the payoffs, weighted by their probabilities [see Equation (2.11) and Table 2.4].

$$\text{expected value} = E(V) = \sum_{i=1}^{k} V_i \cdot P_i \qquad (2.11)$$

The fair price of Jill's bet is about $1.895, but she paid $2.00, so the casino *expects* a net profit of $0.105 on this $2 bet, which is a house edge of about 5.25 cents for every dollar wagered. Not that the casino literally takes 5.25 cents from every dollar; instead, the meaning of the expected value is that over the long run the average profit from the roulette tables will be about 5.25 percent of the money wagered. Jill will make a comparatively small number of bets, which makes her winnings or losses volatile and unpredictable; however, the casino accepts millions if not billions of bets and can count on the *law of averages* to give them something close to the expected 5.25 percent.

The concept of expected value applies to any uncertain quantity, not just money. For example, suppose that Bill tosses a penny until he obtains a head. This could happen on the first toss, or the second, and so on. The probability that Bill will have to toss the coin V times in order to obtain a head is $1/2^V$; consequently the expected number of tosses is the sum of V (the number of tosses) times $1/2^V$ (the probability that V tosses are required). The sum has an infinite number of terms

$$E(V) = \sum_{v=1}^{\infty} v \cdot \frac{1}{2^v} = \frac{1}{2} + \frac{2}{4} + \frac{3}{8} + \frac{4}{16} + \cdots \qquad (2.12)$$

however, it is easy to compute to any degree of accuracy in a spreadsheet. The sum of the first 25 terms is 1.999999, so it appears that the expected value is 2.00. In other words, although it is remotely possibly that Bill might have to toss the penny hundreds of times waiting for a head to appear, the expected number of tosses is only 2.

Table 2.4 shows how to set up the calculation of the expected value of any uncertain quantity that has a finite number of possible values.

The Law of Large Numbers. The mathematical form of the so-called law of averages is the *law of large numbers*. It states that if the same bet is made repeatedly and independently, then the average winnings per bet will approach the expected value as the number of bets approaches infinity. With high probability, the average will deviate from the expected value by less than the maximum allowed bet divided by the square root of the number of independent bets.

For example, a Vegas casino advertises that it has 1500 slot machines that "pay up to 98%", meaning that the casino expects to win more than 2% of every dollar wagered. To simplify the calculation assume these machines accept only a 25-cent bet and that 200 million bets are placed in one year.

The law of large numbers says that the casino's average profit per bet will be 2% of 25 cents, or 0.5 cents *per bet*. Half a cent times 200 million bets comes to an expected profit of $1,000,000 from the slot machines. The actual average profit will deviate somewhat from expectation but the deviation will probably be about 25 cents divided by the square root of 200 million, that is, ±0.002 cents. In other words, the average profit will be in the range 0.5±0.002 cents per bet, somewhere between 0.498 and 0.502 cents per bet. Thus the total profit from

Table 2.4 Expected value of an uncertain quantity.

Sentence S	Value V	Probability P	Product $V \cdot P$
S_1	V_1	P_1	$V_1 \cdot P_1$
S_2	V_2	P_2	$V_2 \cdot P_2$
S_3	V_3	P_3	$V_3 \cdot P_3$
$\vdots$	$\vdots$	$\vdots$	$\vdots$
S_k	V_k	P_k	$V_k \cdot P_k$
		Expected value	$\Sigma V_i \cdot P_i$

200 million bets will be between \$996,000 and \$1,004,000. Not only is there no possibility of losing money, the profits can actually be predicted to an accuracy of ±4 parts per thousand.

On the other hand, a tourist who wagers 1000 times at a 25-cent slot machine expects to *lose* an average of 0.5 cents per bet. However, according to the law of large numbers, his actual average can deviate from expectation by 25 cents divided by the square root of 1000 (the number of bets), which amounts to an uncertainty of about ±0.8 cents. Thus his average loss per bet could be anywhere in the range 0.5±0.8 cents, anywhere between -0.3 cents to +1.3 cents per bet. Of course a negative loss is a gain, so in 1000 bets the gambler could have anything from a \$3 gain to a \$13 loss. That is why insurance companies and gambling casinos are said to be *risk businesses* but for most people gambling is at best a recreation, at worst an addiction, but never a business.

George Bernard Shaw (1956) writing about state-sponsored lotteries, made this point about as well as anyone has,

> A State, being able to make a million bets whilst an individual citizen can afford only one, can tempt him or her to gamble without itself running the slightest risk of losing financially; for, as aforesaid, what will happen in a million case is certain, though no one can foresee what will happen in any one case. Consequently governments, being continually in pressing need of money... are strongly tempted to replenish the Treasury by tempting their citizens to gamble with them. No crime against society could be more wickedly mischievous. (Reproduced with permission of The Society of Authors on behalf of the Bernard Shaw estate.)

An excellent account of the law of large numbers can be found in Freedman et al. (1997).

2.9 EXERCISES

2.1 You roll a six-sided die and toss a coin.The twelve equally probable things that can happen are shown in the matrix below. For example, *T3* stands for

T1	*T2*	*T3*	*T4*	*T5*	*T6*
H1	*H2*	*H3*	*H4*	*H5*	*H6*

the sentence "the coin landed tails and the die landed with three spots showing."

The rules of the game are: you get one point for Heads and one point for every spot showing on the die. So for example if you roll a 3 and toss a Head, you get 4 points. The symbol "4P" stand for the sentence "You got at least 4 points," "H" stands for the sentence "The coin landed heads," and "T" stands for the sentence "The coin landed tails."
(a) Compute P(4P|H) by counting and dividing.
(b) Compute P(4P|H) and P(4P|T) using Equation (2.1) on page 12.

2.2 Referring to exercise 2.1, in the table below fill in the probabilities and compute the expected value of the number of points.

No. of Points V	Probability P
1	
2	
3	
4	
5	
6	
7	

2.3 For each of the $1 bets shown in Fig. 2.1 on page 10:
(a) Compute the fair house odds.
(b) Compute the fair price of a standard bet with total stakes $1.00.

2.4 Use the laws of probability to prove that $P(A \cap B) + P(A \cup B) = P(A) + P(B)$.

2.5 James puts $1 on Odd and $1 on Black at the roulette table shown in Fig. 2.1 on page 10. What is the expected value of his wager?

2.6   Here are two probability tables. In each table compute $P(A \mid B)$, $P(A)$, and
       $P(B)$ and say whether A and B are independent or dependent. In each case
       explain why.

Table I	~B	B
~A	0.20	0.05
A	0.20	0.55

Table II	~B	B
~A	0.06	0.14
A	0.24	0.56

2.7   Three cards are dealt from a standard deck. Use the product rule to
       compute
       (a)   the probability of 3 aces,
       (b)   the probability of AAK in that order.
       (c)   Use the product rule to compute $P(AAK)$, $P(AKA)$, and $P(KAA)$ and
              use the addition rule to compute the probability of two Aces and a
              King.

2.10   REFERENCES

Shaw, G. B. "The Vice of Gambling and the Virtue of Insurance," in *The World of
    Mathematics, Vol 3*, James R. Newman, ed., New York: Simon and Shuster, Inc. 1956.

Freedman, D. Pisani, R., and Purves, R. "The Law of Averages", Chapter 16 of *Statistics,*
    3rd ed. New York: W. W. Norton, Inc., 1997.

3

Subjective Probability

3.1 INTERPRETATIONS OF PROBABILITY

Although there are three so-called *interpretations* or definitions of probability (objective, relative frequency, and subjective), subjective probability is the only definition that is not circular. A *circular definition* is one that includes the term being defined as part of its own definition, as in "A human is an animal with human parents." The circularity of relative-frequency probability is a more subtle but just as fatal.

3.2 DEFINITIONS OF PROBABILITY

The oldest definition of probability is objective probability, which is another name for the "casino" probabilities that we studied in Chapter 3. We made the argument that casino probabilities are actually subjective because they depend on the assumption that there is a set of alternatives that are believed to be equally probable (slots in a roulette wheel, sides on a coin, or faces on a die, for example).

Frequentists take the philosophical position that probability can be defined as *long-run relative frequency.* For example imagine tossing a coin, say, 80 times. It might land heads 48 times in 80 tosses. In that case, the *frequency* of heads would be 48 and the *relative frequency* would be 48/80, or 0.60 or 60%. For example, the eighteenth-century French naturalist Bouffon tossed a coin 4040 times and counted 2048 heads, a relative frequency of 0.5067, or 50.67%. Frequentists propose to define the probability of heads as the limiting value of the relative frequency as the number of tosses goes to infinity.

Unfortunately, long-run relative frequency cannot be the definition of probability, because it assumes that an infinite sequence of relative frequencies converges to some limit (for example, the relative frequency of Heads supposedly converges to 0.50 in an infinite number of tosses). Unfortunately, the axioms of mathematics do not guarantee that such a limit will exist. It is guaranteed that any bounded infinite sequence must contain a subsequence that converges to a limit; however, there is no guarantee that there is only one limit — the sequence could oscillate forever, and therefore the only way to use this limit as the definition of probability is to first assume that it exists. Therein lies the circularity.

Subjective probability is the only definition of probability that survives logical challenges. Subjective probability includes objective (casino) probability as a special case and has a relationship to relative frequencies that we'll explore in the next chapter, where we will make the case that relative frequencies are *data*, not probabilities.

Critics of subjective probability object to any subjectivity in evaluating scientific evidence. They worry that "anything goes" with subjective probability — that it leaves people free to ignore data or free to interpret data in different ways. There are two responses to that criticism: First, scientists frequently do react differently to the same data; second, open-minded and rational people who initially have different subjective opinions will be brought to consensus by large quantities of data. *Open-mindedness* more or less means that if you think a particularly theory is a possible explanation of the facts, then I am willing to consider it (and vice versa). *Rationality* means avoiding pragmatic self-defeat and is the topic of this chapter.

Thus, even though all probability is subjective, a form of objectivity is preserved, not in the sense that everyone must reach exactly the same conclusion from the same data, but in the sense that people who react rationally to data will be brought into practical consensus in the face of compelling evidence.

3.3 OBJECTIVITY AND SCIENCE

This is a good time to tackle the notion that scientific reasoning is objective and therefore the beliefs of individual scientists ought to be immaterial. Here is what a physicist (D'Agostini 1998) had to say on the subject:

> Any experienced physicist knows already that the only "objective" thing in science is the reading of digital scales. When we want to transform this information into scientific knowledge we have to make use of many implicit and explicit beliefs. Nevertheless, the "honest" (but naïve) ideal of objectivity can be recovered if scientific knowledge is considered as a kind of very *solid networks of beliefs*, based on centuries of experimentation, with *fuzzy borders* which correspond to the areas of present research. My preferred motto is that "no one should be allowed to talk about objectivity, unless he has 10 or 20 years of experience in frontier science, economics, or any other real world application." In particular, mathematicians should refrain from using the word objectivity when talking about the physical world. (Reproduced with kind permission of Kluwer Academic Publishers.)

D'Agostini's point is that scientific reasoning is not objective and cannot be made objective. However, scientists universally believe that nature is a unified whole, a belief that is embodied in a self-consistent, interconnecting web of theories. The appearance of objectivity comes from the inherent stability of this highly interconnected system. A new theory not only needs to agree with observations but needs to be consistent with other well-established theories as well. For example, precognition is inconsistent with physics, because it requires that information travel faster than light. This is why claims of evidence for ESP are largely rejected. For more on the subjectivity of scientists, see Press and Tanur (2001).

Objectivity may not be possible, but statistical reasoning does provide coherence and consistency, and we will later demonstrate that these are enough to ensure that evidence will eventually lead rational people to agree on what they think is true. Near consensus on the interpretation of evidence is as close to objectivity as we seem likely to come.

3.4 THOUGHT EXPERIMENTS TO EXPOSE INCONSISTENCY

A thought experiment is a powerful way to test whether ideas are internally consistent. For example, Aristotle thought that heavier objects fall faster than lighter objects. It's a natural mistake because extremely light things like feathers and dandelion fluff really do fall more slowly than heavy things such as rocks and cannonballs due to air resistance. However, we now know that, in a vacuum, feathers fall as fast as rocks.

We've all heard the story that Galileo disproved Aristotle's theory by simultaneously dropping two different sized cannonballs from the leaning tower of Pisa. Actually, this almost certainly never happened — it was a thought experiment. Galileo probably asked himself something like this, "What would happen if I chained the cannonballs together? Reasoning one way, the smaller ball, falling more slowly, would drag on the heavier and slow it down. On the other hand the falling object — two cannonballs and a chain — is heavier than the larger cannonball, so according to Aristotle it ought to fall faster, nor slower than the larger cannonball. The only way to resolve this inconsistency is to conclude that objects must fall at the same speed regardless of their weight." No physical experiment was required — just the conviction that nature is logically consistent and therefore theories about nature must not contain internal contradictions.

Thought experiments of a different sort are fundamental to understanding subjective probability. The subjective definition of probability treats probability as a measure of degree of belief. The idea is that the ultimate test of the strengths of your beliefs is to "put your money where your mouth is," as in this imaginary conversation,

Me: "I think a woman will be elected president by 2024."

You: "How about a friendly little bet on it?"

Me: "OK. I'll bet $300 and give you 3 to 2 odds."

You: "Done. I'll put up $200 and the winner gets $500."

Thinking about bets on sentences is the basic tool of Dutch book thought experiments.

3.5 SUBJECTIVE PROBABILITY AND DEGREE OF BELIEF

As we learned in the previous chapter, probability refers to sentences. A sentence is a declarative statement such as, "230 million years ago Africa and South America were joined," or "A woman will be elected president of the US by 2024." You don't in general know for certain if a sentence is true or false; however, you will hold some degree of belief that the sentence is true. For example, for me there is 90% probability that the second sentence will turn out to be true. Let's make that a little more formal. What does that 90% mean?

We will begin by introducing the idea of a *futures contract*. A futures contract is a bet that a particular sentence will turn out to be true. There are websites that offer futures contracts on elections, apprehending international criminals, whether the stock market will rise 100 points today, and so on. It is easiest to think about bets that pay off quickly (unlike betting on a women president before 2024). For example, we could think about betting on today's closing price of the Dow-Jones average. "The Dow" is published in every newspaper and is a weighted average price of a list of securities traded on the New York stock exchange. The closing price is the value of the average at the end of a trading day. We'll think about betting on the sentence "The Dow will close higher today," which means betting that numerical value of the average at the close of trading today will be higher than the value at the close of the previous trading day. Suppose that John Smith buys a futures contract from a website called OnlineBets.com at 9:00 A.M. today. The contract will look something like this,

> *OnlineBets.com* agrees to pay $1.00 to *John Smith* if the closing value of the Dow-Jones Industrial Average on *today's date* is greater the previous day's closing value.

This contract has value — it could pay Smith $1.00 or it could pay him nothing. Somewhere between these two figures is a price that Smith thinks is fair to buyer and seller — it is a price he would be willing to pay to buy the contract and also the price he would charge to sell it to someone else. Let's say his fair price is $0.37. Smith's *subjective probability* that the sentence S is true is *defined* as his fair price for the contract. We will use the symbol $P(S)$ to stand for the subjective probability of S.

We assumed that Smith's fair price was $0.37, and that *defines* his subjective probability, P(Dow will close higher.) $= 0.37$. In reality, futures contracts have two prices, a bid price, which is an offer to purchase, and an ask price, with is an offer to sell; however, the person doing a Dutch book thought experiment thinks in terms of fair prices — prices at which she believes neither side in a transaction would have an advantage over the other.

Subjective probabilities are personal (Jones' probability can be different from Smith's) and they are conditional (new information changes probabilities).

This rather brief explanation gives an intuitive idea of how subjective probability is defined; however, we have ignored several technical issues, such as the present value of future payoffs and risk aversion. Readers interested in a mathematically rigorous (but difficult to read) axiomatic derivation of subjective probability could consult Savage (1972).

The "fair price" definition of subjective probability operationalizes the idea of "degree of belief." It is considered to be a major philosophical breakthrough of the twentieth century since it makes it possible to analyze mathematically how people change their beliefs in the face of new evidence. According to Talbott (2001), this approach is "...one of the most important developments in epistemology in the 20th century, and one of the most promising avenues for further progress in epistemology in the 21st century."

3.6 BOOKS AND DUTCH BOOKS

A bet is a futures contract. For example, a gambler putting $1.00 on Red at a roulette table in effect buys a contract that pays $2.00 if this sentence is true: "The ball will land in a red slot." A *bookie*, or *bookmaker*, is a person who buys, sells, and settles such contracts. A *book* is a list of the contracts he has bought or sold. A bookie can offer bets for sale or can place his own bets with someone else.

For example, in the Tyson vs. Holyfield fight, Jimmie S. (the bookie) has accepted $1000 in wagers at 1 to 2 odds on Holyfield and $400 in wagers at 1 to 5 on Tyson. These odds are expressed bookmaker style, which means that Jimmie put up $1 for every $2 wagered on Holyfield, for a total of $500, and put up $1 on every $5 wagered on Tyson, for a total of $80. Jimmie's total investment is $580 (see Fig. 3.1).

If Holyfield wins, Jimmie pays $1500 to people who bet on Holyfield, but gets to keep the $480 stakes from the bets on Tyson, so his net profit — his gross winnings minus what he spent — is $480 − $580 = −$100 (a net loss). On the other hand, if Tyson wins, the bookie owes $480 to the people who placed bets on Tyson, but gets to keep the $1500 wagered on Holyfield for a net profit of $920 ($1500 − $580). It's a bit easier to keep track of this book in a balance sheet like Fig. 3.1; the part above the double line shows the stakes that the bookie and betters invested, and the part below the double line shows how the stakes are paid out depending, on who won the fight.

A *Dutch book* is a book that is *guaranteed* to give the bookie a loss. The book in Fig. 3.1 is not a Dutch book, because although there is a chance that the bookie will lose money, there is also a chance that he will gain, so a loss is not guaranteed under all circumstances. To make it into a Dutch book, let's change the example and suppose that Jimmie had given 3:1 odds on Tyson. Jimmie now holds a Dutch

		Stakes (spent)	
	Contract	From bookie	From betters
	Tyson at 1:5	$80	$400
	Holyfield at 1:2	$500	$1000
	Total Stakes	$580	$1400
		Payoffs (received)	
		To bookie	To betters
If Tyson wins	Payoff	$1500	$480
	− Cost	$580	$1400
	= Net profit	$920	-$920
If Holyfield wins	Payoff	$480	$1500
	− Cost	$580	$1400
	= Net profit	-$100	$100

Figure 3.1 Analysis of first book. Net profit is payoff minus amount invested.

book, because he is guaranteed to lose money no matter who wins the fight (see Fig. 3.2). You're probably thinking, "Who would be that stupid?" That is precisely the point — a rational person would not voluntarily put him(her) self in the position of holding a Dutch book. Holding a Dutch book has been described by Talbott (2001) as "pragmatic self-defeat."

Do not confuse a Dutch book with an unfavorable bet. People constantly make unfavorable bets, but that does not constitute a Dutch book. For example, hundreds of thousands of people place bets on Red at roulette. This is an unfavorable bet at house odds, because the average better wins only 18 times out of 38 and therefore loses on average about 5.25 cents of every dollar he bets. But a person betting on Red is not guaranteed to lose, whereas a person holding a Dutch book *is* guaranteed to lose no matter what happens.

3.7 RATIONALITY

Statistical science defines a *rational* person as one who is unwilling to buy and sell a set of futures contracts (i.e., make or cover a series of bets) that would constitute a Dutch book.

Scientific reasoning is founded on the assumption that all aspects of nature are mutually consistent (for example, the laws of physics and chemistry apply to the brain and this means, among other things, that thoughts can't travel faster

	Stakes (spent)	
Contract	From bookie	From betters
Tyson at 3:1	$1200	$400
Holyfield at 1:2	$500	$1000
Total Stakes	$1700	$1400
	Payoffs (received)	
	To bookie	To betters

		To bookie	To betters
If Tyson wins	Payoff	$1500	$1600
	−Cost	$1700	$1400
	= Net profit	−$200	$200
If Holyfield wins	Payoff	$1600	$1500
	−Cost	$1700	$1400
	= Net profit	−$100	$100

Figure 3.2 Analysis of second book. Net profit is payoff minus amount invested.

than light — so we can't have ESP without giving up physics as we understand it). If all parts of nature are mutually consistent, then our beliefs about nature should also be consistent and should be revised in a consistent way in the face of new evidence. The goal of statistical science is to provide that consistency.

What has this to do with Dutch books? Here's an example: Hardrocke, a geologist, thinks there is a 70% chance that the Atlantic Ocean was formed by continental drift (CD) and a 60% chance that it was formed by an asteroid hit (AH). These probabilities are incoherent because they imply that Hardrocke would pay 70¢ for a $1.00 bet on CD and 60¢ for a $1.00 bet on AS. I'll be happy to cover (take the other side of) each of these two bets because when the issue is finally settled to our mutual satisfaction I have to pay him only $1.00 but he paid me $1.30.

The astonishing fact is that it is possible to prove mathematically that if a person is to avoid a Dutch book, his (her) fair prices for bets must obey the laws of probability. For that reason it is legitimate to refer to fair prices as "subjective probabilities." As a corollary, all the machinery of probability theory is now at our disposal for updating our beliefs in the face of new evidence.

3.8 THE LAWS OF PROBABILITY VIA DUTCH BOOK ARGUMENTS

The idea of a *Dutch book argument* is this: If you have serious theories about nature, then (1) you ought to be willing to buy or sell futures contracts contingent on the truth of those theories at prices you judge to be fair, and (2) the contracts you hold or are obligated to pay off should not constitute a Dutch book against you.

In the following sections, a sentence surrounded by braces stands for a futures contract that pays $1.00 if the sentence is true and $0 if it is false. For example, {The author has two brothers} is a futures contract that pays $1 if the sentence in braces is true. A futures contract is a thing that can be bought or sold, like a shirt or a pair of socks; therefore it is at least possible to think about its fair price. My fair price need not be the same as yours, but my fair prices must be internally consistent. For example, if I am willing to sell any quantity of socks at 50 cents a pair and am willing to buy any quantity at $8 for a dozen pairs, then you can make money by buying a twelve separate pairs from me for $6 and then bundling them up as a a dozen pairs and selling them back at $8. Dutch book arguments revolve around the idea that two equivalent bundles ought to have the same fair price — twelve pairs of socks ought to have the same fair price as one dozen pairs of socks. Stores do offer discounts for bundles, but they don't offer to buy them back at the price of the components of the bundle; however, a Dutch book thought experiment is an exercise in exploring internal consistency, not a business model. In thought experiments we work with fair prices, which by definition are prices at which we imagine ourselves equally willing to be a buyer or a seller.

3.8.1 A Subjective Casino

To make things concrete, let's think about a midterm congressional election during a Republican presidency. (In the United States presidents are elected in years divisible by 4: 2004, 2008, ... ; however, congressional elections take place every even year: 2002, 2004, 2006, ... Midterm elections take place in even-numbered years not divisible by 4: 2002, 2006, ...). The sentences of interest are,

H_D: "Democrats will win control of the House."

H_R: "Republicans will win control of the House."

S_D: "Democrats will win control of the Senate."

S_R: "Republicans will win control of the Senate."

Fred is thinking about his subjective probabilities for these sentences — both alone and in various combinations. An interesting combination is $H_D \cup S_D$, which means, "Democrats will win control of the either the House or the Senate or both." Since the president is a Republican, the result would be a divided government. (A divided government occurs when the Senate or House of Representatives is not controlled by the president's party.) Fred's fair prices (subjective probabilities) for bets on sentences and combinations of sentences are shown in Fig. 3.3. For example, his subjective probability that there will be a divided government is 0.86, or 86%.

To determine if his prices are coherent, Fred imagines that he is obliged to cover either side of any of these bets at the stated prices. [In practice this is done by allowing *short sales*; however, in Fred's casino a player who wishes to bet *against* divided government (or any other sentence in the layout) signals his intentions by placing $1 on that square, and Fred gives him the price of the bet, $0.86. If the election results in a divided government, Fred keeps the dollar, earning a net profit of $0.14. If the election goes the other way, the dollar is returned to the player, who nets $0.86.] With the prices in Fig. 3.3, Fred would not be vulnerable to a Dutch book, because, as we will prove below, his fair prices obey the laws of probability. To be specific, the price of each row bet is the sum of the prices of the joint bets in that row, the price of each column bet is the sum of the prices of the joint bets in that column, and the price of any outside bet is consistent with the addition rule [Equation (2.4)].

To see why fair prices must be additive, suppose that Fred had priced S_R at 20¢. This is underpriced, because the price of S_R is not the sum of the prices of its components, $H_R \cap S_R$ and $H_D \cap S_R$. In his thought experiment, Fred imagines what would happen if he set up a casino and agreed to cover either side of any bet at the posted price. This is not the way a real casino operates, but that is not the point; we are trying to identify the characteristics of *fair* prices, and it is in the nature of a fair price that it is fair as a selling price or as a purchase price. Fred therefore thinks that 20¢ is the fair purchase price for a contract on S_R and that 20¢ is also the fair selling price. He would be willing to sell someone the contract at that price, and he would be willing to buy the contract from someone at that price.

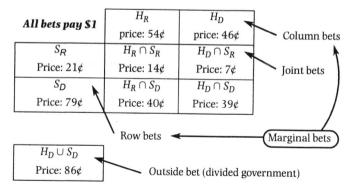

Figure 3.3   Fred's betting layout. To place a bet, player puts the indicated amount of currency on the desired bet. To cover the losing side, player places a dollar on the bet.

I could make Dutch book against Fred this way: I'd bet on S_R, meaning that I'd be the gambler and Fred would be the house. I'd put up 20¢, and Fred would put up 80¢. Then I'd offer to cover Fred's bets on $H_R \cap S_R$ and $H_D \cap S_R$, meaning that for these bets I'd act as the house and Fred would be the gambler placing a bet with me. As the gambler, Fred would have to put up 21¢ (14¢ to bet on $H_R \cap S_R$ plus 7¢ to bet on $H_D \cap S_R$), and I'd have to put up 1.79¢ to cover the other side of these two bets.

Here's the intuition: a store sells shirts a shirt-and-tie special for $20, but separately it charge $14 for the shirt and $7 for the tie. The store accepts returns, no questions asked. I buy the special, and then return the shirt for a $14 refund and the tie for a $7 refund which is $1 more than I spent. Fred's willingness to cover either side is equivalent to a "no questions asked" return policy. The easiest way to see that Fred has a Dutch book is with the balance sheet in Fig. 3.4. Fred's two bets are indicated on the left, and my bet is indicated on the right. In the "Settle the bets" section it is necessary to list all possible outcomes of the election. For example, if the Democrats win the Senate but lose the House, then I lose my bet on S_R, and the stakes (100¢) go to Fred; however, Fred loses both his bets, and the stakes (200¢) go to me. Fred invested 101¢ and won only 100¢, so he has a net loss of 1¢. Examination of all four possible outcomes of the election shows that Fred will lose 1¢ in every case. I have a no-risk profit and Fred has a for-sure loss — that's a Dutch book

Place the bets			
Fred's bets	Fred's stakes	My stakes	My bets
	80¢	20¢	S_R
$H_D \cap S_R$	7¢	93¢	
$H_R \cap S_R$	14¢	86¢	
Total Stakes	101¢	199¢	
Settle the Bets			
If this happens,	Fred wins		and I win
$H_D \cap S_R$	100¢ from $H_D \cap S_R$ Net: −1¢		100¢ from S_R & 100¢ from $H_R \cap S_R$ Net: +1¢
$H_D \cap S_D$	100¢ from S_R Net: −1¢		100¢ from $H_D \cap S_R$ & 100¢ from $H_R \cap S_R$ Net: +1¢
$H_R \cap S_R$	100¢ from $H_R \cap S_R$ Net: −1¢		100¢ from S_R & 100¢ from $H_D \cap S_R$ Net: +1¢
$H_R \cap S_D$	100¢ from S_R Net: −1¢		100¢ from $H_D \cap S_R$ & 100¢ from $H_R \cap S_R$ Net: +1¢

Figure 3.4   Fred's prices are not additive; players can make Dutch book for a no-risk profit.

This argument shows that if Fred is to avoid Dutch book, the price of each row marginal bet needs to equal the sum of the joint bets in that row. Similarly, the price of each column marginal bet must equal the sum of the joint bets in that column.

$$
\begin{array}{ccccc}
P(H_R \cap S_R) & + & P(H_D \cap S_R) & = & P(S_R) \\
+ & & + & & \\
P(H_R \cap S_D) & + & P(H_D \cap S_D) & = & P(S_D) \\
\| & & \| & & \\
P(H_R) & & P(H_D) & &
\end{array}
\tag{3.1}
$$

A *joint* bet is a bet that two sentences such as H_D and S_R are jointly, or simultaneously, true. A row or column bet is a bet that a single sentence such as H_D is true. Row or column bets are called *marginal* bets because they are displayed in the margins of the betting layout. The type of Dutch book argument displayed in Fig. 3.4 proves that if Fred is to avoid a Dutch book, then his fair prices for the joint bets in a given row must add up to his fair price for a marginal bet on that row, and his fair prices for the joint bets in a given column must add up to the fair price for a marginal bet on that column.

We are now poised to argue that subjective probabilities (defined as fair prices for bets) must obey the addition rule

$$
P(H_D \cup S_D) = P(H_D) + P(S_D) - P(H_D \cap S_D)
\tag{3.2}
$$

This rule can be proved using a Dutch book argument. The first step is to move the negative term to the left side of Equation (3.2) to get Equation (3.3).

$$
P(H_D \cup S_D) + P(H_D \cap S_D) = P(H_D) + P(S_D)
\tag{3.3}
$$

The left side of Equation (3.3) is the cost of a bundle of two bets: a fair-priced bet on $H_D \cup S_D$ and another fair-priced bet on $H_D \cap S_D$. The right side is the cost of a bundle containing two other bets: a fair-priced bet on H_D and another fair-priced bet on S_D. To complete the proof, we need to demonstrate that the bundle on the left side of the equation has exactly the same payoffs under all contingencies as the bundle on the right side of the equation; consequently they are identical bundles described differently (just as "a shirt and a tie" is the same thing as "a tie and a shirt").

For Fred avoid a Dutch book, equivalent bundles described differently must have the same fair price. If they do not have the same price, then I'll buy the cheaper bundle from Fred, re-bundle it the more expensive way, and sell it back to him. Since the payoffs to me and to Fred are identical, Fred will lose money on the transaction.

To see this, suppose that Fred's prices don't obey the additive law [Equation (3.2)]. Suppose Fred made a mistake and priced $H_D \cup S_D$ at 90¢ instead of 86¢ (see Fig. 3.3 on page 33). To make a Dutch book against Fred you must "buy low and sell high"; consequently you should buy the contracts (bets) on the right side of Equation (3.3) and sell the contracts on the left side; the details are presented in Fig. 3.5.

This example has taught us that two portfolios of bets can have identical payoffs even though the bets are bundled in different ways. For example, Jane put $1 on each one of the 18 red slots in Fig. 2.1 on page 10 at a total cost of $18. Jill on the other hand put $18 on the Red outside bet. These bets have identical payoffs: Jill and Jane will each win $36 if the ball lands in any red slot, and they will win nothing if the ball lands anywhere else. They have invested the same amount of money and they receive exactly the same payoffs, so their portfolios are

Place the Bets			
Fred's Bets	Fred's Stakes	My Stakes	My Bets
	54¢	46¢	H_D
	21¢	79¢	S_D
$H_D \cap S_D$	39¢	61¢	
$H_D \cup S_D$	90¢	10¢	
Total Costs	204¢	196¢	

Settle the Bets		
If this happens,	Fred wins this,	and I win this,
$H_D \cap S_R$	100¢ from S_D 100¢ from $H_D \cup S_D$ Net: -4¢	100¢ from $H_D \cap S_D$ 100¢ from H_D Net: +4¢
$H_D \cap S_D$	100¢ from $H_D \cap S_D$ 100¢ from $H_D \cup S_D$ Net: -4¢	100¢ from S_D 100¢ from H_D Net: +4¢
$H_R \cap S_R$	100¢ from S_D 100¢ from H_D Net: -4¢	100¢ from $H_D \cap S_D$ 100¢ from $H_D \cup S_D$ Net: +4¢
$H_R \cap S_D$	100¢ from H_D 100¢ from $H_D \cup S_D$ Net: -4¢	100¢ from $H_D \cap S_D$ 100¢ from S_D Net: +4¢

Figure 3.5   Dutch book proof of the addition rule.

identical even though they are bundled differently — 18 separate $1 bets or a single $18 bet.

It seems clear that person's prices for equivalent bundles must be equal; otherwise an opponent could make Dutch book against that person by betting on the cheaper bundle and betting against the more expensive bundle. After establishing some necessary preliminaries, we will use this strategy to prove the central tenet of subjective probability: that fair prices must also obey the product rule [Equation (2.3) on page 13].

3.8.2   Sure Things, Sure losers, Complements

A *sure thing* is a bet that is guaranteed to win. For example betting that "Either it will rain tomorrow or it won't" is a sure winner. Obviously the fair price is $1; otherwise either the better or the person covering the bet will lose money for sure. The symbol for a sure winner is Ω (omega); consequently,

$$P(\Omega) = 1 \qquad (3.4)$$

A *sure loser* is a bet that is guaranteed to lose — for example, betting that $2+2$ equals 5. Obviously the fair price for a sure loser is $0; otherwise the better will lose his stake for sure. The symbol for a sure loser is $\varnothing$ (null); consequently,

$$P(\varnothing) = 0 \qquad (3.5)$$

The *complement* of a sentence is its negation. For example the complement of "It will rain tomorrow" is "It will not rain tomorrow." If S stands for a sentence, the symbol for its complement (negation) is S^C (or $\sim S$), pronounced "S complement" or "not S". Suppose you bet on S and also on S^C. The fair price of this bundle is the sum of its parts: $P(S) + P(S^C)$. But either S is true or its negation is true, so the bundle is a sure winner and its fair price has to be $1; that is, $P(S) + P(S^C) = 1$. Consequently,

$$P(S^c) = 1 - P(S) \qquad (3.6)$$

3.8.3   Fractional and Multiple Bets

So far we've worked with bets that pay $1. For convenience, we're going to use the symbol {$1 if S} to stand for a bet that pays $1 if the sentence S is true. The fair price of this kind of bet is the subjective probability that S is true. For example, the fair price for the roulette bet {$1 if Red} is $18/38$, or about 47¢. What about bets that pay something other than $1, for example, {$2 if Red}? Las Vegas casinos charge $1 to place this bet, but the fair price is obviously double the fair price of {$1 if Red}, in other words, the fair price of {$2 if Red} is $2 \times \$18/38$, or about 95 cents. In general, it seems obvious that the fair price of a bet that pays M dollars must be M times the fair price of a bet that pays $1. Since the fair price of a $1 bet on the sentence S is the subjective probability $P(S)$, we have,

$$\text{fair price of } \{\$M \text{ if } S\} = \$M \cdot P(S) \tag{3.7}$$

This proposition seems obvious (if a gallon of gas costs $1.78, then 10 gallons will cost $17.80); however, here is the idea of how to prove it. Suppose the bet is on sentence S with payoff, let us say, $2.5 or $5/2, instead of the usual $1. The symbol for this bet is {$2.5 if S}, we need to show that the fair price of this bet is 2.5 times the fair price of {$1 if S}. Placing two bets on {$2.5 if S} is equivalent to placing five bets on {$1 if S}, because either bundle pays $5 if S turns out to be true. Let's say your fair price for {$1 if S} is 30¢. Obviously the fair price of making this bet five times is $5 \times 30\textcent = \$1.50$. Since five bets on {$1 if S} is equivalent to two bets on {$2.50 if S}, the second bundle must also have a fair price of $5 \times 30\textcent$; consequently, the two identical parts of the second bundle must be worth $5/2 \times 30\textcent$ each, that is, $2.50 \times 30\textcent$.

3.8.4   House Odds Revisited

Recall that house odds or bookmaker's odds, 7 to 2 for example, represent the bookmaker's stake, $7, vs. the gambler's stake, $2; in other words, the bookmaker agrees to wager $7 for every $2 a gambler wagers, and the total stake, $9, goes to the winner. Our symbol for that contract is {$9 if S}, and the price the gambler paid for it is his stake, $2. The bet {$1 if S} pays nine times less and therefore its price must logically be $2/9 or about $0.22. As with casino bets, this is typically not the fair price, since the bookmaker wants to make a profit. The general rule is

$$\text{bookmaker's price for } \{ \$1 \text{ if } S \} = \frac{\text{gambler's stake}}{\text{gambler's stake} + \text{bookmaker's stake}} \tag{3.8}$$

3.8.5   Conditional Probability and the Product Rule

If D and R are sentences, then the symbol $P(R \mid D)$ stands for the conditional probability of R given D. This purpose of this subsection is to show how to interpret $P(R \mid D)$ as the fair price of a bet, hence as a subjective probability or degree of belief. To set the scene, assume the president is a Republican in his the second year of his first term (as in 2002). We'll work with these two sentences:

D: "The next midterm election will produce a divided government."

R: "The Republican president will be re-elected for a second term."

I'm going to propose a new kind of bet — a *called-off bet*. In this type of gamble you bet on R, but if D doesn't happen, then the bet is called off and your stake is refunded. Here's the idea:

> *You:* "I think a divided government will give the president an advantage in his next election — he'll have someone to blame."
>
> *Me:* "Care to put a little money on the president in the next election?"
>
> *You:* "OK but if the Republicans win the House and Senate, the bet is off."

Your fair price for this bet is different from the fair price for an unconditional bet on R because you are risking your money on R only if D comes true. The subjective conditional probability of R give D, $P(R \mid D)$, is defined as the fair price of the called-off bet.

To get some intuition as to why this is a reasonable definition of subjective conditional probability, suppose that Ed's casino offers this strange wager: The dealer will deal out two cards, one face up and the other face down. The player's decision to bet must be made before any cards are dealt. The player is offered the following wager. If both cards are kings, the player wins a dollar; if the up card is not a king, the bet is called off; and in all other circumstances the player loses whatever he bet. Clearly, this is equivalent to allowing the player to wait to place a bet until after the up card has been dealt. The player would be foolish to place a bet unless the up card is a king, in which case the player has 3 chances that the second card will be a king and 48 chances that it will not, so the conditional probability of winning is $P = 3/51$ and the fair price is $\$P = \$3/51$.

Let R and D stand for two sentences which are not mutually exclusive. A called off bet is a contract with this wording: "If D comes true, the owner of this contract wins \$1 if R is true, but if D is false, the purchase price will be refunded." Example. If it rains tomorrow (Deluge), you will receive \$1.00 if you catch cold (Runny nose). If it does not rain, your purchase price will be refunded. Fig. 3.6 may help clarify how this contract works.

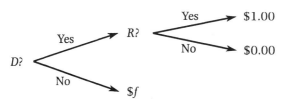

Figure 3.6 A called-off bet with fair price $\$f$. The gambler's stake is $\$f$. If D is false, her stake is refunded; if D and R are true, she wins \$1.00; otherwise she wins nothing.

We're trying to work out a fair price for this complicated contract. For now, we'll call the fair price f. Whatever that price turns out to be, it is, by definition, the conditional probability, so $f = P(R \mid D)$.

To prove that the fair price must obey the product rule [Equation (2.3) on page 13], we're going to construct a bundle of bets that is equivalent to the called-off bet in Fig. 3.6 but is easy to price. The equivalent bundle consists of a bet that pays $1 if $R \cap D$ comes true plus a bet that pays f if D turns out to be false; we'll argue that the called-off bet and this bundle have identical payoffs under all contingencies, and therefore they must have the same fair price.

The fair price of the called-off bet is symbolized f. To compute the fair price of the equivalent bundle we add up the fair prices of its two components. The fair price of the bet that pays $1 on the truth of $R \cap D$ is, by definition, the subjective probability, $P(R \cap D)$. The fair price of second part of the bundle, the bet that D is false, requires a little thought, since the payoff is f instead $1. We learned in the previous section that the fair price of this fractional bet is f times the fair price of a bet that pays $1. A bet that D is false is the same as a bet that $\sim D$ is true. By definition the fair price of the standard bet {$1 if $\sim D$} is the subjective probability $1 \cdot P(\sim D)$, that is $1 \cdot [1 - P(D)]$. Consequently, the fair price of the non-standard bet {f if $\sim D$} is $f \cdot [1 - P(D)]$.

The fair price of the bundle is the sum of its parts, $P(R \cap D) + f \cdot [1 - P(D)]$. But the bundle is equivalent to the called-off bet, which has fair price f. Consequently,

$$f = P(R \cap D) + f \cdot [1 - P(D)] \qquad (3.9)$$

Solving for f we get

$$P(R \cap D) = f \cdot P(D) \qquad (3.10)$$

But f is the fair price of the called-off bet, and is the definition of the conditional subjective probability, $P(R \mid D)$; consequently, substituting $P(R \mid D)$ for f, we conclude that subjective probabilities must obey the product rule:

$$P(R \cap D) = P(R \mid D) \cdot P(D) \qquad (3.11)$$

Dividing both sides of Equation (3.11) by $P(D)$ gives an equation for computing conditional probability when joint and marginal probabilities are available,

$$P(R \mid D) = \frac{P(R \cap D)}{P(D)} \qquad (3.12)$$

3.8.6   Independence Revisited

Recall that sentences R and D are said to be *statistically independent* if learning that D is true does not change the subjective probability of R. In other words, if D and R are independent, learning that D is true would not change the fair price of a bet on R.

There are two ways to verify that two sentences are independent. One is to check whether the numerical values of the subjective probabilities (fair prices) satisfy the independence condition: $P(R) = P(R|D)$. The second method is to persuade yourself that learning that D is true would not change your fair price for a bet on R. Let's apply the first method to Fred's subjective probabilities in Fig. 3.3 on page 33. Is H_D independent of S_D? You'd think not, but let's check the independence condition, that is, lets check to see if $P(H_D|S_D) = P(H_D)$.

The first step is to compute the conditional probability using Equation (3.12) on page 40: $P(H_D|S_D) = 0.39/0.79 = 0.4937$, and the second step is to compare it to $P(H_D)$. Since $P(H_D) = 0.46$ is not equal to $P(H_D|S_D) = 0.4937$, Fred's fair price of a bet on H_D is changed by learning that S_D is true. Consequently the knowledge that S_D is true increases Fred's probability that H_D is true, and we conclude that the two sentences are dependent.

To illustrate the second method for checking independence, let $T =$ "It will rain tomorrow" and $W =$ "A woman will be elected president in 2024." I can assure you that whether or not it rains tomorrow will not change my opinion about W. Consequently, for me, rain tomorrow and electing a woman president in 2024 are independent of each other.

When sentences T and W are independent, there is a special form of the product rule:

If sentences T and W are independent,

$$P(T \cap W) = P(T) \cdot P(W) \qquad (3.13)$$

To prove this special product rule, insert sentences W and T in place of R and D in the general product rule [Equation (3.11) on page 40], and then apply the condition for independence, $P(W \mid T) = P(W)$.

It is important to understand that the special product formula is not a *law* — it is not necessarily true. In order to apply Equation (3.13) you must first verify that sentences R and W are independent, and if they are not independent you must use the general form of the product rule [Equation (3.11)]

3.9   EXPECTED VALUE REVISITED

The expected value of an uncertain quantity — for example, the payoff of a complicated bet — was introduced in Section 2.8. Here we present a short proof that expected value is interpretable as fair price. Suppose that S_1, S_2,..., S_k is a mutually exclusive and exhaustive list of sentences. In other words, only one of the sentences can be true, and at least one must be true. For example, referring to a single spin of a roulette wheel, S_1=Red, S_2=Black, and S_3=Green are mutually exclusive, because each of the 38 compartments has only one color (red, black, or green), and exhaustive, because the ball has to land somewhere.

Suppose further that each of these sentences has an associated payoff value: V_1 is the value associated with sentence S_1, V_2 is the value associated with sentence S_2, etc. Prior to learning which sentence is true, the expected value, EV, of the future payoff is defined as

$$EV = \sum_{i=1}^{k} V_i \cdot P(S_i) \tag{3.14}$$

Suppose that the future payoff, whatever it turns out to be, will be paid to me, but I need to sell my right to the payoff now, before its exact value is known. If the probabilities in Equation (3.14) are *my* subjective probabilities, then EV is the fair price at which I would sell my right to receive the future payoff. To prove this, note that owning the right to receive the payoff is equivalent to owning this bundle of contracts: {$\$V_1$ if S_1}, {$\$V_2$ if S_2},..., {$\$V_k$ if S_k}. According to Equation (3.7), the fair price of {$\$V$ if S} is $\$V \cdot P(S)$; consequently, the fair price of the bundle of contracts is the sum of their individual fair prices, which is Equation (3.14).

For example, suppose that there are two drugs available to treat a mild fungal infection: an inexpensive drug (A) costing $5 per dose, and an expensive drug (B) costing $200 per dose. A physician thinks that drug A has a 60% chance of curing the patient and drug B has a 100% chance. Since the infection is slow-growing and not life-threatening, it is ethical for the physician to try drug A first and, if that fails, to prescribe drug B. The cost of the treatment depends on these sentences: S_1 = "Drug A cures the patient" and S_2 = "Drug A does not cure the patient." The associated values (costs) are V_1 = $5 (because if drug A works there is no need to prescribe drug B) and V_2 = $205 (the cost of drug A plus the cost of drug B). The expected cost of treatment is therefore $V_1 \times P(S_1)$ + $V_2 \times P(S_2)$ = $5 \times 0.6 + \$205 \times 0.4 = \$85$.

Now suppose that a new drug, C, becomes available. The physician believes that drug C has an 80% chance of curing the patient, but it costs $30 per dose. The expected cost of treating with drug C followed by drug B if the first drug fails is, $30 \times 0.8 + \$230 \times 0.2 = \$70$. Thus, even though drug C is more expensive than drug A, the expected cost of treatment with C (followed by B if necessary) is $15 lower than the expected cost of treatment with A (followed by B if necessary). The higher cost of C is offset by its higher probability of curing the infection. This rather contrived example illustrates how expected values form the basis of rational choices among different courses of action. We will return to this topic in Chapter 13.

3.10 EXERCISES

3.1 Suppose Sam's fair prices are as shown below. Make a Dutch book against him. Which bet is incoherently priced?

All bets pay $1	H_R price: 54¢	H_D price: 48¢
S_R price: 21¢	$H_R \cap S_R$ price: 14¢	$H_D \cap S_R$ price: 7¢
S_D price: 79¢	$H_R \cap S_D$ price: 40¢	$H_D \cap S_D$ price: 39¢

3.2 Suppose Ted's fair prices are as shown below. Assuming Ted is coherent, use the laws of probability to fill in his fair prices for $H_D \cap S_D$, and for the four marginal bets, and use the addition rule, Equation (3.2) on page 35, to compute the Ted's fair price for the outside bet $H_D \cup S_D$.

All bets pay $1	H_R price: ?	H_D price: ?
S_R price: ?	$H_R \cap S_R$ price: 23¢	$H_D \cap S_R$ price: 9¢
S_D price: ?	$H_R \cap S_D$ price: 55¢	$H_D \cap S_D$ price: ?

3.3 A bookie offered the following odds on a race involving four horses (you may refer to the horses as D, H, M, and L). The odds are expressed bookmaker style. (See Section 2.7 on page 18) Use Equation (3.8) on page 38 to convert the bookmaker's odds into his prices for standard bets of the form {$1 if D}. Suppose, for the sake of illustration that the fair prices are $0.12, $0.45, $0.11, and $0.41. It would cost $1.09 to "bet the field", that is, to place all four standard bets.

(a) Demonstrate that you can make a Dutch book against the bookie by betting the field at his prices.

(b) Change the price on Dog Biscuit to avoid a Dutch book and convert that price back to bookmaker's odds.

Dog Biscuit	Horse Collar	Mama Mia	Lead Foot
3 to 2	3 to 1	4 to 1	7 to 1

3.4   In Fred's betting layout, Fig. 3.3 on page 33, determine whether the
       sentences H_R and S_D are independent.

3.5   Suppose that the sentence D implies the sentence M. For example, D: "The
       next nonhuman animal I see will be a dog," and M: "The next nonhuman
       animal I see will be a mammal." Obviously if D is true than M is true, but
       the reverse is not true, because, for example, I might see a cat. Make an
       argument that $P(D) \leq P(M)$.

       Hint: You can prove this in either of two ways. Let C = "A non dog
       mammal" (cow, camel, ...). Obviously, mammals (M) are either dogs (D) or
       some non-dog mammal (C); consequently $M = D \cup C$. Since a mammal can't
       be a dog and a noonday at the same time, $D \cap C = \varnothing$. Apply the addition rule
       to $P(D \cup C)$.

       Another approach is to make a Dutch book argument: Suppose Chuck's
       subjective probabilities are $P(D) = 20\cent$ and $P(M) = 15\cent$. Let B = "A non
       mammal" (bird, butterfly, ...). Construct a Dutch book against Chuck using a
       diagram along the lines of Fig. 3.4 on page 34 or Fig. 3.5 on page 36 (in the
       "Settle the bets" section you will need to consider what happens if D or C, or
       B is true).

3.11 REFERENCES AND ADDITIONAL READINGS

D'Agostini, G., "Bayesian Reasoning Versus Conventional Statistics in High Energy Physics," *Maximum Entropy and Bayesian Methods: Garching, Germany 1998*, ed. Linden von der. et al., Kluwer Academic Publishers, Dordrecht, 1999.

Press, S.J. and Tanur, J.M., *The subjectivity of Scientists and the Bayesian Approach*, New York: John Wiley & Sons, Inc., 2001.

Savage, Leonard J., *The Foundations of Statistics*, New York: Dover Publications.

Steup, Matthias, "The Analysis of Knowledge," in *Stanford Encyclopedia of Philosophy*, Edward N. Zalta, ed. The Metaphysics Research Lab Center for the Study of Language and Information, Ventura Hall, Stanford University Stanford, CA 94305-4115, http://plato.stanford.edu/contents.html, 2002.

Talbott, William J., "Bayesian Epistemology," in *Stanford Encyclopedia of Philosophy*, Edward N. Zalta, ed. The Metaphysics Research Lab Center for the Study of Language and Information, Ventura Hall, Stanford University Stanford, CA 94305-4115, http://plato.stanford.edu/contents.html, 2001.

4

Distributions and Descriptive Statistics

4.1 INTRODUCTION

The purpose of this chapter is to introduce the reader to three basic statistical concepts: *distributions, quantiles, and moments*. Here they are introduced as ways of describing and summarizing observed data; however, the most important use of distributions, moments, and quantiles is in describing and summarizing probabilities, that is, degrees of belief. We start with describing data because that has its uses and is a quick and easy way to introduce the ideas; however, the rest of the book is devoted to learning how to describe and summarize degrees of belief.

4.2 CASES AND VARIABLES

The data in Table 4.1 on page 48 were generated by students in the fall and spring sections of an engineering course. The students worked in pairs to measure each other's visual reaction times. Each pair was equipped with an aluminum rod marked in centimeters and millimeters. One student (the investigator) dropped the rod without warning, and the other student (the subject) caught it between thumb and forefinger. After several practice drops, a final drop was made and the distance that the rod fell was recorded.

Each row in Table 4.1 contains data about one student. Each column contains data for one *variable*. The variables are: semester (fall or spring), subjects' initials, genders, and ages; the distances the rod fell in centimeters; and the equivalent reaction times in seconds based on the law for freely falling bodies, $t = \sqrt{2 \cdot d/g}$, where d is the distance, t is the free-fall time, and $g = 981$ cm/s^2 is the acceleration due to gravity at sea level. Table 4.1 is an example of a *cases by variables data matrix*. The cases (rows) are individual subjects, and the variables (columns) are things we know about the subjects. The last variable, reaction time in seconds, is a *transformation* of the distance variable. The data table contains variables that represent quantities (age, distance, time) and non-quantitative, or *categorical*, variables that label or classify subjects in some way (semester, initials, gender).

Table 4.1   Reaction time data[a].

Semester	Initials	Gender	Age	Distance (cm)	Time (s)
Fall	AEI	M	24	11.0	0.153
Fall	SUN	F	22	28.0	0.241
Fall	SLA	M	23	9.0	0.139
Fall	AHR	F	23	29.0	0.245
Fall	RWJ	M	23	6.0	0.111
Fall	VNW	F	28	38.0	0.277
Fall	DAT	F	23	60.0	0.351
Fall	COR	F	21	40.0	0.286
Fall	PIB	F	27	71.0	0.380
Fall	TOH	M	29	32.0	0.254
Fall	PTP	M	27	23.0	0.217
Fall	JLA	M	20	29.0	0.244
Fall	WJE	M	29	24.0	0.221
Fall	MAM	F	21	31.0	0.252
Fall	COI	F	20	26.0	0.230
Fall	GAW	M	22	25.0	0.226
Fall	MSN	M	20	18.0	0.192
Fall	AIO	M	20	26.0	0.230
Fall	AFW	M	21	11.0	0.150
Fall	PRU	M	47	42.0	0.291
Fall	TCC	F	32	26.0	0.232
Fall	BCK	M	28	11.0	0.150
Fall	KIH	F	26	113.0	0.479
Spring	ZJM	M	21	18.0	0.191
Spring	KRB	M	21	13.0	0.162
Spring	PTY	F	21	14.0	0.168
Spring	DGL	F	26	10.0	0.142
Spring	CAQ	M	21	15.0	0.174
Spring	NVK	M	21	19.5	0.199
Spring	BDE	F	23	15.0	0.174
Spring	NLV	M	22	20.0	0.201
Spring	OMP	F	21	23.5	0.218
Spring	DNG	F	21	30.0	0.247
Spring	SPN	F	22	19.0	0.196
Spring	JTR	M	23	21.5	0.209
Spring	DCB	F	22	2.0	0.063
Spring	ART	M	25	7.0	0.119
Spring	RAD	F	23	15.0	0.174
Spring	GTN	M	22	8.7	0.133
Spring	IRD	F	21	24.0	0.221
Spring	BHB	F	21	20.0	0.201

a. Source: Professor Thomas J Armstrong, personal communication. Semester and initials are fictional.

4.3 DISTRIBUTIONS

Just looking over the reaction times in Table 4.1 is not particularly revealing. For example, it is not easy to spot the fastest and slowest reaction times, nor is it apparent whether men or women tend to have quicker reflexes. The purpose of a distribution is to arrange the data in a way that makes it easier to see these things and others. For a quantitative variable (age, distance, time) a distribution is a picture or a table showing the positions of the values of a variable (reaction time for example) on the number line. There are several ways to do this including *stemplots* (stem and leaf diagrams), *histograms*, and *frequency tabulations*.

4.3.1 Stemplots

There are 20 female students in Table 4.1. Their reaction times, rounded to two significant digits, are 0.24, 0.25, 0.28, 0.35, 0.29, 0.38, 0.25, 0.23, 0.23, 0.48, 0.17, 0.14, 0.17, 0.22, 0.25, 0.20, 0.06, 0.17, 0.22, and 0.20. The first step in constructing a stemplot of these data is to determine the largest and smallest values of the most significant digit (called the *stem*). The most significant digit for the reaction times is that of tenths of a second. The stem of the shortest reaction time, 0.06 seconds, is 0 (0 tenths of a second), and the stem of the longest reaction time, 0.48 seconds, is 4 (4 tenths of a second). In the stemplot, the stem values are listed in a column to the left of a vertical line as shown in Panel A of Fig. 4.1.

Sometimes a particular stem does not actually appear in the data; for example if the longest time had been 0.58 instead of 0.48, then the stems would be 0, 1, 2, 3, and 5, and there would be no observations with stem 4. However, stem 4 should still be included in the diagram to give a correct impression of the distribution.

A. Stems Only		B. First Four Females	
0		0	
1		1	
2		2	458
3		3	5
4		4	

C. All Females		D. Sorted	
0	6	0	6
1	7477	1	4777
2	458953325020	2	002233455589
3	58	3	58
4	8	4	8

Figure 4.1   Steps in constructing a stemplot.

The next step is to record all 20 female reaction times as *leaves* on the right side of the vertical line. The second digit is used instead of an "X" or a " | " as a tally mark. For example the first reaction time is 0.24, so it is displayed as a "4" on stem 2. The first four reaction times in the list (0.24, 0.25, 0.28, and 0.35) are shown in Fig. 4.1 B. After the rest of the data have been entered, there will be one leaf for each observation, as shown in Fig. 4.1 C. As a final step the leaves on each stem are sorted from smallest to largest, as shown in Fig. 4.1 D. Notice that decimal points are never entered and that some leaves occur several times.

The completed stemplot in Fig. 4.1 D makes two data features instantly visible: the range of reaction times and the most common reaction times. The stemplot shows that the female reaction times ranged from 0.06 to 0.48 seconds but were most often in the range 0.20 to 0.29. In addition, the stemplot also makes it easy to compute percentiles simply by counting. For example, the 50th percentile, or *median*, is the middle reaction time, 0.23 seconds. Half the subjects had shorter reaction times, half had longer. Sorting the leaves makes it easy to determine the median by counting — no calculations are needed. There are 20 reaction times, so the median is between the 10th and 11th reaction times in the sorted stemplot.

Back-to-Back Stemplots. Back-to-back stemplots are occasionally used to make it easier to compare two groups. For example, Fig. 4.2 is a back-to-back stemplot designed to permit a visual comparison of the distributions of female and male reaction times. It is easy to see that the male reaction times tend to be somewhat shorter than the female reaction times (nearly half of the male reaction times are between 0 and 19 seconds but a much smaller proportion of female times are in that range). It is also apparent that in this group of students, the female reaction times are more variable, ranging from 0.06 to 0.84 seconds whereas the male reaction times range from 0.11 to 0.29 seconds.

In a back-to-back stemplot the stems are in a central gutter and two distributions of leaves are presented, one to the left of the gutter and one to the right. The leaves on the left side of the gutter are sorted from right to left and the leaves on the right side of the gutter are sorted in the usual way from left to right. The principle is that the leaves are sorted outward from the gutter, with the smallest values nearest the gutter.

Females		Males
6	0	
7774	1	1234556799
985544332200	2	00112233459
85	3	
8	4	

Figure 4.2   Back-to-back stemplots of male and female reaction times (stem = 0.1 s).

Two-Digit Stems and Other Split Stems. James and Knuiman (1987) displayed a stemplot of the flush volumes of 147 toilets ranging from 4.0 to 20.0 liters. Their stemplot, Fig. 4.3 A, is an extremely clear depiction of the data; however, it does not strictly follow the rules of stemplot construction that were presented above, since according to the rules a stemplot is a display of data rounded to two digits. In the case of the flush volume data the standard stemplot would be based on data rounded to the nearest liter (so that the smallest flush volume would be 04 liters and the largest would be 20 liters); however, strict adherence to the rules produces the ugly, useless stemplot in Fig. 4.3 B, and for these data it was better to bend the rules and use two-digit stems.

In some cases two-digit stems make the stemplot too tall and narrow. A solution is to split stems into two or five parts, as is illustrated with artificial data in Fig. 4.4. The standard stemplot (Panel A) is too compressed, but a stemplot based on two-digit stems (not shown) is equally unsatisfactory. In panel B, each stem has been split between two rows; so for example, stem 6 in panel A is split

A. stem unit = 1 liter

```
04 | 005
05 | 4
06 | 001145569
07 | 000000023455678
08 | 00000012222455677899
09 | 0000022234556777899
10 | 00000000000000111455566666789
11 | 00000000000012244444555568889
12 | 0000023455
13 | 00145
14 | 16
15 | 00
16 | 0
17 |
18 |
19 |
20 | 0
```

B. stem unit = 10 liters

```
0|4445666666666777777777777777788888888888888888888888999999999999999999999
1|00000000000000000000000000000000111111111111111111111111111111122222222222333344556
2|0
```

Figure 4.3 Stemplots of toilet flush volumes. Double-digit stems (A) give a clearer picture of the shape of the distribution than do single-digit stems (B). Source of Panel A: James and Knuiman (1987), Figure 2, reprinted with permission from *The Journal of the American Statistical Association*, Copyright 1993, by the American Statistical Association, All rights reserved.

A. One-digit stems are too compressed.

```
6 | 2335689
7 | 00000111122233333444445555666677778889
8 | 01123
```

Two- or five-way splits improve the stemplot:

B. Stems split in two		C. Stems split in five	
6·	233	6T	233
6*	5689	6F	5
7·	000001111222333334444	6S	6
7*	5555666677778889	6*	89
8·	01123	7·	000001111
		7T	22233333
		7F	44445555
		7S	66667777
		7*	8889
		8·	011
		8T	23

Figure 4.4   Splitting stems in two or five parts (B or C) improves the appearance of a compressed stemplot (A).

into stems 6· and 6*. Leaves 0 through 4 are recorded on stem 6· and leaves 5 through 9 on stem 6*. The reader is invited to work out the reason for labeling the stems "T," "F" and "S" in the five-way split (panel C).

4.3.2   Frequency Tabulations

A *frequency tabulation* displays the distribution of data values in the form of a table of numbers rather than in a graphic form such as a stemplot. Table 4.2, for example, displays the reaction time data of Table 4.1 on page 48 in the form of a frequency tabulation. The reaction times have been grouped into convenient *class intervals* (which need not have equal lengths), and the number, or frequency, of people with reaction times in each of the class intervals is listed. For example, times between 0.05 and 0.09 seconds (rounded to the nearest 0.01 second) are grouped together in the class interval 0.05 – 0.09.

The columns labeled "Frequency" display the number of people in each class interval. For example, there were 8 females and 9 males with reaction times between 0.20 and 0.24 seconds (rounded to the nearest 0.01 second). The columns labeled "Relative frequency" display the frequencies expressed as percentages of the total; so, for example, the table indicates that 40% of females (8 out of 20) and 42.9% of males (9 out of 21) had reaction times between 0.20 and 0.24 seconds.

Table 4.2 Frequency Tabulation of Male and Female Reaction Times

Reaction time	Frequency		Relative frequency (%)	
(s)	F	M	F	M
0.05 – 0.09	1	0	5.0	0.0
0.10 – 0.14	1	4	5.0	19.0
0.15 – 0.19	3	6	15.0	28.6
0.20 – 0.24	8	9	40.0	42.9
0.25 – 0.29	4	2	20.0	9.5
0.30 – 0.34	0	0	0.0	0.0
0.35 – 0.39	2	0	10.0	0.0
0.40 – 0.44	0	0	0.0	0.0
0.45 – 0.49	1	0	5.0	0.0

4.3.3 Histograms

A *histogram* is a type of bar chart that displays the information in a relative frequency tabulation. For example, Fig. 4.5 is a histogram that depicts the relative frequencies of female reaction times tabulated in Table 4.2. Notice that in a histogram, unlike a standard bar chart, the base of each bar runs from the lower endpoint of its class interval to the lower endpoint of the next. Histograms are bulky compared to frequency tables and are used only when a strong visual impact is necessary to make a point. For example, back-to-back histograms,

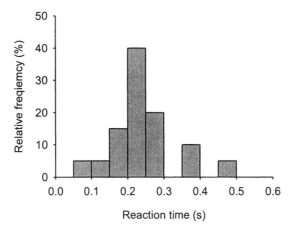

Figure 4.5 Histogram of female reaction times.

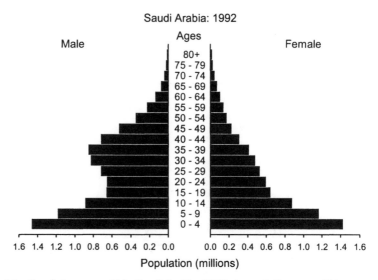

Figure 4.6 Population pyramid for Saudi Arabia, 1992. A population pyramid is a back-to-back raw frequency histogram. Source: U.S. Census Bureau, International Data Base.

called *population pyramids*, are commonly used to show the distribution of a population by age and sex. Fig. 4.6 is the 1992 population pyramid of all residents of Saudi Arabia. The reader is invited to account for what appears to be an excess of males aged 25 to 65.

4.4   DESCRIPTIVE STATISTICS

The word *statistics* can either refer to the field of statistical science, or mean numerical summaries, such as averages. Descriptive statistics fall into two broad categories: *quantiles* and *moments*. Quantiles are statistics that divide the distribution in fixed proportions; for example, the median divides the distribution into halves, quartiles divide the distribution into quarters, and the 90th percentile divides the lower 90% from the upper 10% of the distribution. The word quantile was apparently coined from "percentile." A percentile cuts off a given percentage of the observations; a quantile cuts off a particularly *quantity* of the distribution. The term *fractile* is also used and has the same meaning as quantile — a point on the horizontal axis that cuts off a particular *fraction* of the distribution. Thus the 90th percentile is the 0.90th quantile or fractile.

4.4.1 Quantiles

It is easy to compute quantiles such as medians and quartiles from a stemplot. The median is either the middle number in an odd number of observations, or the average of the two numbers on either side of the middle of an even number of observations. For example, there are 21 males in Fig. 4.2 on page 50, so the median reaction time is the 11th from the bottom (or top) of the list, that is, 0.20 seconds. There are 20 females, so their median is the average of the 10th and 11th reaction times from the bottom (or top) of the list, that is, 0.23 seconds.

There are three quartiles: the median and the first (lower) and third (upper) quartiles. We will define the first quartile as the median of the cases below the median, and the third quartile as the median of the cases above the median. For example, the first quartile of the female data is the median of the ten observations below the overall median, that is, it is the average of the 5th and 6th observations. As it happens, the first quartile of the male data is also the average of the 5th and 6th observations because there are 10 observations below the median.

Note that the observation at the median is not included in either quartile. The rule for calculating the median and quartiles is diagrammed in Fig. 4.7 for 20, 21, 22, or 23 observations only; however, the pattern repeats itself, so that, for example, the rule for 24 observations is similar to the rule for 20, and so on. In the diagram x's stand for observations, and vertical bars | stand for the average of two adjacent observations. In cases where a quartile or median falls on an observation, the letter x is replaced by the letter M or Q. In cases where a median or quartile is the average of two adjacent observations, M or Q is written above the line of x's.

For example, when there are 21 observations, the median is the 11th observation (which is therefore replaced by the letter M), the first quartile is the average of the 5th and 6th observations counting in from the left (and is therefore represented by a vertical line between those observations), and the third quartile is the average of the 5th and 6th observations counting in from the right. The

No. of cases	Rule for median and quartiles
20	**Q** **M** **Q** xxxxx | xxxxx | xxxxx | xxxxx
21	**Q** **Q** xxxxx | xxxxx**M**xxxxx | xxxxx
22	**M** xxxxx**Q**xxxxx | xxxxx**Q**xxxxx
23	xxxxx**Q**xxxxx**M**xxxxx**Q**xxxxx

Figure 4.7 Templates for counting in to the median and quartiles. Each x represents an observation; a line between x's indicates the average of the two adjacent observations.

Table 4.3   Five-number summaries of male and female reaction Times

	Min	Q1	Med	Q3	Max
Females	0.06	0.185	0.23	0.265	0.48
Males	0.11	0.150	0.20	0.225	0.29

counting-in process quite easy in a stemplot since the final sorted stemplot arranges the observations from smallest to largest.

Five-Number Summaries.   The *five-number summary* of a distribution consists of the minimum, maximum, quartiles, and median. Table 4.3 presents five-number summaries of reaction times for men and women. Not only is the median reaction time shorter for males, but the reaction times are less variable among males (ranging from 0.11 to 0.29), than among females (ranging from 0.06 to 0.48). The quartile range (QR) — the distance between the quartiles — is a measure of spread or variability. The QR of the distribution of female reaction times is 0.080 seconds (0.265 minus 0.185), almost the same as the 0.075 second QR of the distribution of male reaction times.

Outliers.   Outliers are deviant observations. They can be interesting for a variety of reasons; the presence of one or more outliers could indicate something as simple as data entry errors, or could provide useful insights about previously unnoticed phenomena. Outliers are defined as observations that deviate from the nearest quartile by at least one and a half times the quartile range (see Fig. 4.8). For example, the quartile range of female reaction times is 0.08 seconds (Q3 − Q1 = 0.265 − 0.185 = 0.08); consequently any female reaction time greater than 0.385 seconds (0.265 + 1.5 · 0.08) or less than 0.065 seconds (0.185 − 1.5 · 0.08) would be classified as an outlier. There are two outliers among the female reaction times: one unusually short, 0.06 seconds, and one unusually long, 0.48 seconds.

Extreme outliers deviate more that twice as much as ordinary outliers; in other words, extreme outliers deviate from the nearest quartile by at least three times the quartile range. For example, male reaction times greater than 0.500

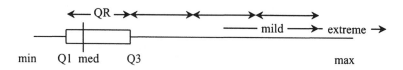

Figure 4.8   Mild and extreme positive outliers. Negative outliers are defined similarly.

(0.225 + 3.0 · 0.075) or less than -0.075 (0.150 − 3.0 · 0.075) would have been classified as extreme outliers.

Boxplots. Five-number summaries are sometimes displayed as *boxplots*, or box-and-whisker diagrams) such as Fig. 4.9. The long vertical line (the *whisker*) is the range of the non-outlying reaction times; in other words, it extends from the lowest non-outlier (0.14 for the females) to the highest (0.38). Outliers, if there are any, are plotted as open circles (ordinary outliers) or asterisks (extreme outliers). Boxplots are most useful when there are two or more groups to be contrasted. For example, side-by-side boxplots of male and female reaction times (Fig. 4.9) make two things immediately apparent: The males in this data set have a slightly shorter median reaction time, and there is more variation among the female reaction times (the whisker is longer, and there are outliers). Since we know nothing about the characteristics of the males and females in this particular sample, it is not possible to make any generalization from this set of data.

4.4.2 Moments

Moments are statistics computed from sums of observations or sums of powers of observations. The first moment, or *mean*, is the sum of the observations divided by the number of observations, and the second moment, or *standard deviation*, is computed from the sum of squares of observations. The term *moment* was probably borrowed from physics; for example, if the number line is thought of as a horizontal rod pivoted at zero, and the observations are one-gram loads, then the

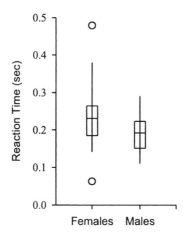

Figure 4.9 Boxplot of male and female reaction times.

mean of the observations is the proportional to the torque that they collectively exert on the rod. Of such fanciful analogies is terminology born.

Formulas for the mean (symbol: $\bar{x}$) and standard deviation (symbol: s) are

$$\bar{x} = \frac{\sum_{i=1}^{n} x_i}{n} = \frac{Sum}{n} \tag{4.1}$$

and

$$s = \sqrt{\frac{\sum_{i=1}^{n}(x_i - \bar{x})^2}{n-1}} = \sqrt{\frac{CSS}{n-1}} = \sqrt{\frac{Centered\ sum\ of\ squares}{n-1}} \tag{4.2}$$

Expressions like equations (4.1) and (4.2) are recipes for setting up a calculation in a spreadsheet. For a simple example, suppose that there were only $n = 5$ females in the reaction time data table (the women with initials AHR, DAT, DGL, BDE, and BHB). The reaction times of these women were 0.245, 0.351, 0.142, 0.174, and 0.201, respectively. The generic symbol x_1 stands for AHR's reaction time, x_2 stands for DAT's reaction time, and so on.

The algebraic expression in Equation (4.2) is a complete set of instructions for computing the standard deviation, s. Algebraic expressions such as this one are interpreted from the inside out; therefore, the instructions are: "calculate the mean, calculate the deviations by subtracting the mean from each observation, square each deviation, add up the squared deviations, divide the sum of squared deviations by $n-1$, and take the square root."

Table 4.4 shows how to set up Equations (4.1) and (4.2) in a spreadsheet format and Fig. 4.10 how to calculate the mean and standard deviation of the five

Table 4.4   Generic spreadsheet for the mean and standard deviation.

Subjects	Data	Deviations	Squared deviations
1	x_1	$(x_1 - \bar{x})$	$(x_1 - \bar{x})^2$
⋮	⋮	⋮	⋮
n	x_n	$(x_n - \bar{x})$	$(x_n - \bar{x})^2$
Totals:	$\sum_{i=1}^{n} x_i$		$\sum_{i=1}^{n}(x_i - \bar{x})^2$
Divided by n:	$\bar{x}$	Divided by n-1:	s^2
		Square root:	s

IDs	Reaction times	Deviations	Squares
AHR	0.245	0.0224	0.0005
DAT	0.351	0.1284	0.0165
DGL	0.142	-0.0806	0.0065
BDE	0.174	-0.0486	0.0024
BHB	0.201	-0.0216	0.0005
Sum:	1.113	CSS:	0.0263
$\bar{x} = \text{Sum}/n$:	0.2226	$s^2 = \text{CSS}/(n-1)$:	0.0065
		$s = \sqrt{s^2}$:	0.0811

Figure 4.10   Mean and standard deviation of the reaction times of five women. Sum and CSS are defined in Equations (4.1), and (4.2).

female reaction times. The mean and standard deviation of the reaction times of the full set of 21 female students in Table 4.1 on page 48 are $\bar{x} = 0.239$ seconds (the average female reaction time), and $s = 0.089$ (the standard deviation of the female reaction times). The standard deviation is meant to quantify the "typical" or "average" deviation of individual reaction time values from the mean — some deviations are larger than the standard deviation, some are smaller. We will later learn that for *normal* data about a third of the deviations are larger than the standard deviation.

4.5   COMPUTING DESCRIPTIVE STATISTICS

The SAS System statistical package, has two procedures for computing descriptive statistics: PROC MEANS and PROC UNIVARIATE. Instructions for running these procedures are in Appendix C. PROC UNIVARIATE produces moments, quantiles, stemplots, and boxplots and is the more versatile of the two; however, PROC MEANS is preferable if what is needed is a compact table of means and standard deviations.

4.6 EXERCISES

4.1   Table 4.5 lists consonant recognition scores for 42 patients wearing
      second- (S) or third-generation (T) cochlear implants. Third generation
      devices incorporate changes and enhancements suggested by experience
      with second generation devices.
      (a) Make back-to-back stemplots (with S on the left).
      (b) Compute the five-number summary for each device, and draw the
           boxplots. What do you conclude?

**Table 4.5   Consonant recognition scores[a]
for cochlear implant patients.**

S		T	
21	53	78	86
47	70	64	78
24	36	81	54
13	31	45	86
27	31	21	40
12	52	19	73
32	37	38	92
72	43	32	70
35	11	45	81
46	23	67	69
72	08		

a. Iowa Cochlear Implant Project, unpublished
data.

4.2   Compute the mean $\bar{x}$ and standard deviation s for these data: 74, 113,
      136, 119, 144. Do the work with a calculator, not a spreadsheet program,
      and present your work as in Table 4.4.

4.3   Repeat exercise 4.2 using Microsoft Excel® or some other spreadsheet.
      Readers unfamiliar with spreadsheets should read, "Introduction to
      Microsoft Excel," on the textbook website.

4.4   Use SAS PROC UNIVARIATE to produce moments, quantiles, and
      stemplots for the two brands of cochlear implants. See Appendix C,
      "Introduction to SAS System Software," for instructions.

4.5   Use SAS PROC MEANS to compute $\bar{x}$ and s broken down by gender for the
      reaction time data in Table 4.1 on page 48. The data file, ReacTime.txt,
      can be downloaded from the textbook website. See Appendix C for
      instructions.

4.7 REFERENCES

James, I. R. and Knuiman, M. W., "An application of Bayes Methodology to the Analysis of Diary Records from a Water use Study," *Journal of the American Statistical Association*, Vol. 82 (1987), pp. 705-711.

5

Statistical Inference

5.1 MODELS

A model is a representation of some part of reality. A model contains the essential characteristics of the thing modeled but is not the thing itself. A model is sufficiently detailed to be useful but does not contain extraneous detail. For example, panel A of Fig. 5.1 is Crick and Watson's original model of DNA made out of metal rods and plates. Its purpose was to establish the three-dimensional structure of DNA consistent with Chargaff's ratios, Rosalind Franklin's x-ray diffraction images, and the physics of chemical bonds (see, among many sources, Watson, 1980). Panel B is also a model of DNA, but in the form of a list of the sequence of bases (thymine, adenine, cytosine, guanine) in a segment of the cystic fibrosis gene. Its purposes are to predict the amino acid sequence in the protein coded by the gene and to locate mutations within the gene. Each of these models refers to the same thing, DNA, but clearly neither is the thing itself and neither incorporates all the characteristics of DNA.

A. Crick and Watson's DNA Model   B. Segment of CFTR Gene

.... GAA
AAT ATC
ATC TTT
GGT GTT
TCC ....

Figure 5.1   Two DNA models. Panel A credit: Science Museum/Science & Society Picture Library, used with permission.

63

Modeling is an essential tool of scientific reasoning; a biologist is reported to have said, "Models are what get you through the day." A model is a theory, or *working hypothesis*, about some part of nature. Making models is the difference between doing science and just fooling around. In 1897 the Geologist Thomas Chamberlain wrote,

> The method of multiple working hypotheses is the surest way to ... detect imperfections of our knowledge ... The [idea] is to bring up into view every rational explanation of new phenomena, and to develop every tenable hypothesis respecting their cause and history. The investigator thus becomes the parent of a family of hypotheses: and, by his parental relation to all, he is forbidden to fasten his affections unduly upon any one. ... Having thus neutralized the partialities of his emotional nature, he proceeds with a certain natural and enforced erectness of mental attitude to the investigation, knowing well that some of his intellectual children will die before maturity, yet feeling several of them may survive the results of final investigation.

What Chamberlain meant is that to develop an understanding of some part of reality it is useful to think of all possible models (theories, hypotheses) that could explain the data you have observed, and then, guided by the models, seek additional data to distinguish among the working hypotheses, and reevaluate the plausibility of each model in the light of the new data.

Models are essential to scientists for two reasons: first, they give a unified explanation of diverse observations, and second, they suggest new data to look for. An example is Einstein's theory (model) of general relativity which *explained*, among many other things, a long-standing, previously unexplained discrepancy in the orbit of Mercury (O'Conner and Robertson 1996) and *predicted*, among other things, that light "feels" gravity. This prediction lead to Eddington's 1919 eclipse expedition to Principe Island to attempt to observe the bending of light by the sun's gravity. It is hard to imagine why anyone would undertake such observations without Einstein's model.

Statistical inference is the science of evidence, which means that it provides the quantitative tools for evaluating the evidence for and against each of Chamberlain's working hypotheses. In the Bayesian statistical paradigm, an investigator begins with some prior degree of belief (subjective probability) attached to each of her working models. She gathers new data and updates her subjective probabilities based on that data. We'll illustrate the details of how this is done with a simple, very artificial example involving three possible models of a very small part of nature.

Example: A Cup of Marbles.   I have in my hand a cup containing six marbles. Some are white and some are black, but the exact numbers are not known. The cup got into my hands this way: someone filled three opaque plastic cups with six

M_1 M_2 M_3

Figure 5.2 Three working hypotheses (models) for a cup of marbles.

black or white marbles, as shown in Fig. 5.2 and put lids on the cups. Someone else selected one of the cups, handed it to me, and discarded the other two. Consequently, Fig. 5.2 is a complete list of all possible working hypotheses about the cup of marbles in my hand; only one of these models is true, and the other two are false. The model is not the cup itself, but it has all the necessary elements to accurately represent the characteristics of the cup that are relevant. We don't care whether the cup is made of foam or waxed cardboard, or whether the marbles are made of glass or plastic or are really poker chips. We are not interested in how much coffee the cup could hold. We are interested only in discovering the number of black and white marbles in the cup.

What makes a model *scientific* is that it makes predictions. For example, model M_1 predicts that a randomly selected marble will most likely be white, whereas model M_3 predicts that it will most likely be black. Consequently, drawing a white marble would give more support to model M_1, whereas drawing a black marble would give more support to model M_3. Models that don't make predictions cannot be verified or refuted by data.

A model is an abstraction, a statement about things not observed. In order for us to work out which model is true (or at least which is most believable), there must be some connection between the unseen world of the model (the interior of the cup) and things we can actually observe (the color of one marble selected from the cup). The quantitative link between model and data is called the *likelihood function*. Likelihood is defined as the probability that a given model could produce the observed data. Typically, an observation, say drawing a white marble out of the cup, will have different likelihood under each working hypothesis, which is how we are able to distinguish among the models.

Suppose for example the marble I randomly selected from the cup has turned out to be white. The likelihood of observing a white marble depends on which model is correct, and although we don't know which model is correct, we can compute the likelihood of observing a white marble for each of the models. The likelihood calculation is quite simple in this case. For example model M_1 is a cup containing 5 white and 1 black marble, so if this model is correct the probability (likelihood) of randomly drawing a white marble is 5/6. Similarly, the likelihood of a white marble is 3/6 under model M_2 and is only 2/6 under M_3. Observing a white marble is more consistent with model M_1, but it does not absolutely rule out either of the other two models.

Prior to observing any marbles from the cup, I had no basis to favor one model of the cup over another. In my state of ignorance, I believed equally in each of the models; in other words, before I had observed any data my probability of each model was $1/3$. These probabilities represented my knowledge about the contents of the cup prior to my observing any data and are therefore called *prior probabilities*. The question is how to rationally revise prior probabilities after obtaining relevant data. The revised probabilities are called *posterior probabilities*. Since model M_1 is most consistent with the data, the posterior probability of M_1 must be bigger than its prior probability but how much bigger? Bayes' rule gives the answer.

5.2   UPDATING BELIEFS

5.2.1   Bayes' Rule

Although Bayes' rule is the fundamental law of statistical reasoning, it is mathematically nothing but a rearrangement of the product rule; hence, it is arguable that the product rule is the fundamental principle of statistical reasoning.

Bayes' rule is used to compute how the data change our opinions about the truth or falsity of each model. Equation (5.1) is the formal, algebraic statement of Bayes' rule, although it is not the most practical way to arrange the calculations. The symbols M_1, M_2, M_3,... M_k stand for the multiple working hypotheses (models) that we think might be true. Each hypothesis, or model, can be thought of as a (possibly very long) sentence, such as "The cup contains 5 white and 1 black marble." Only one of these sentences *can* be true and we think one of them (although we don't know which) *must* be true ("bring up into view *every* rational explanation of new phenomena," wrote Chamberlain).

$$P\left(M_j\right) = \frac{P\left(M_j \cap D\right)}{P(D)} = \frac{P\left(D|M_j\right) \cdot P\left(M_j\right)}{\underset{\text{all models}}{\sum} P\left(D|M_i\right) \cdot P\left(M_i\right)} \tag{5.1}$$

The symbol D represents a sentence describing what we observed, that is, D represents the *data*. For example, "I selected a marble at random and it was white." We can observe data but we cannot directly observe which model is true; however, Bayes' rule permits us to compute how the data (D) rationally changes our beliefs about which of the models is true. First we will sketch a proof of the rule and then learn how to use it in artificial and practical examples.

The first equality in Equation (5.1) is a rearrangement of the product rule, Equation (3.11). The numerator of the third term *is* the product rule; however, the denominator of the third term requires explanation. The idea, illustrated in

Fig. 5.3, is to slice region D into parts that can be computed using the product rule. For example, $P(D \cap M_1) = P(D \mid M_1) \cdot P(M_1)$, which is computable because $P(M_1)$ is the fair price of a bet on M_1 prior to observing data D and $P(D \mid M_1)$ is the likelihood of data D assuming M_1 to be true.

It bears repeating that Equation (5.1) is not the most convenient way to carry out Bayes' rule calculations; it is much easier to set up the calculations in spreadsheet form. We'll illustrate that with the marbles in a cup example and then proceed to serious applications.

Example: Which Cup? I am holding a covered cup containing six marbles, some black, some white. My goal is to figure out what's in the cup. There are three working hypotheses about the composition of the cup (see Fig. 5.2); however, I don't know which is correct, and I'm not allowed to look in the cup. In the real world, our ability to gather data is limited by time and funding. In the tiny world of a cup of marbles, we'll imitate this sort of restriction by allowing me to randomly select only two marbles from the cup. The symbol D will stand for the sentence that describes whatever data I observe. From this point on we'll assume that I observed two white marbles. So $D =$ "I drew two white marbles." The data suggest that the marbles in the cup are predominantly white and would support (but not prove) the statement, "The cup in my hand is M_1." However, we must use Bayes' rule to compute the exact amount of support that the data D gives to model M_1 and to the other two working hypotheses.

The inputs to Bayes' rule are prior probabilities and likelihoods. The investigator's *prior probabilities* of the working hypotheses, $P(M_1)$, $P(M_2)$, and $P(M_3)$, are his or her beliefs, expressed as fair prices based on current information but prior to seeing any new data. Likelihoods are conditional probabilities of the data given each particular model. For example $P(D \mid M_1)$ is the probability of drawing two white marbles from a cup containing four white and two black marbles and $P(D \mid M_2)$ and $P(D \mid M_3)$ are the corresponding probabilities for the other two models. The prior probabilities are subjective, but the likelihoods are in most cases "objective" in the way casino probabilities are objective: people agree

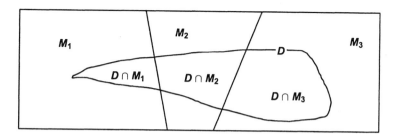

Figure 5.3 Denominator of Bayes' rule illustrated for three working hypotheses. The probability of the data, $P(D)$, is the sum of $P(D \cap M_1)$, $P(D \cap M_2)$, and $P(D \cap M_3)$.

$$P(WW|M_1) = \frac{5}{6}\cdot\frac{4}{5} = 0.6667$$

$$P(WW|M_2) = \frac{3}{6}\cdot\frac{2}{5} = 0.2000$$

$$P(WW|M_2) = \frac{2}{6}\cdot\frac{1}{5} = 0.0667$$

Figure 5.4   Likelihoods of observing two white marbles. For a given set of data, likelihood is a *function* of the model.

on them. For example, most people would agree that the probability of drawing a white marble from model M_1 of Fig. 5.2 on page 65 is 5/6 or about 83%.

There is often less consensus on prior probabilities, presumably because each person comes to the situation with different prior experience and knowledge. In this example, I'm going to assume that the investigator started with a completely open mind about the working hypotheses and in the absence of any data thinks that they are equally probable. In other words, the prior subjective probabilities are $P(M_1) = 1/3$, $P(M_2) = 1/3$, and $P(M_3) = 1/3$.

Likelihoods are based on the assumption that any given marble is as likely to be drawn as any other. So for example, the probability of drawing two white marbles from model M_1 is $(5/6) \times (4/5)$ (there are 5 chances out of 6 that the first marble is white, which leaves 4 whites in the cup, so there are 4 chances in 5 that the second marble is white). Likelihoods for this and the other two working hypotheses are listed in Fig. 5.4. It is important to understand that if the data had been different — for example, a white followed by a black marble — then the likelihoods would change (see Fig. 5.5). This is entirely reasonable, since different evidence supports different conclusions.

$$P(WB|M_1) = \frac{5}{6}\cdot\frac{1}{5} = 0.1667$$

$$P(WB|M_2) = \frac{3}{6}\cdot\frac{3}{5} = 0.3000$$

$$P(WB|M_2) = \frac{2}{6}\cdot\frac{4}{5} = 0.2667$$

Figure 5.5   Likelihoods of observing one white and one black marble. The likelihood of each model is different from the corresponding likelihood in Fig. 5.4 because the data are different.

Computing Posterior Probabilities with Bayes' Rule. Posterior probabilities are the investigator's revised or updated beliefs after taking new data into account. Prior probabilities, which represent degrees of belief in the various working hypotheses prior to obtaining new data, are combined with likelihoods to produce revised (posterior) probabilities that represent degree of belief in each working hypothesis after taking the new data into account. We have all the necessary numbers to do the calculation — prior probabilities and likelihoods — so the only thing required is to apply Bayes' rule.

Fig. 5.6 shows the Bayes' rule calculations done the hard way by plugging numbers into Equation (5.1); this computing layout is inefficient and practically impossible to scale up when there are more than a few working hypotheses. It is much more efficient to arrange the calculations in a spreadsheet similar to Fig. 5.7. Spreadsheet calculations are comparatively easy to scale up to hundreds of working hypotheses simply by appending more rows. Prior probabilities and likelihoods are entered in the second and third columns of the spreadsheet. The

$$P\left(M_1 \mid WW\right) = \frac{0.6667 \times 0.3333}{0.6667 \times 0.3333 + 0.2000 \times 0.3333 + 0.0667 \times 0.3333} = 0.7142$$

$$P\left(M_2 \mid WW\right) = \frac{0.2000 \cdot 0.3333}{0.6667 \times 0.3333 + 0.2000 \times 0.3333 + 0.0667 \times 0.3333} = 0.2144$$

$$P\left(M_3 \mid WW\right) = \frac{0.0667 \cdot 0.3333}{0.6667 \times 0.3333 + 0.2000 \times 0.3333 + 0.0667 \times 0.3333} = 0.0714$$

Figure 5.6 Bayes' rule the hard way. This layout fails to exploit repetitive elements such as 0.6667×0.3333. The spreadsheet layout in Fig. 5.7 is more efficient.

Bayes' rule spreadsheet in symbols

Model	Prior	×	Likelihood	=	Joint	Posterior
M	$P(M)$	×	$P(D \mid M)$	=	$P(D \cap M)$	$P(M \mid D)$
M_1	$P(M_1)$	×	$P(D \mid M_1)$	=	$P(D \cap M_1)$	$P(D \cap M_1)/P(D)$
M_2	$P(M_2)$	×	$P(D \mid M_2)$	=	$P(D \cap M_2)$	$P(D \cap M_2)/P(D)$
M_3	$P(M_3)$	×	$P(D \mid M_3)$	=	$P(D \cap M_3)$	$P(D \cap M_3)/P(D)$
			$P(D)$	=	$\Sigma P(D \cap M)$	

Figure 5.7 An efficient spreadsheet layout for Bayes' rule calculations. The "Joint" column lists the joint probabilities of data and each model: the product of the prior and the likelihood. $P(D)$, the marginal probability of the data, is the sum of the joint probabilities. Posterior probabilities are joint probabilities divided by their total.

Model	Prior	Likelihood	Joint (product)	Posterior (joint/total)
M	$P(M)$	$P(D \mid M)$	$P(D \cap M)$	$P(M \mid D)$
M_1	0.3333	0.6667	0.2222	0.7142
M_2	0.3333	0.2000	0.0667	0.2144
M_3	0.3333	0.0667	0.0222	0.0714
		$P(D)$	0.3111	

Figure 5.8   Bayes rule spreadsheet for the three-cup example with equal prior probabilities after observing two white marbles.

calculations are: First compute the joint probabilities (prior × likelihood), next compute the probability of the data (total the joint probabilities), and finally, compute the posterior probabilities (divide each joint probability by their total). Fig. 5.8 is the spreadsheet version of the Bayes' rule calculation shown in Fig. 5.6; the spreadsheet computes exactly the same posterior probabilities, but in a more efficient format.

The calculations, done the hard way or the spreadsheet way, confirm that observing two white marbles does indeed support model M_1 and makes model M_3 rather unlikely. The data changed my beliefs: Prior to observing the data I thought the three models were equally probable but after observing two white marbles, I think that there is a 71.4% chance that model M_1 is correct and only a 7.1% chance that model M_3 is correct. However, model M_2 is still a contender with 21.44% posterior probability of being correct. These posterior probabilities of course depend on the data; if different data had been observed, then the posterior probabilities would have been different, but once the data are observed the calculation must obviously be based on those data and not some imaginary data that might have been observed.

On the other hand, it *is* good practice to reanalyze the data using a variety of prior distributions to investigate the sensitivity of the conclusions to prior differences of opinion. For example, if Jill thinks the half-black-half-white model M_2 has prior probability 0.5 and the other models have prior probabilities 0.25 each, then her posterior probabilities based on the same data that I observed will be somewhat different: $P(M_1 \mid WW) = 0.59$, $P(M_2 \mid WW) = 0.35$, and $P(M_3 \mid WW) = 0.06$. Her prior belief that M_2 was more probable than M_1 has been reversed by the data, and she and I now agree that M_1 is the most probable of the three working hypotheses, although she is less certain about this (59%) than I (71%) because she was initially more skeptical about M_1. Calculation of Jill's posterior probabilities is left as an exercise for the reader.

5.2.2 Bayes' Rule in Diagnostic Testing

Diagnostic testing was first introduced in Chapter 1. The flowchart method of calculating what we now recognize to be posterior probabilities (Fig. 1.1 on page 2) is Bayes' rule in disguise. We will rework the calculation here as a Bayes' rule spreadsheet.

In the terminology of this chapter, Nate's working models are, M_1: Nate has HBV and M_2: Nate does not have HBV. His prior subjective probabilities are $P(M_1) = 3/100,000 = 0.00003$ and $P(M_2) = 0.99997$; however, the likelihoods require some thought.

Recall that a likelihood is the probability of the observed data assuming a particular model to be correct. There are only two possible kinds of data that Nate could observe: a positive test result or a negative test result. Let's first assume that Nate tested positive, so that the sentence (D) describing the data is "Nate tested positive." In this case, we need to compute the likelihood of a positive test result under each of the two models and then use Bayes' rule to compute Nate's posterior probabilities.

If model M_1 is true (Nate does have HBV, but he doesn't know it) then there is a 99% probability that his test will come back positive. We know this because the sensitivity of the test is 99%, which means that 99% of HBV-infected individuals will test positive. Consequently the likelihood is $P(D \mid M_1) = 0.99$ (the true positive rate). On the other hand, if M_2 is true (Nate does not have HBV), then there is 0.995 probability that he will test negative and therefore only 0.005 that he will test positive. Consequently the likelihood of a positive test under model M_2 is $P(D \mid M_2) = 0.005$ (the false positive rate).

Equation (5.2) shows how to compute the likelihood of a positive test result, and Equation (5.3) below shows how to compute the likelihood of a negative test result:

If D = "The patient tested positive," the likelihood is

$$P(D|M_1) = P(+ \mid \text{Infected}) = \text{sensitivity} \qquad (5.2)$$
$$P(D|M_2) = P(+ \mid \text{Not infected}) = 1 - \text{specificity}$$

If D = "The patient tested negative," the likelihood is

$$P(D|M_1) = P(- \mid \text{Infected}) = 1 - \text{sensitivity} \qquad (5.3)$$
$$P(D|M_2) = P(- \mid \text{Not infected}) = \text{specificity}$$

Fig. 5.9 shows how to calculate Ned's posterior probabilities under two scenarios. In panel A it is assumed that Nate tested positive, so the likelihoods are calculated using Equation (5.2). In this case Nate's posterior probability that he is

A. Nate's posterior probabilities assuming he tested positive.

Model	Prior	Likelihood	Joint	Posterior
M	$P(M)$	$P(D \mid M)$	$P(D \cap M)$	$P(M \mid D)$
M_1	0.00003	0.990	0.00002970	0.0059
M_2	0.99997	0.005	0.00499985	0.9941
		$P(D)$	0.00502955	

B. Nate's posterior probabilities assuming he tested negative.

Model	Prior	Likelihood	Joint	Posterior
M	$P(M)$	$P(D \mid M)$	$P(D \cap M)$	$P(M \mid D)$
M_1	.00003	.010	0.00000030	0.0000003
M_2	.99997	.995	0.99497015	0.9999997
		$P(D)$	0.99497045	

Figure 5.9   Nate's Bayes' rule calculations under two data scenarios. If he tests positive (A), there are about six chances in a thousand he is infected (0.0059). If he tests negative (B) there are about 3 chances in ten million he is infected (0.000,000,3).

infected, given that he tested positive, is 0.0059, or about 6 in 1,000, much higher than his prior probability.

In panel B we assume that Nate tested negative. The prior probabilities are the same because they represent Nate's beliefs before seeing the test result. However, the data (a negative test) are different and therefore the likelihoods are different: the likelihood of a negative test under model M_1 is the probability that an infected person tests negative (a false negative), which is $1 -$ sensitivity $=$ 0.01. The likelihood of the data under model M_2 is the probability of a negative test from an uninflected person (a true negative), which is the specificity, 0.995. When these likelihood values are inserted in the Bayes' rule spreadsheet in panel B of Fig. 5.9, we see that Nate's posterior probability that he is infected drops to about three chances in ten million. It is important to understand that prior probabilities do not, and logically cannot, depend on the data, because prior probabilities represent Nate's beliefs before seeing the data. The impact of the data on Nate's beliefs is entirely contained in the likelihoods.

5.2.3   Bayes' Rule in DNA Evidence

DNA fingerprinting is used to identify the source of a trace of human tissue found in association with a crime scene or found in association with the alleged perpetrator. In 1996 the celebrity athlete O. J. Simpson (OJS) was tried for the alleged murder of his ex-wife Nicole Brown Simpson (NBS) and her friend Ronald Goldman. The prosecution case involved two kinds of trace evidence: item 52 and the bloody sock. Item 52 was a drop of blood found at the scene of the crime not belonging to either victim. The bloody sock was found in the defendant's bedroom shortly after the crime. The DNA fingerprint of the crime-scene blood, item 52, matched a DNA sample from OJS, and the probability of a random match was reported to be 1 in 170 million. The DNA fingerprint of blood on the sock matched that of the female victim (NBS), and the probability of a random match was reported to be 1 in 6.8 billion (People vs. Simpson, May 11, 1995). To help benchmark these probabilities, the population of greater Los Angeles is about 16 million, the population of the United States is about 300 million, and the population of the world is about 5 billion.

We will analyze the crime-scene blood here, and the bloody sock analysis will be left as an exercise. The prosecution seems to have had in mind two models of the origin of the crime-scene blood: model M_1: "The blood came from OJS," and model M_2: "The blood came from someone else." However, the defense appears to have suggested a third model, M_3: "The evidence was planted." Ordinarily this model is not proposed by defense lawyers because the prosecution is able to provide a well-documented chain of possession of the physical evidence that precludes tampering. We will initially use the conventional, two-model analysis implicit in the prosecution case, and then see what happens if a juror finds the third model credible (evidence tampering).

Likelihoods in DNA Testimony.   The data, with respect to the crime scene blood, is D ="The DNA in the crime scene blood matched a DNA sample taken from OJS." The Bayes' rule spreadsheet requires us to enter the likelihood of observing D given model M_1 and the likelihood given model M_2. A complete statement of model M_1 is "Two DNA samples, one from the crime scene, the other from OJS, both came from OJS." Obviously, if M_1 is true then there must be a perfect DNA match, consequently the likelihood of D given model M_1 (assuming no laboratory errors) is 1.0. A full statement of Model M_2 is "The crime-scene DNA came from a random person who, by chance, has the same DNA fingerprint as OJS." Clearly, the likelihood of a DNA match under model M_2 is the probability of a random match, 1 in 170 million, or 0.588×10^{-8}. These numbers have entered in the likelihood column of the Bayes' rule spreadsheet in Fig. 5.10; however, it is not clear how to fill in the prior probabilities.

Prior Probabilities.   Prior probabilities are subjective, and the question is whose priors should be used in the calculation. The purpose of Bayes' rule is to rationally update prior belief. Clearly, the only relevant beliefs in a criminal trial are the beliefs of the members of the jury. After the trial is underway, it is not possible to interrogate jury members to discover their prior beliefs; therefore, the prosecution might propose a "worst case" analysis using a prior probability they think is smaller than that entertained by the most skeptical juror. For example, the prosecutor might argue, "Let's assume that before hearing the DNA evidence you thought that the blood on the sidewalk could have come from anyone in Los Angeles and that you have no other evidence to link it to this defendant." Since OJS lived in Los Angeles, it could have come from him, but he is only one among 16 million, so with no other evidence, the prior probability that the blood came from him was 1 in 16 million, or 6.25×10^{-8}. The resulting posterior probability that the crime scene bloodstain came from OJS is about 91% (Fig. 5.11). This is certainly not evidence beyond a reasonable doubt. The prosecution would have to reinforce the DNA evidence with other forms of evidence.

The prosecution's DNA expert witness, Dr. Robin Cotton, also gave evidence about the bloody sock found in the defendant's bedroom and in this testimony there appears to be an implicit worst case analysis. The evidence was that DNA on a bloody sock found in the defendant's bedroom after the crime was an exact match to the DNA of the female victim (NBS). The probability of a random match was reported as one in 6.8 billion western hispanics, one in 9.8 billion Caucasians, or one in 530 billion African Americans (People vs. Simpson, May 11, 1995). One of the prosecuting attorneys, Deputy District Attorney George Clarke, apparently invited the jury to use the most skeptical prior imaginable — to suppose that the blood was equally likely to have come from anyone on Earth.

Model	Prior	Likelihood	Joint	Posterior
M	$P(M)$	$P(D \mid M)$	$P(D \cap M)$	$P(M \mid D)$
M_1: OJS	$P(M_1)$	1		
M_2: other	$1-P(M_1)$	0.5882×10^{-8}		
		$P(D)$		

Figure 5.10 Bayes' rule spreadsheet for crime-scene bloodstain. Prior probabilities are required in order to complete the calculation.

Model	Prior	Likelihood	Joint	Posterior
M	$P(M)$	$P(D \mid M)$	$P(D \cap M)$	$P(M \mid D)$
M_1: OJS	6.25×10^{-8}	1	6.25×10^{-8}	0.91
M_2: other	≈ 1	0.5882×10^{-8}	$\approx 0.5882 \times 10^{-8}$	0.09
		$P(D)$	$\approx 6.84 \times 10^{-8}$	

Figure 5.11 Bayes' rule spreadsheet for crime scene bloodstain. Prior probabilities are based on the assumption that the blood is equally likely to have come from anyone in greater Los Angeles. Approximations are good to the seventh decimal place.

Mr. Clarke: All right. Dr. Cotton, you have written in under frequencies one in 6.8 billion, one in 530 billion; is that right?

Dr. Cotton: Yes.

Mr. Clarke: How many people are on earth?

Dr. Cotton: Well, I don't personally know, but the figure I've been quoted is about five billion, I think.

Perhaps Mr. Clarke intended to suggest that, even if a juror's prior opinion was that the blood on the sock could have come from any human on Earth, the data would overwhelm even that degree of skepticism. It seems that Mr. Clarke is inviting the jurors to run a Bayes' rule calculation with prior probability one in 5 billion for the hypothesis that the blood came from NBS. It seems plausible enough, one person in 6.8 billion has the DNA fingerprint, and there are only 5 billion people in the world, so the fingerprint must be unique. We know that NBS had the fingerprint, so the blood must be hers. In fact the intuition, plausible as it seems, is completely wrong. The flaw in that informal reasoning comes from confusing the *expected* number of people with the fingerprint (less than one) with the actual number. It is left as an exercise for the reader to find the flaw in that line of reasoning by computing the posterior probability that the blood on the sock came from NBS assuming prior probability of one in 5 billion.

The defense could have suggested a third model to account for the DNA matches, M_3: "The evidence was planted." Fig. 5.12 shows a Bayes' rule analysis with the third model included. The likelihood of the data under model M_3 is the also 1.00 because the theory of M_3 is that someone obtained a sample of the defendant's blood and either physically planted a drop at the crime scene or substituted the defendant's blood for the purported crime scene sample, either of which guarantees a DNA match. For this hypothetical defense strategy to work it would be necessary to persuade evan a single juror that there was a non-zero prior probability of model M_3. In Fig. 5.12 we assume this hypothetical juror assigned a one in a million prior probability for model M_3. The result is that the

Model	Prior	Likelihood	Joint	Posterior
M	$P(M)$	$P(D \mid M)$	$P(D \cap M)$	$P(M \mid D)$
M_1: OJS	6.25×10^{-8}	1	6.3×10^{-8}	0.06
M_2: other	≈ 1	0.5882×10^{-8}	$\approx 0.6 \times 10^{-8}$	0.00
M_3: planted	$\approx 10^{-6}$	1	$\approx 100 \times 10^{-8}$	0.94
		$P(D)$	$\approx 106.9 \times 10^{-8}$	

Figure 5.12 Alternate Bayes' rule spreadsheet for crime scene bloodstain. The defense is assumed to have created in the juror's mind a one in a million prior probability that the evidence was planted.

evidence tampering hypothesis, M_3, would have the highest posterior probability of the three models.

This example illustrates Chamberlain's dictum: bring up into view every rational explanation of new phenomena. In this case introducing a third explanation of the DNA match had a striking effect on the weight of the evidence. This is sometimes taken to mean that any alternate theory however far-fetched weakens the evidence in favor of other theories. However, by his use of the word "rational", Chamberlain rejects that interpretation. In terms of Bayes' rule his point is this: a preposterous or irrational explanation, or theory, will have vanishingly small prior probability and will therefore have vanishingly small posterior probability. On the other hand the history of science makes it clear that what appears to us at this moment to be a list of every rational explanation of a new phenomenon might not include the true explanation. A case in point is the etiology of peptic ulcers: it was long thought that the major causes were spicy food, acid, and stress; however, it is now known that the bacterium *H. pylori*, discovered in 1982, is the cause of most ulcers. In 1981 a list of all working hypotheses regarding the cause of peptic ulcers would certainly not have included *H. pylori*. A working hypothesis nobody has thought of has zero prior probability and therefore zero posterior probability.

5.3   BAYES' RULE IN THE ANALYSIS OF RATES

5.3.1   The Importance of Studying Rates

Epidemiology is the statistical study of morbidity and mortality (sickness and death). Epidemiological studies often involve making comparisons of the rates of a particular disease or cause of death in different groups of people. For example, pack-a-day smokers reportedly have a lifetime lung cancer rate about 9 times higher than the rate for non-smokers.

Rates are expressed as probability per unit of time. For example, everyone dies, but not at the same age, so although the lifetime probability of dying is 100% the one-year death rate for 20-year-olds in the United States is low, about 0.0013, or 13 per 10,000 per year. The lifetime risk of breast cancer for women in the United States is reported to be 6% to 10%; however, the one-year breast cancer rate for women in their forties is reported to be about 4 per 10,000 per year or 0.04% per year.

Medical discoveries, as well as policy decisions such as banning smoking in restaurants, are often based on comparisons of rates. For example, the EPA report on second-hand smoke (Environmental Protection Agency 1992) reported that nonsmoking spouses of cigarette smokers had a 19% higher lung cancer rate than nonsmoking spouses of nonsmokers. This has been interpreted as evidence that long-term exposure to second-hand smoke in restaurants and bars probably also raises the risk of lung cancer.

Many experimental studies, such as the polio vaccine field trial of 1954 involve the comparison of two rates. In this study, 200,745 children received active vaccine and 201,229 children received a placebo injection. During the follow-up period (July–December of 1955), 57 of the vaccinated children developed polio (about 28 per 100,000), as did 142 of the placebo children (about 71 per 100,000). Vaccination appeared to cut the polio rate in half; however, these figures are only estimates of the rates that would have been observed in the whole population. The research question was whether these estimated rates were sufficiently reliable to support the conclusion that the vaccine would be more effective than the placebo in the wider population of all U.S. children. We will learn how to answer this sort of question using Bayes' rule.

We begin by learning how to compute posterior probabilities and precision (margin of error, credible interval) of a single rate using Bayes' rule. Later, we will learn how to compare rates in two groups (vaccinated vs. placebo or smokers vs. non-smokers, for example) using Bayes' rule.

5.3.2 Box Models

We're going to start with a real research project, estimating the rate of Heads in coin tossing. The procedure, as in the "three cups of marbles" example, will be to make a list of working hypotheses (models) and collect data in order to try to discover the correct model.

What does it mean to model a coin toss? An individual coin toss is governed by Newton's laws of motion, thus an engineer could develop a deterministic model such as Fig. 5.13 or equations (1), (2), and (3) in Cusumano and Hecht (1995). Given very precise information about the initial impulse that sent the coin spinning in the air this mathematical model is capable of predicting, while the coin is in the air, exactly how it will land. However, to use the model, we would need to be able to measure the vertical velocity, height, and spin rate while the coin was in flight. Since most of us don't have the equipment to make such measurements, a coin toss appears to us to be unpredictable. However, the unpredictability is in us, not in the coin; it is the result of our lack of information about the initial conditions of the toss (height, vertical velocity, spin rate) and is not intrinsic to the coin-tossing process, which is deterministic and in theory perfectly predictable.

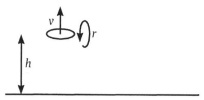

Figure 5.13   Deterministic model of a coin toss. The final state of the coin (Heads or Tails) is determined by the initial height, vertical velocity, and spin rate of the coin.

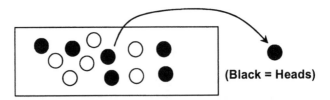

Figure 5.14   Box model of a coin toss. The box contains an equal number of black marbles
(representing heads) and white marbles (representing tails). The box is shaken, a marble is
selected blindly, its color is recorded, and the marble is returned to the box.

In our state of ignorance of the initial conditions of the toss, we perceive a
coin toss as a 50 : 50 gamble. So our model of a coin toss is a box of marbles like
Fig. 5.14. The box contains an equal number of black marbles and white marbles.
The box model of a coin toss is a three-step *process*: first, shake the box really well;
second, close your eyes and select one marble out of the box; third, write down its
color (black = Head, white = Tail); and finally, put the marble back in the box.
This model incorporates all the features that we associate with a coin toss: there is
an equal chance of getting heads or tails, one toss is independent of the next
(because we shake the box and replace the marble), and we can make as many
tosses as we please.

5.3.3   A Toy Research Project: Spinning a Penny

Instead of tossing a penny, we will hold it vertical on a hard surface and flick it so
that its spins like a top. A perfectly smooth, symmetrical disk would be equally
likely to fall with either side up; however, if the disk is even slightly bowl-shaped, it
will somewhat favor the concave side. Consequently, we have no intuition about
how often the coin will land heads, it could be more than 50% of the time, or it
could be less. The *rate* of heads is an unknown quantity.

We can gather data by spinning a real penny. However, in order to process the
data, we need to list all possible working hypotheses about the rate at which the
coin will land heads. There is actually an infinite number of possible models,
because the heads rate could be any real number between zero (never lands
heads) and 1 (always lands heads). The heads rate could be 0.25, it could be
0.3333..., it could be 0.721; we simply do not know what it is, and must gather
real world data to learn about the rate.

To make the problem manageable, we will model the heads rate to three
decimal places, which means that our working hypotheses will be boxes
containing 1000 marbles. For example, the model for "The heads rate is 0.25" is
a box containing 250 black marbles and 750 white marbles, and the model for
"The heads rate is 0.333" is a box containing 333 black marbles and 667 white
marbles. We'll call these models M.250, M.333, and so on.

Since we don't know which model best describes penny spinning, we must consider *multiple working hypotheses* (i.e., all plausible models) and gather data to help us identify which of these is the correct model. We can eliminate two models immediately: model M.000 can't be right, because it predicts that the spun coin will never land Heads. Likewise M1.000 can't be correct because it predicts that the coin will always land Heads. That leaves 999 other working models, any one of which *could* be correct.

Prior Probabilities. The are 999 possible working models, ranging from M.001 (heads rate 0.001), to M.999 (heads rate 0.999). The prior subjective probability of each model is defined as the fair price for a bet that the model is correct. Nobody can tell you what *your* prior probabilities *should* be, although to avoid Dutch book your probabilities must add to 1.

My prior probabilities are based on the facts that I have no idea which side of the penny is more concave, nor do I have any idea how much it is biased in favor of the concave side. Keeping an open mind, not favoring any model before I have data, I'll assign each of the 999 working hypotheses a prior probability of 1/999. This appears to be about as unopinionated as a prior can be and is sometimes called a *flat prior.* In general, there is no such thing as correct prior probabilities — it is completely rational for you to entertain prior probabilities different from mine. As long as neither of us puts zero prior probability on a model that the other person puts positive prior probability on, the data will eventually bring us into agreement.

Data and Likelihood. Gathering data for our research project consists in observing the results of many independent spins of the penny ("independent" means that one spin does not physically influence another). For example, I just did $n=10$ spins and observed TTTTHTTTHH (seven tails and three heads).

The likelihood of these data depends on the model. A likelihood is the conditional probability of the observed data assuming that a specific model is true. Each likelihood is a "what if" calculation: if this model is true, what is the probability of the data? The *likelihood function* is a table listing the likelihood of each of the 999 working hypotheses; thus, we do not assume any particular model is true; instead we open-mindedly make a list of the likelihood of each and every model. For example, for model M.200 the likelihood of this particular set of data is

$$
\begin{aligned}
\text{likelihood} \;&=\; P\big(\text{TTTTHTTTHH}|\text{M.200}\big) \\
&=\; 0.8 \cdot 0.8 \cdot 0.8 \cdot 0.8 \cdot 0.2 \cdot 0.8 \cdot 0.8 \cdot 0.8 \cdot 0.2 \cdot 0.2 \\
&=\; 0.001678
\end{aligned}
$$

The general formula for the likelihood is

$$\text{likelihood} \quad = \quad P\big(x \text{ heads in } n \text{ spins} \mid \text{heads rate} = p\big)$$

$$= \quad p^x \cdot (1-p)^{(n-x)}$$

(5.4)

In this expression the symbol p stands for the heads rate, that is, the proportion of black marbles in the box. For example, if there are 333 black marbles in the box, then p = 0.333. Symbol n stands for the number of spins and symbol x for the observed number of heads among those spins. The use of symbols p, n, and x for these quantities is traditional and should not be over interpreted: there is no connection between lowercase x meaning the number of heads and an uppercase X meaning a generic unknown quantity: in fact x is a *known* quantity — the observed number of heads. Likewise, lowercase p, meaning a rate, should not be confused with uppercase P standing for subjective probability in general.

For example, the data TTTTHTTTHH is the result of $n = 10$ spins and contains $x = 3$ heads (black marbles). The likelihood function of the data is $p^3 \cdot (1-p)^7$. So, for model M.389, for example, the likelihood would be $0.389^3 \times 0.611^7 = 0.0018713$. Of course, to complete the Bayes' rule spreadsheet the likelihood has to be computed for each of the 999 working hypotheses.

We now have all the necessary inputs for the Bayes' rule spreadsheet: the prior probability of each model, and a way to calculate the likelihood of the data given each model. Computing the posterior probabilities is simply a matter of "plug and chug" — plug the priors and likelihoods into Bayes' rule and chug out the answer. With 999 working hypotheses, this is not feasible by hand and we must resort to computers. We will later learn how to use a general purpose statistical computer package called WinBUGS to compute posterior probabilities; however, the spinning penny calculation is fairly easy to set up in a computer spreadsheet similar to Fig. 5.15. (Readers not familiar with computer spreadsheets should read, "Introduction to Microsoft Excel" on the textbook website.) The prior probabilities are inserted in column C, the likelihood formula is inserted in column D, and the posterior probabilities are automatically computed by the formulas in column E, cell E1001, and column F.

Fig. 5.16 is the numerical view of the same spreadsheet. For example, the posterior probability of model M.300 is 0.002935. Before observing any data I thought that the probability of this model was about 0.001, so the data have almost tripled my belief in this particular model. On the other hand the data left me virtually certain that model M.001 is incorrect (posterior probability 3.6×10^{-7}).

	A	B	C	D	E	F
1	Model	p	Prior	Likelihood	Joint	Posterior
2	M.001	.001	=1/999	=B2^3*(1-B2)^7	=C2*D2	=E2/E1001
3	M.002	.002	=1/999	=B3^3*(1-B3)^7	=C3*D3	=E3/E1001
...	...	...	...	...	...	...
501	M.500	.500	=1/999	=B501^3*(1-B501)^7	=C501*D501	=E501/E1001
...	...	...	...	...	...	...
1000	M.999	.999	=1/999	=B1000^3*(1-B1000)^7	=C1000*D1000	=E1000/E1001
1001				P(D)	=SUM(E2:E1000)	

Figure 5.15   Formula view of Bayes' rule spreadsheet for the spinning penny investigation. Data are three heads and seven tails. The spreadsheet computes the likelihoods using the formulas in column D and computes the posterior distributions using formulas in cell E1001 and column F.

5.3.4   Posterior Distributions

My posterior probability is *distributed* over the 999 working hypotheses, and although the Bayes' rule spreadsheet (Fig. 5.16) lists all 999 posterior probabilities in column F, it is difficult to get a general impression of which models have been favored by the data and which have become less probable. A picture is often the best way to display large amounts of information, and that is the case here. The distribution of my posterior probability, or degree of belief, for each of the 999 models, can be shown as a line graph (Fig. 5.17).

The 999 models are represented on the horizontal axis (for example, model M.333 is represented by the point 0.333 on the horizontal axis, model M.500 is represented by 0.500, and so on). The posterior probability of each model is

	A	B	C	D	E	F
1	Model	p	Prior	Likelihood	Joint	Posterior
2	M001	0.001	0.001001	9.93021E-10	9.94015E-13	1.31079E-09
...			... models M002 through M298 ...			
300	M299	0.299	0.001001	0.00222351	2.22574E-06	0.00293504
301	M300	0.300	0.001001	0.00222357	2.22579E-06	0.00293511
302	M301	0.301	0.001001	0.00222351	2.22574E-06	0.00293504
...			... models M302 through M998 ...			
1000	M999	0.999	0.001001	9.97003E-22	9.98001E-25	1.31604E-21
1001					7.58334E-04	

Figure 5.16   Numeric view of Bayes' rule spreadsheet for the spinning penny investigation. Data are three heads and seven tails.

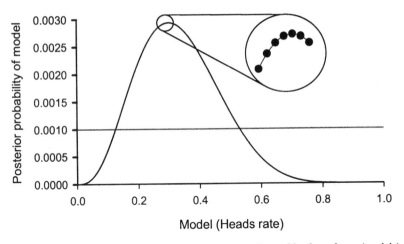

Figure 5.17   Distribution of posterior probability over the working hypotheses (models) after observing three heads and seven tails. The horizontal gray line shows the prior probabilities. The graph consists of 999 points representing posterior probabilities of the working hypotheses; a few of these are shown in the inset.

plotted on the vertical axis. For example, the posterior probability of model M.300 is 0.00294, so that model is represented by the point with coordinates (0.300, 0.00294). The graph makes it clear that the data (3 heads in 10 spins) give the greatest support to model M.300 but that the data also support nearby models such as M.299 and M.301 almost strongly. On the other hand, the data have virtually ruled out rates above 0.800.

The textbook website contains a link to a spreadsheet (called Bernoulli.xls for reasons that will be explained shortly) programmed to do these Bayes' rule computations and graph the posterior distribution — the user enters the data (number of spins, number of heads), and the spreadsheet displays posterior probabilities; no programming is required. In addition, the spreadsheet computes a 95% credible set comprising the most probable working hypotheses. A screen shot of the spreadsheet is shown in Fig. 5.18.

5.4   CREDIBLE SETS

The word *credible* means "worthy of belief"; posterior probability is degree of belief; a *set* is a collection of things (in this case models); so a *credible set* is a collection of models with high posterior probability. A 95% credible set is a set of working hypotheses (models) that, taken together, has 95% posterior probability of including the correct model. A *highest-density* (HD) credible set is constructed by beginning with the most probable model and adding models to the set in order of their posterior probability until 95% of the posterior probability has been accumulated in the set. In the analysis reported in Fig. 5.18 the 95% credible set comprises all models between and including M.093 and M.589. In other words after observing 3 heads and 7 tails I think there is 95% posterior probability that the rate of Heads in penny-spinning is between 0.093 and 0.589. A 95% HD credible set has two defining characteristics: it contains 95% of the posterior probability, and any model included in the credible set is more probable than any model not in the set. Sets with other levels of credibility (90% or 80%, etc.) are also sometimes used.

After only 10 spins of the coin, many models are still credible to me; however, as I gather more data, fewer models will remain credible. For example, if I observed 23 heads in 100 spins, the 95% credible set would shrink to encompass only those models between M.155 and M.319. I would, in other words, believe that there is 95% probability that the rate of Heads in penny-spinning is somewhere between 0.155 and 0.319. The shorter credible set and more sharply peaked shape of the posterior distribution reflect the fact that the additional data have made my opinions sharper and more concentrated in the neighborhood of the most probable model.

5.5 BERNOULLI PROCESSES

Coin spinning is an example of a Bernoulli process, that is, a procedure that has only two possible outcomes. In the coin-spinning process, the only possibilities are Heads or Tails. The two outcomes are traditionally called success and failure, although these terms are arbitrary. Typically, the outcome we are investigating, heads for example, is labeled success and the other outcome is labeled failure. This can seem a bit bizarre if we are studying, say, postsurgical mortality. In that case a death would be termed a "success" and survival a "failure". Not all binary (two outcome) processes are Bernoulli. To put it simply, a binary process is a Bernoulli process if the box-of-marbles model is appropriate. The formal definition involves four conditions,

- Each repetition (trial) of the process produces a binary observation — either a success or a failure.
- There is no limit (other than time and budget) to the number (n) of trials that can be made.
- Trials of the process are independent of each other.
- The underlying success rate (p) does not change over time.

The word "trial" is a generic term for making an observation (drawing a marble from the box). Examples of binary processes that can reasonably be modeled as Bernoulli processes are betting on Red at a roulette table, spinning a penny, monitoring the five-year survival of patients who were given a specific treatment, and randomly dialing people to ask a yes-no question such as "Have you ever failed to fill a prescription because it was too expensive?" In the roulette example each spin is a trial, winning is a success and the success rate is $p = 18/38 = 0.47368$.... In the penny-spinning example each spin is a trial, heads is a success, and the success rate is an unknown quantity. In the medical follow-up example, each treated patient is a trial, five-year survival is a success, and the purpose of the study is to learn something about the unknown success rate. In the random dialing example, a completed phone call is a trial, a yes answer is a success and the purpose of the investigation is to learn about the proportion of yeses in the entire population.

Data from any data-generating process that satisfies the Bernoulli assumptions is analyzed the same way (provided the investigator starts from a state of ignorance represented by a flat prior distribution). A spreadsheet implementing this analysis, Bernoulli.xls, is available for downloading from the textbook website. Fig. 5.18 shows the appearance of the spreadsheet.

The user is required to enter only two data items, the number of trials (n) and the number of successes (x). Fig. 5.18 displays a portion of the spreadsheet computation of the posterior distribution of the success rate after observing 3 successes in 10 trials. The most probable value of the success rate is 0.300 and the 95% posterior credible set includes all models between M.093 and M.589. In other words, there is 95% posterior probability that the success rate is between 0.093 and 0.589. Bernoulli.xls performs all the calculations, presents a graph of the posterior distribution, and reports the 95% credible set.

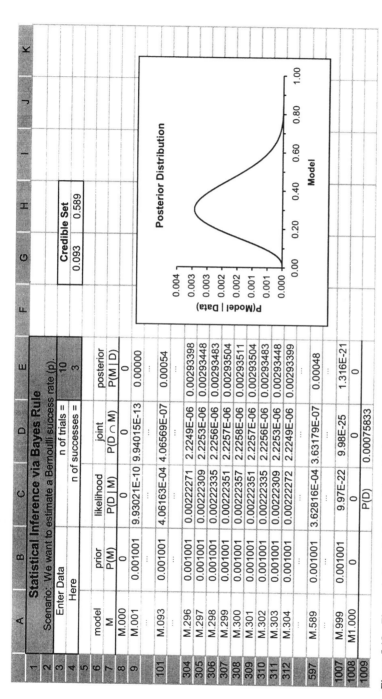

Figure 5.18 Electronic Bayes' rule spreadsheet for the posterior distribution of a Bernoulli rate (p). The spreadsheet uses a flat prior distribution. Here the data are three successes and seven failures in ten trials. The posterior distribution peaks at model M.300, and the credible set comprises models M.093 through M.589, that is, there is 95% probability that the success rate is somewhere between 0093 and 0.589.

85

5.6 EXERCISES

5.1 Repeat the three-cup example of Fig. 5.2 on page 65 but assume the observed data is D: "you drew one black marble and one white marble." Use a spreadsheet layout and calculate the posterior probabilities of the three working hypotheses. Which model has the greatest posterior probability?

5.2 A physician ordered a diagnostic test for her patient. The test has 99% sensitivity and 98% specificity. The physician knows that the incidence of the disease is 1 in 30,000 and uses this as her prior probability that this patient has the disease. Calculate the posterior probability that the patient has the disease given (a) that the test was positive, and (b) that the test was negative.

5.3 The bloody sock DNA mentioned in Section 5.2.3 matched the DNA of the female victim (NBS). The probability of a random match was reported to be 1 in 6.8 billion. Assume a prior probability of 1 in 5 billion that the blood came from NBS. Calculate the posterior probability that the DNA on the sock came from NBS. Comment on the prosecutor's apparent choice of 1 in 6.8 billion as a worst case prior. Was it necessary to be that skeptical? Suggest a less skeptical, but persuasive worst-case prior, and redo the calculations.

5.4 Download the spreadsheet Bernoulli.xls from the textbook website. Spin a penny 5 times. (Use a fairly new penny with Lincoln on one side and the Lincoln Memorial on the other.) Enter your data, and write down the 95% credible interval as shown on the spreadsheet. Now spin the coin 25 times, enter the new data, and again write down the 95% credible interval. Does more data make you more certain? In what way? What does this say about your knowledge? What do you think would happen if you had the time and patience to spin the penny 1000 times? Of course you don't know what the results would be, but just try a few examples, like $n = 1000$ (spins), $x = 350$ (heads), and generalize.

5.5 Make a box model each of these processes, write down an expression for the likelihood of observing SSFSFFFSFF, compute the numerical value of the likelihood.
   (a) Tossing a fair coin.
   (b) Betting on Red at a roulette table.

5.7   REFERENCES AND FURTHER READING

Chamberlin, T.C., The Method of Multiple Working Hypotheses: *Journal of Geology,* Vol. 5 (1887), pp. 837-848. For a modern "translation" see http://www.gly.uga.edu/railsback/railsback_chamberlin.html.

Cusumano, J.P. and Hecht, N.K., "On the Predictability of a Tossed Coin," Department of Engineering Science and Mechanics, The Pennsylvania State University, http://www.esm.psu.edu/nld/images/coin/coinpred.pdf, 1995.

Environmental Protection Agency, *Respiratory Health Effects of Passive Smoking: Lung Cancer and Other Disorders,* EPA/600/6-90/006F, December 1992, Table 5-9.

O'Connor, J.J. and Robertson, E.F., "Urbain Jean Joseph Le Verrier," The MacTutor History of Mathematics Archive, School of Mathematics and Computer Science, University of St. Andrews, http://www-history.mcs.st-andrews.ac.uk/history/Mathematicians/ Le_Verrier.html, 1996.

O'Connor, J.J. and E F Robertson, E.F., "Arthur Stanley Eddington," The MacTutor History of Mathematics Archive, School of Mathematics and Computer Science, University of St. Andrews, http://turnbull.mcs.st-and.ac.uk/history/Mathematicians/ Eddington.html, 2002.

People vs. Orrenthal James Simpson, Superior Court of the State of California in and for the County of Los Angeles, Case # BA097211, 1995. At the time of publication of this book, the complete trial transcript was available online at http:// simpson.walraven.org/. Testimony of the People's DNA expert, Dr. Robin Cotton can be found in the May 8–15 transcripts.

Watson, James D., *The double helix: A Personal Account of the Discovery of the Structure of DNA,* edited by Gunther S. Stent, New York: W. W. Norton, 1980.

6

Continuous Probability Distributions

6.1 INTRODUCTION

Distributions of observed data on the number line were introduced in Chapter 4. Distributions of posterior probability on the number line were introduced in Section 5.3.4. Those probabilities were distributed over 999 discrete (separated) points on the horizontal axis (0.001, 0.002, etc.). Not surprisingly, a distribution of probability on discrete spots on the number line is called a *discrete probability distribution*. In this chapter we will learn how to deal with probability that is distributed continuously over all or part of the number line.

6.2 CONTINUOUSLY DISTRIBUTED PROBABILITY

We'll start with a simple example. Imagine spinning the pointer depicted in Fig. 6.1. The pointer can come to rest at any angle or fraction of an angle between 0° and 360° (measured clockwise from "12 o'clock"). (For example, in Fig. 6.1, the pointer stopped at about 40°.) The imaginary pointer is balanced, spins smoothly, and is equally likely to stop at any angle around the circle. Imagine that the angle can be measured to any degree of accuracy, which means that there are an infinite number of places the pointer could stop. This thought experiment leads to a paradox: the pointer is as likely to stop in one place as another, yet no place can have positive probability (otherwise the total probability would be infinite).

Figure 6.1   Spin the pointer.

As a matter of fact, this is an old familiar paradox. Here's another version (Zeno's Arrow paradox): a car traveling at 75 mile per hour, moves 34 meters every second. In an instant of time (0 seconds) the car will move 0 meters; yet when the instants in one second are "added up," the car will have traveled 34 meters. How can any number of zeros add to 34? Not to trivialize the deep mathematical and philosophical issues raised by this paradox, the working answer to the puzzle is integral calculus: the distance traveled in a given time interval is the area (integral) under a graph of the velocity over the time interval.

Probability density is analogous to velocity. Velocity is the rate at which distance accumulates over time. Probability density is the rate at which probability accumulates from left to right over the number line. For the imaginary spinner (Fig. 6.1) the probability density is a constant $1/360$ (0.002778 or 0.2778%) per degree, because the spinner has 1 chance in 360 of landing within any $1°$-wide sector of the circle.

We'll use the symbol X to stand for the stopping point (in degrees) of the next spin of the pointer. The probability density function of X, $f(x)$, is graphed in Fig. 6.2. The density is a constant 0.002778 over the interval $0°$ to $360°$, which means that the pointer is as likely to stop at one spot as any other around the circle. For example, suppose we want to compute the probability that the pointer will come to rest somewhere between $20°$ and $56°$. The probability that the pointer stops in that interval is the area under the graph of the probability density function over that interval (the black shaded area in Fig. 6.1). In this case the shaded area is a rectangle, so it is easy to compute the area as the product of the base times the height. The height is $1/360$ (probability per degree) and the width of the base is $36°$, so the area is 0.10. Thus $P(20 < X \leq 56) = 0.10$.

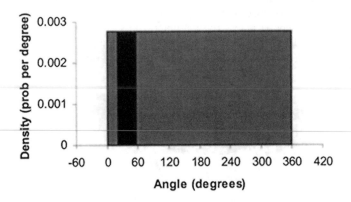

Figure 6.2   Probability distribution for the spinner. (The black area is the probability that the pointer stops somewhere between $20°$ and $56°$ clockwise from "12 o'clock.")

6.3   PROBABILITY DENSITY FUNCTIONS

6.3.1   Probability as Area

In general, probability refers to any declarative sentence, such as "It will rain tomorrow." However, in this chapter we will confine our attention to sentences that refer to a quantity. Examples of such sentences are "The mass of the neutrino is less than 5.1 eV," "The Dow will top 20,000 by the year 2005," or "Humans acquired the language gene between 50 and 100 thousand years ago." Each of these sentences refers to a quantity that we do not know for certain; the mass of the neutrino, the value of the Dow-Jones Industrial Average in 2005, and the exact instant in time that a particular mutation entered the human genome. Each of these things is a quantity, each is unknown, and each is *continuous*, in the sense that it can take on any numerical value in an interval (a quantity which is always an integer is said to be *discrete*[4]). Each of the above sentences predicts that the unknown quantity will turn out to be in some interval; the first sentence states that the mass of the neutrino is in the interval [0, 5.1) eV, the second sentence states that in 2005 the Dow-Jones Industrial Average will be in the half-infinite interval [20,000, +∞), the third sentence places the time of the language mutation in the interval [100,000, 50,000] ybp, or years before the present.

We will use the generic symbol X to stand for an as yet unknown continuous quantity, for example, while the pointer is still spinning, we don't yet know where it will stop (see Fig. 6.1 on page 89). When X is continuous, a person's beliefs about the actual value of X must be expressed as a probability density function, $f(x)$; consequently the probability that a statement about X is true is equal to the area under the density curve over the interval where the sentence is true (see Fig. 6.3). The area under a curve over an interval is a *definite integral*; in simple cases it can be calculated using geometry (see Fig. 6.2, and exercise 6.1), but in

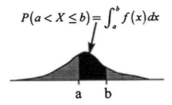

$$P(a < X \le b) = \int_a^b f(x)\,dx$$

a   b

Figure 6.3   The area (black-shaded) over an interval under a density curve is the probability that X is in the interval.

---

4. To be precise, continuous quantities can take any of an uncountably infinite number of values (an interval of real numbers, for example) and discrete quantities can take on only a finite or countable number of values (the positive integers, for example).

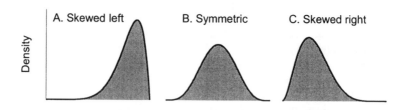

Figure 6.4   Distribution shapes. Left or right skew refers to the direction of the longer tail.

many cases, including the normal distribution, areas under the density graph cannot be hand calculated.

6.3.2   Normal Density Functions

Density curves come in a variety of shapes, as illustrated in Fig. 6.4; by far the most important of these, however, is the normal probability density function, the famous *bell curve*. The normal density function has two *parameters*, the mean μ (mu) and standard deviation σ (sigma); μ is the horizontal position of the peak of the density curve, and σ determines whether the curve is short and wide (indicating a comparatively high degree of uncertainty about the exact value of *X*) or tall and narrow (indicating comparatively precise knowledge of the exact value of *X*); see Fig. 6.5. A normal density curve is symmetrical around a vertical line through μ, and σ is the horizontal distance from μ to the two points of inflection of the curve; consequently, small values of σ correspond to

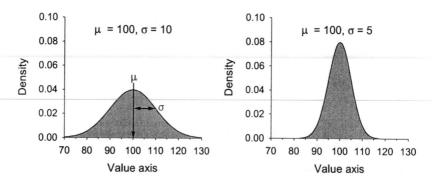

Figure 6.5   Normal density functions. μ is the center of symmetry, and σ is the distance to the point of inflection. A smaller σ value (right) makes the distribution more

comparatively concentrated probability distributions as in the right panel of Fig. 6.5, and large values correspond to comparatively diffuse probability distributions as in the left panel. The reciprocal of σ^2 is called the *precision* and is symbolized by the Greek letter tau, $\tau = 1/\sigma^2$, so for example $\sigma = 10$ corresponds to precision $\tau = 0.01$.

6.3.3 Computing Areas under the Normal Density

It is generally true that if the distribution of probability is expressed in terms of a density function, probabilities can be computed as areas under the density curve. It is particularly important to be able to compute areas under a normal density function. For example, Fig. 6.6 shows the probability distribution of X, John Smith's resting systolic blood pressure 90 days after he starts the a new medication his physician prescribed today. X is at present unknown because it is a number that will not be known until 90 days from today. The density function represents John's physician's beliefs about what X is likely to be. The black-shaded area is the probability that X will turn out to be somewhere in the interval [115, 130].

Unfortunately, there is no general formula for an area under the normal curve, which gives us two options: Use a computer or, when a computer is not handy, use a standard normal tail area table like Table A.1 on page 300. The *standard normal distribution* has parameters $\mu = 0$ and $\sigma = 1$; however, it can be used to compute the probability of any finite or half-infinite interval for any normal distribution. First we'll learn how to use Table A.1 to compute areas (probabilities) for the standard normal distribution, then how to compute

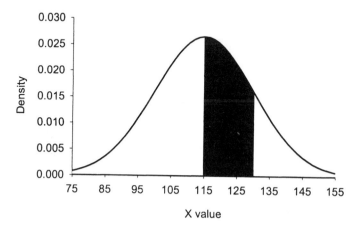

Figure 6.6 Normal density with μ=115, σ=15. The black-shaded area is the probability that X is in the interval [115, 130].

Z	0.00	0.01	0.02	0.03	0.04	0.05	0.06	0.07	0.08	0.09
...	...	...	...	...	...	...	...	...	...	...
1.3	0.9032	0.9049	(0.9066)	0.9082	0.9099	0.9115	0.9131	0.9147	0.9162	0.9177
...	...	...	...	...	...	...	...	...	...	...

Figure 6.7   Fragment of Table A.1 with $P(Z \leq 1.32) = 0.9066$ circled.

probabilities for any normal distribution. We will adopt the almost universal convention of using the symbol Z to stand for an unknown quantity that has a standard normal distribution. Table A.1 is a list of *left tail areas* under the standard normal density curve; in other words, it presents values of the probability that Z is less than or equal to a given value. For example the probability $P(Z \leq 1.32)$ is the number 0.9066 found at the intersection of row 1.3 and column 0.02 (see Fig. 6.7).

    Fig. 6.8 shows how to compute probability that Z is in the interval [0.50, 1.50] by subtracting left tail areas obtained from Table A.1 on page 300. The general computing formula is

$$P(a < Z \leq b) = P(Z \leq b) - P(Z \leq a) \qquad (6.1)$$

Thus, as illustrated in Fig. 6.8, the probability that Z is between 0.50 and 1.50 is the tail area to the left of 1.50 minus the tail area to the left of 0.50, that is

$$P(0.50 < Z \leq 1.50) \; = \; P(Z \leq 1.50) - P(Z \leq 0.50)$$
$$= \; 0.9332 - 0.6915$$
$$= \; 0.2417$$

The left tail areas, 0.9332 and 0.6915, were obtained from Table A.1.

The upper tail formula is a special case of Equation (6.1) (set $b = +\infty$):

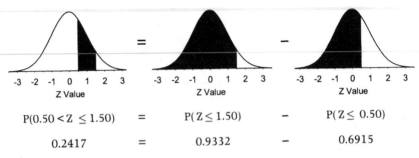

P(0.50 < Z ≤ 1.50)	=	P(Z ≤ 1.50)	−	P(Z ≤ 0.50)
0.2417	=	0.9332	−	0.6915

Figure 6.8   Computing an interval probability as the difference of two tail probabilities.

$$P(a < Z) = 1 - P(Z \leq a) \tag{6.2}$$

For example, the probability that Z is greater than 0.50 is $1 - 0.6915 = 0.3085$.

Computing probabilities for general normal distributions requires an initial transformation from the X scale (value scale) onto the Z scale

$$Z = \frac{X - \mu}{\sigma} \tag{6.3}$$

To compute the probability that an approximately normally distributed unknown quantity X (e.g., a Bernoulli success rate) is in an interval (A, B), first use Equation (6.3) to convert the endpoints A and B into Z values (let's call them a and b) using Equation (6.3), and then apply Equation (6.1) to compute the interval probability using the transformed endpoints a and b.

$$
\begin{aligned}
P(A < X \leq B) &= P\left(\frac{A - \mu}{\sigma} < Z \leq \frac{B - \mu}{\sigma}\right) \\
&= P(a < Z \leq b) \\
&= P(Z \leq b) - P(Z \leq a)
\end{aligned}
\tag{6.4}
$$

To illustrate how the calculation works, suppose that the distribution of X is normal with $\mu = 115$ and $\sigma = 10$. Suppose that we want to calculate the probability that X is in the interval [120, 130]. The first step is to use Equation (6.3), to convert the endpoints of this interval into Z values: for the lower endpoint, A = 120, the Z value is $a = (120 - 115)/10 = 0.50$, and for the upper endpoint B = 130, the Z value is $b = (130 - 115)/10 = 1.50$. The final step is to compute the probability that Z is between 0.50 and 1.50, which we demonstrated in Fig. 6.8.

Formulas for half-infinite intervals are

$$
\begin{aligned}
P(X \leq B) &= P\left(Z \leq \frac{B - \mu}{\sigma}\right) = P(Z \leq b) \\
P(X > A) &= P\left(Z > \frac{A - \mu}{\sigma}\right) = P(Z > a) = 1 - P(Z \leq a)
\end{aligned}
\tag{6.5}
$$

6.4   POSTERIOR DISTRIBUTION OF A BERNOULLI SUCCESS RATE

In the previous section we used X as a generic symbol for an approximately normally distributed unknown quantity. This section is concerned with a specific unknown quantity: the success rate (p) of a Bernoulli process such as coin spinning. In the previous chapter we used a spreadsheet to compute the posterior distribution of p, and it appeared that it sometimes resembles a normal distribution (see Fig. 5.18 on page 85). We now state the conditions under which the posterior distribution of a Bernoulli rate is approximately normal.

Normal Approximation.   If an investigator's prior knowledge about a Bernoulli success rate (p) is expressed as a flat or nearly flat density function, and if she has observed s successes in n trials (observations) of the Bernoulli process, then the posterior distribution of her (updated) knowledge about the success rate p is approximately normal with parameters

$$\mu = \hat{p} \ = \frac{s}{n} \text{ and } \ \sigma = \text{sep} = \sqrt{\frac{\hat{p} \cdot (1 - \hat{p})}{n}} \tag{6.6}$$

Rule of Thumb.   This approximation is accurate for practical purposes if the analyst has observed at least 10 successes (s) and at least 10 failures ($n - s$).

Example: A Medical Device.   50 patients were implanted with an experimental medical device, and 19 of them experienced one or more device-related complications in a 5-year follow-up period. Obtain the approximate posterior distribution of the 5-year complication rate p, and compute the posterior probability that the rate would be less than 0.50 (50%) in the population from which the sample was obtained. Note that in this case a complication is a "success."

The data satisfy the rule of thumb, so we can use the normal approximation with parameters $\mu = \hat{p} = 19/50 = 0.38$ and $\sigma = \text{sep} = \sqrt{0.38 \times 0.62/50}$ $= 0.0686$, computed with Equation (6.6). In other words, the posterior distribution is approximately normal with $\mu = 0.38$ and $\sigma = 0.0686$.

The investigator wanted to compute the probability that the population complication rate p is less than 0.50, in other words to compute $P(p \le 0.50 \mid \text{data})$. This expression is of the form $P(X \le B)$, which can be computed using Equation (6.5). The calculation is diagrammed in Fig. 6.9; the first step is to convert 0.50 into a Z value: $Z = (B - \mu)/\sigma = (0.50 - 0.38)/0.0686 = 1.75$. The second and final step is to look up the tail area in Table A.1 on page 300, where we find $P(Z \le 1.75) = 0.9599$ (see Fig. 6.9). Thus the

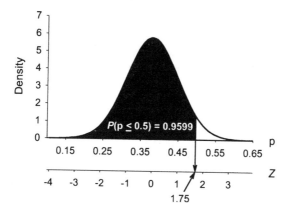

Figure 6.9   Approximate posterior distribution of a Bernoulli success rate p based on observing 19 successes in 50 trials. The transformation of the value 0.50 on the p scale to the corresponding value on the Z scale is illustrated. The black-shaded area is $P(p \le 0.50 \mid \text{data})$.

investigator would conclude that the data have persuaded her that there is a 96% probability that the complication rate is less than 0.50. Someone else viewing the same data, but with a strong prior opinion based perhaps on experience with similar devices, would have computed a somewhat different posterior probability. We will return to that point in Section 6.6.3.

Fig. 6.10 on page 97 explores the accuracy of the normal approximation when the rule of thumb is satisfied (10 successes, 15 failures) and when it is

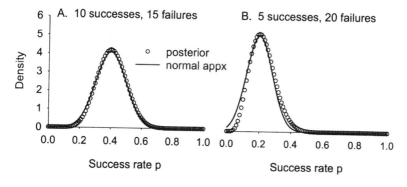

Figure 6.10   Posterior density functions (circles) and normal approximations (lines) of a Bernoulli success rate p for two data configurations. In panel B, the approximation is inaccurate because the rule of thumb is violated.

violated (5 successes, 20 failures). The approximation is better when the numbers of successes and failures are equal or nearly so and when there are many observations. When the rule of thumb is violated and the normal approximation is not accurate, it is necessary to use the spreadsheet Bernoulli.xls or a software package such as WinBUGS to make an accurate calculation (see Appendix B).

6.5   APPROXIMATE CREDIBLE INTERVALS

A 95% credible interval (CI) is an interval of possible values of the success rate (p) that has 95% posterior probability of containing the true rate. The CI is called an HD (highest density) interval if any value of p not in the interval is less probable than any value of p in the interval. HD CI's are preferred because every model in the interval is more probable than any model outside, which means that the interval comprises the 95% *most credible models*. However, in practice it is somewhat difficult to compute HD CI's and equal-tail credible intervals are used instead. An equal-tail 95% CI is the interval between the 2.5th and 97.5th percentiles of the posterior distribution. This interval contains 95% probability but may not contain the 95% most credible models.

An approximate CI can easily be obtained with the normal approximation, Equation (6.6). The procedure is to obtain a 95% CI on the Z scale and then translate that interval back to the X axis. In Table A.1 on page 300 we find that the interval between −1.96 and 1.96 has 95% probability, because

$$P(-1.96 < Z \le 1.96) = P(Z \le 1.96) - P(Z \le -1.96)$$
$$= .9750 - .0250$$
$$= .9500$$

To obtain the credible interval for an unknown quantity X (in this case the success rate p), it is necessary to translate the endpoints of the Z confidence interval 1.96 and -1.96 onto the X axis by running the Z transformation in reverse

$$X = \mu + Z \cdot \sigma \tag{6.7}$$

Consequently, an approximate 95% credible interval (CI) for an unknown quantity X in general and an unknown success rate p in particular, is,

$$95\% \text{CI: } [\mu - 1.96 \cdot \sigma \text{ to } \mu + 1.96 \cdot \sigma] \tag{6.8}$$

Fig. 6.11 illustrates the computation of the approximate 95% CI in the medical device failure rate example on page 96. The data consisted of 19 successes (patients with complications) in 50 trials (patients). We calculated $\mu = 0.38$ and $\sigma = 0.0686$ on page 96; consequently the approximate credible

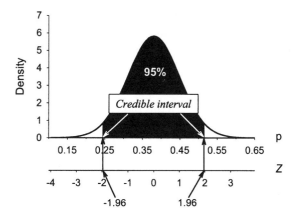

Figure 6.11 Approximate 95% credible interval for rate p after observing 19 successes in 50 trials ($\mu = 0.38$, $\sigma = 0.0684$). The standard normal (Z) 95% interval -1.96 to 1.96 is reverse-transformed to the X scale using.

interval is $0.38 \pm 1.96 \cdot 0.0686$, or $[0.246, 0.514]$. Based on the evidence provided by the data, an investigator who had little prior knowledge of the complication rate for this device would conclude that there is 95% probability that the true complication rate is somewhere in that interval. Another person with different prior opinions would arrive at a somewhat different CI (see Section 6.6.3).

Credible intervals are in many cases numerically very similar to what conventional, frequentist statisticians call *confidence intervals*; however, they are conceptually distinct, and only credible intervals describe the knowledge (beliefs) supported by the data that were actually observed. The fair price of a bet that the unknown quantity (in this case a Bernoulli success rate) is inside the credible interval is $0.95; the fair odds are 1 to 19 (bookmaker style). Confidence intervals, it is claimed, do not refer to the data actually observed, but instead refer to hypothetical sets of data that might be obtained. A confidence interval is actually not a single interval, but a formula for computing an interval once the data are obtained. The claim is that this formula will make a correct statement in 95 data sets out of 100. The more relevant question of whether *this specific interval* for this specific data set makes a correct statement is not answerable by a confidence interval. Generally this distinction makes a difference only when the analyst has relevant auxiliary information that the formulaic confidence interval fails to take into account.

Here is an analogy that may help clarify the distinction. The 3:30 P.M. Bison Wing Airlines flight into Denver is known to land within 10 minutes of the scheduled 3:30 arrival time in 95 flights out of 100. Today, at 2:30, the TV monitors at the airport are still showing the flight "on time." A frequentist statistician would, we presume, argue that a confidence interval for today's arrival time is 3:20 to 3:40 PM. He claims to mean that if he bet on that interval, he'd win on 95 days out of 100. However, suppose I happen to notice that as of 2:30 most other airlines are posting 15 minute delays. This would be a good day for me to bet against the frequentist.

The point of this example is that a frequentist claims to be confident only about the long run "batting average" of his interval-producing formula, not in whether it will "hit" today, and even this rather weak claim to confidence relies on the assumption that people who might bet against him will refrain from using any relevant information not used by his formula. Although this was a frivolous example, there are a number of widely used confidence intervals and other conventional statistical procedures that have this flaw (see Olshen 1973 for one example).

6.6 THE BETA FAMILY OF DISTRIBUTIONS

6.6.1 Beta Densities

We found that the normal approximation to the posterior distribution of a rate is not accurate when the number of successes or the number of failures is less than 10. In that case a different approximation must be used involving a member of the beta family of distributions. Examples of beta distributions are displayed in Fig. 6.12.

In general, a beta distribution can be used to model the distribution of an unknown quantity that is known to be a number between 0 and 1. For example, the rate (p) at which a spun penny lands heads is such a quantity. That rate could be 0.68, for example, meaning that the coin lands heads 68% of the time; but the rate could not be negative, nor could it exceed 1.0.

The beta family includes a variety of distribution shapes: from nearly normal to highly skewed in either direction or even bathtub-shaped. The shape, spread, and peak of the beta density are controlled by two *parameters* called α and β (or a and b). A specific beta density is denoted beta(α, β); for example, beta(10,20). One important special case is the beta(1,1) density, which is uniform (flat) over its entire range. Some nonflat beta densities are depicted in Fig. 6.12. Each point on the horizontal axis represents a possible value v of the unknown quantity (e.g., the Heads rate p in penny-spinning). The height of the curve, as usual, is the probability density at that value, and, as usual, the probability of a statement about the unknown quantity is an area under the curve (see Fig. 6.3 on page 91).

Notice that when the parameters α and β are between 0 and 1, the curve is bathtub-shaped (beta(0.5,0.5) in Fig. 6.12). If $1 < \alpha \le 2$ or $1 < \beta \le 2$ then the curve is *unimodal* (has one peak) but lacks one or both points of inflection and is not bell-shaped [beta1(1.5,1.5) Fig. 6.12].

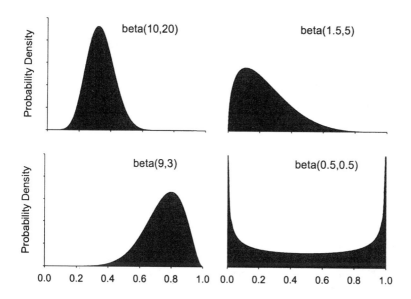

Figure 6.12 Examples of beta(a,b) distributions. Parameters a and b determine the shape.

If both parameters exceed 2 ($\alpha > 2$ and $\beta > 2$), then the curve is unimodal with mode (most probable value).

$$\text{mode } = \tilde{\mu} = \frac{\alpha - 1}{\alpha + \beta - 2} \qquad (6.9)$$

and points of inflection located symmetrically around the mode at

$$\text{inflections } = \tilde{\mu} \pm \sqrt{\frac{\tilde{\mu} \cdot (1 - \tilde{\mu})}{\alpha + \beta - 3}} = \tilde{\mu} \pm \tilde{\sigma} \qquad (6.10)$$

For example, the mode of the beta(9,3) distribution is at 0.80, and the points of inflection are at 0.80 ± 0.13. As a rule of thumb, if both a and b exceed 10, then the curve is approximately normal with the same mode and inflection points; in other words, normal with parameters $\mu = \tilde{\mu}$ and $\sigma = \tilde{\sigma}$.

6.6.2 Beta Posterior Distribution of a Rate

For even moderate numbers of observations, the posterior distribution of a Bernoulli rate is adequately approximated by the normal distribution (see Fig. 6.10); however, for any number of observations the posterior distribution can be exactly represented by a member of the beta family. Specifically, if the investigator has a flat prior distribution for the rate, then his/her posterior distribution is exactly beta with parameters $\alpha = s+1$ and $\beta = n-s+1$, where n is the number of observations of the Bernoulli process and s is the number of observed successes.

$$\text{Flat prior: beta}(1,1)$$
$$\text{Data: } s \text{ successes and } n-s \text{ failures} \qquad\qquad (6.11)$$
$$\text{Posterior: beta with parameters } \alpha = s+1 \text{ and } \beta = n-s+1$$

[For example, if n = 25 and s = 18, then the posterior distribution is beta(19, 8).] This is very useful, because in this case the normal approximation is not accurate (see Fig. 6.10 on page 97). However, in order to compute posterior probabilities and credible intervals it is necessary to have a tail area table for the beta(19,8) distribution similar to Table A.1 and that table would be relevant only for 18 successes in 25 observations and could not be used to assess the evidence provided by a different number of successes or observations. Printing all possible beta distribution tables is impossible — it would require 26 pages of tables just to cover all the posterior distributions that could result from only 25 observations. For that reason, the beta distribution table has been made available as a spreadsheet, BetaTailArea.xls, which can be downloaded from the textbook website.

Fig. 6.13 shows the layout of the spreadsheet set up to calculate tail areas of the beta(19, 8) distribution (the posterior for 18 successes in 25 trials). The parameters α=19 and β=8 have been entered in the gray boxes under α and β, after which, left tail areas for the beta(19, 8) distribution appear at the bottom of the spreadsheet. The table of left tail areas can be used to compute posterior probabilities; for example, the probability that the Bernoulli rate p is between 0.75 and 0.85 is

$$P(p \leq 0.85) - P(p \leq 0.75) = 0.968 - 0.685 = 0.283$$

The "Custom Values" area of the spreadsheet can be used to compute the tail area for a value not listed in the table, or it can be used to find a quantile — a value that cuts off a given tail area. For example, Fig. 6.13 shows how to compute the 0.975th quantile (the 97.5th percentile) of the posterior distribution. To obtain the 95% posterior credible interval for the Bernoulli rate we need to find a lower credible limit (LCL) that cuts off 2.5% (0.025) in the lower tail and an upper credible limit (UCL) that cuts off 2.5% in the upper tail or 97.5% in the lower tail.

Beta Distribution Tail Area and Quantiles

Instructions: 1) Enter parameters α and β in the shaded boxes only.
2) Look up required tail area in the Left Tail Area Table
3) Optional: enter a custom value (v) and get a tail area OR enter a tail area and get a quantile.

1) Parameters	
α	19
β	8

3) Custom Values		
Value	Area	
0.5	0.0145	<-- Read tail area here
0.8567	0.975	<-- Enter tail area here
Enter value here -->		
Read quantile here -->		

2) Left Tail Area Table

Value	0.00	0.01	0.02	0.03	0.04	0.05	0.06	0.07	0.08	0.09
0.0	0.000	0.000	0.000	0.000	0.000	0.000	0.000	0.000	0.000	0.000
0.1	0.000	0.000	0.000	0.000	0.000	0.000	0.000	0.000	0.000	0.000
0.2	0.000	0.000	0.000	0.000	0.000	0.000	0.000	0.000	0.000	0.000
0.3	0.000	0.000	0.000	0.000	0.000	0.000	0.000	0.000	0.000	0.000
0.4	0.001	0.001	0.001	0.002	0.003	0.004	0.005	0.006	0.008	0.011
0.5	0.014	0.019	0.024	0.030	0.038	0.047	0.057	0.070	0.085	0.102
0.6	0.122	0.144	0.168	0.196	0.226	0.260	0.295	0.334	0.374	0.417
0.7	0.460	0.505	0.551	0.596	0.641	0.685	0.727	0.767	0.804	0.838
0.8	0.869	0.896	0.919	0.939	0.955	0.968	0.978	0.986	0.991	0.995
0.9	0.997	0.998	0.999	1.000	1.000	1.000	1.000	1.000	1.000	1.000
1.00	1.000	na	na	na	na	na	na	na	na	na

Figure 6.13 BetaTailArea.xls spreadsheet for the beta(19,8) distribution. Enter data only in the shaded boxes. Entering a custom value of 0.975 for the tail area produces the upper confidence limit, 0.8567.

To compute the upper credible limit, enter 0.975 in gray box under "Custom Area" and read UCL = 0.8567 in the white box to the left, as shown in Fig. 6.13. To compute the lower credible limit, enter 0.025 in the gray box and read LCL=0.5221 (not illustrated). These two quantiles, the 2.5th and 97.5th percentiles, delimit the 95% equal-tail posterior credible interval, 0.52 to 0.86. In other words, after observing 18 successes in 25 observations, the investigator is 95% certain that the true success rate is somewhere between 0.52 and 0.86. On the other hand, the normal approximation to the credible interval [see Equation (6.8) on page 98] is 0.54 to 0.90. The normal credible interval is inaccurate, because the posterior distribution is not approximately normal.

6.6.3 Informative Beta Priors

Up to this point we have presumed that the investigator uses a flat prior distribution, either as a "neutral" *reference prior* likely to be persuasive to a wider audience, or as a reflection of a genuine lack of relevant prior information. Using a beta distribution as the prior, the investigator can introduce relevant prior information into the analysis when it is available. Prior information about a rate (the potential cure rate for an experimental drug, for example) is based on prior experience with similar phenomena (perhaps the cure rate in animal studies, or the cure rate for similar drugs). The process of discovering the investigator's prior beliefs about the rate under investigation is called *prior elicitation*. Various methods for accurate prior elicitation have been proposed, most of them based on structured interviews with graphic or numeric feedback, and some have been used in biomedical studies (see Stangl and Berry 2003 for an excellent review and references).

Here we will give a simplified example. Suppose that Kate is investigating the rate p at which a spinning penny lands Heads. Kate has been told that a spinning disk favors the more concave side, and it appears to her that the tails side of her penny is somewhat more concave than heads side. Therefore, even before making a single spin, she has an opinion about the unknown quantity — she believes that the Heads rate likely to be less than 0.50.

Let's suppose that after a bit of introspection, Kate was willing to offer 2 : 1 odds that the heads rate is less than 0.50 but thinks there is only a 1% chance that the rate is less than 0.10. In other words for Kate, $P(p < 0.50) = 0.67$, and $P(p < 0.10) = 0.01$. The reader is invited to use BetaTailAreaTable.xls to verify that the beta(3.4,4.6) distribution comes close to these tail areas, so Kate can use it as her prior distribution. In other words, Kate's prior knowledge is equivalent to starting with a flat prior, beta(1,1), and "observing" $s = 2.4$ successes and $n - s = 3.6$ failures [see Equation (6.11) on page 102]. These are of course not real observations but are *virtual* observations that would carry the same amount of information as Kate's prior knowledge.

Generically the symbols α_0 and β_0 stand for the parameters of an informative prior distribution for an unknown rate, p, so Kate's prior parameters are $\alpha_0 = 3.4$

and $\beta_0 = 4.6$, whereas the parameters of a flat prior distribution would be $\alpha_0 = 1$ and $\beta_0 = 1$. After observing s successes in n spins of the penny, there are, in effect, $2.4 + s$ successes (virtual prior successes plus actual observed successes) and $3.6 + n - s$ failures. Thus, by Equation (6.11) on page 102, Kate's posterior distribution will be beta with parameters $\alpha = 2.4 + s + 1 = \alpha_0 + s$, and $\beta = 3.6 + n - s + 1 = \beta_0 + n - s$:

> Informative prior: beta(α_0, β_0)
>
> Data: s successes and $n - s$ failures $\hspace{2cm}$ (6.12)
>
> Posterior: beta with parameters $\alpha = s + \alpha_0$ and $\beta = n - s + \beta_0$

To finish the example, suppose that Kate spins the penny 10 times and observes 7 heads (7 successes and 3 failures). Her prior distribution for the heads rate p was beta(4.8, 5) and, according to Equation (6.11) her posterior distribution is beta(10.4, 7.6). Kate's posterior 95% equal tail credible interval can be obtained by finding the 2.5th and 97.5th percentiles of the posterior distribution. To get the 97.5th percentile, enter $\alpha = 10.4$ and $\beta = 7.6$ in BetaTailArea.xls (see Fig. 6.13) and enter 0.975 in the shaded box labeled "Enter tail area here." The percentile, 0.7887, appears in the box labeled "Read quantile here." To get the 2.5th percentile enter 0.025 in the "Enter tail area here" box and read 0.3505 for the percentile. Therefore Kate's posterior 95% credible interval for the heads rate is 0.35 to 0.79.

Kate's posterior probability that the heads rate is greater than 50% is calculated by entering 0.50 in the box labeled "Enter v here" and reading the tail area in the box labeled "Read tail area here." The left tail area is 0.2492, but we want the right tail area, which is $1 - 0.2492 = 0.7508$. Thus, after seeing the data, Kate thinks there is 75% probability (3 to 1 odds) that the heads rate is *greater* than 0.50; whereas prior to collecting data, Kate would have bet 2 to 1 that the heads rate was *less* than 0.50.

6.7   EXERCISES

6.1   You have measured the lengths of two dried beans to the nearest 0.1 mm.
and calculated the average of the two measurements. The unknown
quantity X is the round-off error in the average. The probability density of
X is shown below. Use geometry to verify that the total area under the
density graph is 1.00. Explain why the black-shaded area is the probability
that the round off error is between -0.01 and +0.01. Use geometry to
calculate the black-shaded area.

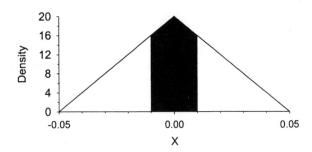

6.2   The best currently approved treatment for a particular disease has a 63%
cure rate (success rate = 0.63). A new, experimental treatment has been
given to $n = 85$ patients and $s = 68$ of them were cured. Use the normal
approximation to obtain an approximate 95% credible interval for the
success rate p of the new treatment, and compute the probability that the
success rate exceeds 63%, that is, compute $P(p > 0.63 \mid \text{data})$.

6.3   The unknown quantity X has a normal distribution with parameters
$\mu = 25$ and $\sigma = 3$. Compute the following probabilities,
(a) $P(22 < X \leq 28)$,
(b) $P(X > 19)$,
(c) $P(X \leq 29.5)$.

6.4   Suppose that U is an unknown quantity with a beta(3, 7) distribution.
Compute the mode and points of inflection for this distribution. Use the
normal approximation to compute $P(U \leq 0.20)$. Compute the exact value
using the BetaTailArea.xls spreadsheet.

6.5   Repeat exercise 6.4 but assume that U has a beta(12, 28) distribution.
What do you conclude about the normal approximation?

6.6   The unknown quantity X has a beta distribution with parameters $\alpha = 11$
and $\beta = 13$. Use the BetaTailArea.xls spreadsheet to obtain an exact 95%
equal-tail credible interval. Hint: Find the 2.5th and 97.5th percentiles.

6.7   A sample of n=200 bolts made to attach bumpers to sport utility vehicles contained 15 defectives.

   (a)   What is the normal approximation to the posterior distribution of the rate of defectives in the population?

   (b)   Justify using the normal approximation.

   (c)   Obtain a 95% CI for the population proportion.

6.8   A sample of n=200 cardiac pacemaker leads contained 2 defectives.

   (a)   Why can I not use a normal approximation for the posterior distribution of the rate of defectives in the population?

   (b)   Assuming a flat beta(1, 1) prior, what is the posterior distribution of the rate of defectives?

   (c)   Use BetaTailArea.xls to look up the 2.5th and 97.5th percentiles and use them to make a 95% posterior CI.

   (d)   Use BetaTailArea.xls to compute the posterior probability that the rate of defectives is less than 4%.

6.8   REFERENCES

Olshen, R., "The Conditional Level of the F-test", *Journal of the American Statistical Association*, 68, (1973), pp. 692–698.

Stangl, D. and Berry, D. (2003) "Elicitation of Priors," in *Bayesian Statistics in Medicine: Where are We and Where Should We be Going?* http://citeseer.nj.nec.com/401035.html (consult the course web page for an updated citation).

7

Comparing Two Rates

7.1 POSTERIOR DISTRIBUTION OF A DIFFERENCE

Many studies of rates involve comparing rates in two or more different groups of people (smokers and nonsmokers, for example) or in the same group of people under different conditions. For example, van Balen et al. (1996) reported a double-blind placebo-controlled study of amoxicillin versus placebo for persistent otitis media with effusion (OME) in general practice. The research question was whether antibiotic treatment is any better than watchful waiting. In this study 162 children were randomized to receive amoxicillin or placebo. The two treatments, amoxicillin and placebo, are referred to as *arms* of the experiment.

The outcome measure was the absence of persistent OME after two weeks of treatment. The outcome measure in an experimental study such as this is often called the *endpoint*. Endpoint data for the two arms for 149 children who completed the two-week follow-up period are shown in Table 7.1. The observed rate of OME is lower among the 79 children in the amoxicillin arm of the study; the question is whether this is convincing evidence that the rate of OME would be lower in the general population of children in need of treatment for middle ear infection. We will use Bayes' rule to answer this question.

7.2 MODELS, LIKELIHOODS, AND BAYES' RULE

The first step in applying Bayes' rule is to list all working hypotheses and their prior probabilities, in other words, to list all models that might account for the

Table 7.1 Relative frequency of absence of bilateral OME[a]

Arm	Number of children	Number without OME	Observed rate (relative frequency)
Amoxicillin	79	37	0.468
Placebo	70	11	0.157

a. Source: Derived from van Balen et al. (1996), Table 4. Reprinted with permission from Elsevier (The *Lancet*, 1996, Vol 374, pp713-716).

data. The purpose of Bayes' rule is to compute the extent to which the data increase our belief in some models (confirmation) and decrease our belief in others (disconfirmation). For example, most people would agree that the data in Table 7.1 ought to increase our belief that amoxicillin lowers the rate of OME in children. The purpose of Bayes' rule is to quantify that belief.

The previous chapter showed how to analyze a single rate such as the Heads rate in penny spinning. In this chapter there are two rates under investigation: the OME rate in children treated with amoxicillin, and the OME rate in children "treated" by watchful waiting. Not surprisingly, two boxes of marbles are needed to model the two rates; the marbles in each box represent children with OME — not the 149 children in this particular study, but all children who will experience OME sometime in the near future, probably several hundred thousand. Box T (the treatment population) represents what would happen if all of these children received amoxicillin. A black marble represents a child who would not have OME at the end of the two-week treatment period, and a white marble a child who would still have OME at the end of the treatment period. Box C (the control population) represents what would happen if all children received the placebo.

The proportions of black marbles in the two boxes are unknown quantities, which we will designate p_T and p_C. Thus p_T represents the success rate of amoxicillin treatment. (Success means the absence of OME two weeks after treatment began.) Similarly, p_C is the proportion of black marbles in box C and represents the success rate of watchful waiting. The two rates p_C and p_T refer to hypothetical futures and are therefore both unknown, which of course is the reason for doing the experiment.

Although there are two separate rates, the principal investigator, Dr. van Balen, and his collaborators were primarily interested in learning about the *difference* between the two rates, which we will designate by the Greek letter delta:

$$\Delta = p_T - p_C \qquad (7.1)$$

The purpose of van Balen's research was to answer the question, "How much does amoxicillin reduces the rate of OME?" In other words, how big is the difference Δ between the two rates?

Models.   Any application of Bayes' rule begins with a list of all possible models (working hypotheses). In this case, each working hypothesis is a *pair* of boxes; for example, the working hypothesis designated {T.487, C.189} is: "The success rate would be 0.487 if all children with OME were treated with amoxicillin and it would be 0.189 if the same children were treated with placebo." There are an infinite number of working hypotheses; however, knowing rates to three digits is probably sufficient for most purposes. Even at that resolution there are over one million working hypotheses (1001 models for box T times 1001 models for box

C); consequently, any thought of using a spreadsheet for the calculations has to be abandoned. So we will illustrate the ideas with a low-resolution spreadsheet, but will learn how to do the calculations with other software and with approximations.

Prior Probabilities.   Inputs to Bayes' rule are prior probabilities of models and likelihoods of observed data under each model. Prior probabilities represent beliefs that were held prior to observing the data in Table 7.1 on page 109. We have to assume that an experienced physician would not be entirely ignorant about success rates for antibiotic treatments and watchful waiting; however, it is useful to present an analysis using a *reference prior*. A reference prior (a *noninformative* or *flat prior*) is meant to represent the beliefs of a neutral person who does not believe that either treatment is necessarily better than the other. The purpose of using a reference prior in Bayes' rule is to present an analysis that in some sense lets the data speak and keeps the influence of prior opinion to a minimum. Many, but not all conventional statistical analyses produce results similar to Bayes' rule analyses with a reference prior.

The power of Bayesian statistical analyses is its capacity to separate the influences of data and prior opinion, which conventional analysis is incapable of doing. The prudent analyst will run analyses with a variety of *informative* prior probabilities representing the prior opinions of readers ranging from "skeptics" to "believers." We will later learn how to construct and use informative priors, for now we will use a flat prior that assigns equal probability to every working hypothesis.

Likelihood.   In the two-box model of a *two-arm experiment with a binary endpoint*, data are generated by selecting given numbers of marbles from box T and box C. The symbols used to represent data are

n_T = number of marbles selected from box T

x_T = number of black marbles selected from box T

n_C = number of marbles selected from box C

x_C = number of black marbles selected from box C

$\hspace{10cm}$ (7.2)

For example, the observed data in the OME study (Table 7.1 on page 109) are $n_T = 79$, $x_T = 37$, $n_C = 70$, $x_C = 11$.

The likelihood of the data is the probability of drawing 37 black marbles and 42 white marbles from box T and 11 black marbles and 59 white marbles from box C. The likelihood of the data for each box is given by Equation (5.4) on page 80, and since the boxes are sampled independently, the joint likelihood of the two boxes is given by the product rule for independent sentences [Equation (3.13) on page 41]:

$$\text{Two-box likelihood} = \left(p_T\right)^{x_T} \cdot \left(1 - p_T\right)^{n_T - x_T} \cdot \left(p_C\right)^{x_C} \cdot \left(1 - p_C\right)^{n_C - x_C} \qquad (7.3)$$

For example, the likelihood of the OME data for a specific model, say $\{T.475, C.115\}$, is $0.475^{37} \times 0.525^{42} \times 0.115^{11} \times 0.885^{59} = 6.6326 \times 10^{-38}$.

The Bayes' rule calculation requires computing the likelihood of the data for each one of over one million working hypotheses. Spreadsheet computations at this resolution are infeasible, and we will shortly resort to other computational strategies; however, to clarify the ideas, we will do a coarser-grained analysis using boxes of 100 marbles, which allows us to represent success rates to two digits only. At two-digit resolution, there are 101 models for box T and 101 models for box C, and therefore $101 \times 101 = 10,201$ working hypotheses of the two rates.

The Bayes' rule calculation is shown in Table 7.2. The working hypotheses are listed in the p_T and p_C columns under the heading "Model." The prior probability of each working model is listed under the heading "Prior." (In this case we have assigned equal prior probabilities to all 10,201 models.) Likelihoods of the OME study data in Table 7.1 are computed under the heading "Likelihood." Equation (7.3) was used to compute the likelihoods. Thus, for example, the likelihood of the data for working model $\{p_T = 0.39, p_C = 0.19\}$ is,

$$\text{likelihood} = \left(0.39\right)^{37} \times \left(1 - .39\right)^{42} \times \left(.19\right)^{11} \times \left(1 - .19\right)^{59} = 3.3124 \times 10^{-38}$$

This calculation was repeated for each of the other 10,200 working models, and the results were entered in the likelihood column. At that point Bayes' rule takes over to compute the posterior probability of each working model. For example, the posterior probability of the model $\{T.37, C.27\}$ is 0.0001883415.

The spreadsheet computes probabilities of sentences such as "The success rate of amoxicillin treatment is 0.28 and the success rate of placebo treatment is 0.11"; however, the investigator wants to know the probability of statements about the difference Δ between the rates. For example the sentence "Amoxicillin has a higher success rate than placebo," is equivalent to the sentence "$\Delta > 0$". The strategy for computing the probability of that sentence is to add up the probabilities of the working models for which that sentence is true.

To illustrate this point, suppose, for example, we want to compute the probability that amoxicillin is exactly 20 percentage points better than placebo, that is, $P(\Delta = 0.20 \mid \text{data})$. There are 81 different models for which the sentence "$\Delta = 0.20$" is true; for example, it is true if $p_T = 1.00$ and $p_C = 0.80$, or if $p_T = 0.99$ and $p_C = 0.79$, and so on through $p_T = 0.20$ and $p_C = 0.00$. Those working hypotheses are included in the oval in Table 7.2; their combined probability, 0.0193, is the probability that $\Delta = 0.20$. This sort of calculation was repeated for all other possible two-digit values of Δ, and the results are graphed at the bottom of Table 7.2. This graph is the posterior distribution of Δ and can be used to compute any desired posterior probability as an area under the curve.

If more than two digits of resolution are required, the spreadsheet approach quickly becomes infeasible; however, the computation can be carried out to any desired degree of accuracy with software specifically designed for Bayes' rule calculations (see Appendix B). However, the shape of the posterior distribution

Table 7.2   Spreadsheet for two-box model (11 of 10201 models)

Model (M)		Prior	Likelihood	Joint	Posterior
p_T	p_C	$P(M)$	$P(\text{data} \mid M)$	$P(M \cap \text{data})$	$P(M \mid \text{data})$
— Not shown: models with Δ values less than 0.20 —					
0.20	0.00	0.000098030	0	0	0
0.21	0.01	0.000098030	2.317E−52	2.2714E−56	1.3174E−17
0.22	0.02	0.000098030	8.5365E−49	8.3683E−53	4.8538E−14
— Not shown:18 models with $\Delta = 0.20$ —					
0.39	0.19	0.000098030	3.3124E−38	3.2472E−42	0.001883415
0.40	0.20	0.000098030	3.5662E−38	3.496E−42	0.002027720
0.41	0.21	0.000098030	3.5744E−38	3.5039E−42	0.002032331
0.42	0.22	0.000098030	3.3453E−38	3.2794E−42	0.001902073
0.43	0.23	0.000098030	2.9311E−38	2.8734E−42	0.001666604
— Not shown: 56 models with $\Delta = 0.20$ —					
0.98	0.78	0.000098030	2.1607E−112	2.1182E−116	1.2286E−77
0.99	0.79	0.000098030	5.2886E−126	5.1844E−130	3.0070E−91
1.00	0.80	0.000098030	0	0	0
— Not shown: models with Δ values greater than 0.20 —					

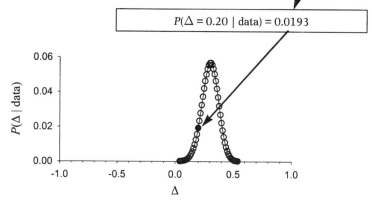

$P(\Delta = 0.20 \mid \text{data}) = 0.0193$

graph at the bottom of Table 7.2 looks like a normal curve, and that impression is correct — for moderately large samples there is a simple and reasonably accurate normal approximation that can be computed without special software.

7.3   AN APPROXIMATE POSTERIOR DISTRIBUTION OF Δ

The normal distribution is determined by two parameters, μ and σ; so what we need in order to make use of the normal approximation is some way to calculate the μ and σ parameters of the posterior distribution of Δ. The procedure for doing that is laid out in Fig. 7.1. Notice that the μ and σ parameters of the posterior distribution of the rate difference Δ are derived from corresponding parameters of the posterior distributions of the success rates p_T and p_C [see Equation (6.6) on page 96]. The normal approximation is reasonably accurate if there are at least 10 successes and 10 failures in each arm of the study, as was the case in the OME study.

The parameter μ of the posterior distribution of Δ is called $\hat{\Delta}$ ("delta-hat"). It is the most probable value of Δ and consequently is the best estimate of Δ. The parameter σ is called sed (pronounced "s-e-d," or "standard error of the difference"). For the OME study, the posterior distribution of Δ is approximately normal with parameters $\mu = 0.311$ and $\sigma = 0.071$. Consequently, the approximate 95% credible interval (CI) for Δ is $\mu \pm 1.96 \cdot \sigma$ $= 0.311 \pm 1.96 \cdot 0.071$, or 0.17 to 0.45. In other words, the analyst (or anyone

	Arms	
Data	Amoxicillin (T)	Placebo (C)
n	79	70
x	37	11
μ and σ of the posterior distributions of rates in each arm:		
$\mu = \hat{p} = x/n$	0.4684	0.1571
$\sigma = \text{sep} = \sqrt{\hat{p} \cdot (1 - \hat{p})/n}$	0.0561	0.0435
μ and σ of the posterior distribution of the difference of the rates (Δ):		
$\mu = \hat{\Delta} = \hat{p}_T - \hat{p}_C$	0.3113	
$\sigma = \text{sed} = \sqrt{\text{sep}_T^2 + \text{sep}_C^2}$	0.0710	
95% credible interval for the difference:	(0.17, 0.45)	

Figure 7.1   Normal approximation to the posterior distribution of Δ, the difference of two Bernoulli rates. The approximation is acceptable when there are at least 10 successes and 10 failures in each arm of the experiment.

else who puts flat priors on the two rates p_T and p_C) would be 95% certain that the true difference between the OME rates in the two populations is somewhere between 0.17 and 0.45.

CI's are useful in a generic sort of way, but the clinically relevant question is, "What is the probability that amoxicillin is better than placebo?" Literally, the sentence "Amoxicillin is better than placebo," means "$\Delta > 0$"; however, *clinical* superiority requires that the treatment be substantially better than placebo. There is no standard of clinical superiority, but something like a 5- to 20-percentage-point improvement is often used as a standard. For the sake of illustration, let us suppose that physicians would regard an 8-percentage-point improvement (or better) over placebo as clinically useful. So the clinically interesting sentence is "$\Delta > 0.08$," and the relevant calculation is the posterior probability that sentence is true, $P(\Delta > 0.08 \mid \text{data})$. The normal approximation makes it easy to approximately compute both the posterior probability that amoxicillin is better than placebo ($\Delta > 0$) and the posterior probability that the improvement is clinically useful ($\Delta > 0.08$).

The posterior probability that amoxicillin is better than placebo (possibly by a trivial amount) is computed using the methods of Section 6.3.3 on page 93.

$$
\begin{aligned}
P(\Delta > 0 | \text{data}) &= P\left(Z > \frac{0-\mu}{\sigma}\right) = P\left(Z > \frac{0-0.3112}{0.0710}\right) \\
&= P(Z > -4.38) = 1 - P(Z \leq -4.38) \\
&\cong 1.0000
\end{aligned}
\tag{7.4}
$$

The posterior probability that amoxicillin is a clinically useful improvement over placebo is,

$$
\begin{aligned}
P(\Delta > 0.08 | \text{data}) &= P\left(Z > \frac{0.08-0.3112}{0.0710}\right) = P(Z > -3.26) \\
&= 1 - P(Z \leq -3.26) \cong 0.9994
\end{aligned}
\tag{7.5}
$$

Assuming that there were no hidden flaws in the conduct of the study, the investigator can be virtually 100% certain that amoxicillin treatment is better than watchful waiting, and 99.94% certain that amoxicillin treatment is a clinically useful improvement over watchful waiting.

7.4 AUXILIARY AND HIDDEN HYPOTHESES

As strong as these results appear to be, they rest on numerous *auxiliary hypotheses*, which are assumed to be true. Generally speaking, auxiliary hypotheses are assumptions that allow the data to be taken at "face value." Investigators, for example, are required to disclose any financial interest in the manufacturer of the drug under investigation, and as we will see later, other strict rules of experimental design must be followed to ensure (among other things) that:

- Assignment of subjects to amoxicillin or placebo was not influenced by the severity of the infection,
- Assessment of a child's OME status after two weeks of treatment was not influenced by knowing which treatment the child had received (in a borderline case, a physician who believed in "watchful waiting" might unconsciously judge amoxicillin patients more negatively than placebo patients)

Any suspicion that these assumptions might not be true might introduce additional working hypotheses and could substantially change the computed posterior probabilities.

7.5   CONVENTIONAL STATISTICAL PRACTICES

7.5.1   Significance Testing and p Values

Significance testing is a central tool of conventional, frequentist statistical practice that has no counterpart in Bayesian analysis. The reporting of statistical significance has been strongly criticized even by practitioners of conventional statistics (see Parkhurst 1997). We will not attempt to present the theory of significance testing in detail, and will merely sketch a version of it here and explain how it can in some cases be reinterpreted in terms of posterior probabilities. Goodman (1999) is an excellent discussion the p value fallacy, written for biomedical scientists rather than statistical experts.

A classical account of significance testing is R.A. Fisher's "Lady Tasting Tea" (Fisher, 1935). The episode took place in England at a departmental tea break. A lady claimed to be able to determine by taste alone whether milk had been added to the cup before or after the tea. According to Fisher's account, she was presented with eight cups of tea (four of them were poured milk first, the other four tea first) in random order and was able to identify every one correctly. Fisher calculated that the probability of the data (getting all eight correct) assuming that she was "just guessing" was $1/70$ or 0.0143. Conventional statisticians call this number the "p value" or the "significance level".

The "just guessing" hypothesis is called the *null hypothesis*. The word "null" means zero, nothing. In other words the null hypothesis is that there is nothing new here; the lady does not have any special powers.

In our terms, Fisher's p value is the likelihood of the data under the null hypothesis. What is missing is any consideration of other working hypotheses. The logic of significance testing, as we understand it, is that the data are so improbable under the null hypothesis (1 chance in 70) that there must be some other explanation than "She's just guessing". There are three errors in treating a small p value as evidence against the "just guessing" hypothesis: First, it does not take into account the *prior* probability of that or any other working hypothesis, yet (as the almost universal rejection of paranormal research attests) prior probabilities can strongly influence how people react to data. Second, it does not take into account the likelihood of the data under any *other* working hypothesis; yet, as improbable as the data are under the null hypothesis, they could be even less probable under any other plausible hypothesis. Finally, Fisher's p value is the probability of the data given one particular model rather than the probability of that model given the data. Carver (1978) made the latter point with particular clarity (think of Carver's D as observed data and his H as a working hypothesis or model):

> What is the probability of obtaining a dead person (D) given that the person was hanged (H); that is, in symbol form, what is P(D | H)? Obviously, it will be very high, perhaps 0.97 or higher. Now, let us reverse the question: What is the probability that a person has been hanged (H) given that the person is dead (D); that is, what is P(H | D)? This time the probability will undoubtedly be very low, perhaps 0.01 or lower. No one would be likely to make the mistake of substituting the first estimate (.97) for the second (.01); that is, to accept 0.97 as the probability that a person has been hanged given that the person is dead. Even thought this seems to be an unlikely mistake, it is exactly the kind of mistake that is made with the interpretation of statistical significance testing — by analogy, calculated estimates of P(H | D) are interpreted as if they were estimates of P(D | H), when they are clearly not the same. (Carver 1978, pp 384-385, quoted with permission of the *Harvard Educational Review*)

In general the concept of p value (or significance level) is even stranger than Carver's analogy suggests. Consider the OME study comparing amoxicillin and placebo as an example. In this study, the null hypothesis (H_0) is that there is no difference between amoxicillin and placebo, that is, $\Delta = 0$. The relevant datum (D) is the observed difference between the two rates, $\hat{\Delta} = 0.3112$, which of course is the best estimate of the true difference Δ. The p value is not the probability of the data that were actually observed; instead it is the probability of hypothetical data at *least as unfavorable* to the null hypothesis as the observed data:

$$p = P\left(\hat{\Delta} \geq 0.3112 | \Delta = 0\right) \tag{7.6}$$

The story gets even stranger. Equation (7.6) defines the *one-sided p value*; however, the almost universal practice is to compute the *two-sided p value*:

$$p = P\left(\hat{\Delta} \geq 0.3112 | \Delta = 0\right) + P\left(\hat{\Delta} \leq -0.3112 | \Delta = 0\right) \qquad (7.7)$$

In words, assuming there is no true difference ($\Delta = 0$), the two-sided p value is the probability of *hypothetical data* that is more unfavorable *in either direction* than the data actually observed. The relevance of entirely fictional data directly contradictory to what was observed is difficult for most people to understand. Nevertheless, and mostly by coincidence, it is sometimes possible to reinterpret p values as posterior probabilities, as we will explain in the next section. Do not confuse lower case italic p, meaning a p value, with lower case p, meaning a Bernoulli rate.

7.5.2 Confidence Intervals

A 95% *confidence interval* for the true difference Δ is not an specific interval such as $(0.17, 0.45)$; instead, it is a pair of formulas for computing the endpoints of an interval. The upper and lower endpoints of the most commonly used (approximate) confidence interval for the unknown quantity Δ are

$$\text{LCL} = \text{lower confidence limit} = \hat{\Delta} - 1.96 \cdot \text{sed}$$
$$\text{UCL} = \text{upper confidence limit} = \hat{\Delta} + 1.96 \cdot \text{sed} \qquad (7.8)$$

The confidence is supposed to be the probability that samples, as yet not observed, will produce intervals that contain the true value of Δ. In other words, it is claimed that over the long run, Equation (7.8) will produce intervals that truly contain Δ in 95 *studies* out of 100. This might sound like a credible interval but there is a difference. Confidence is a sort of lifetime batting average; over the "working life" of Equation (7.8), during which it will be applied to thousands of studies, it will produce intervals that contains the true difference Δ in 95 studies out of 100.

On the other hand, the credible interval for the difference between the cure rates of amoxicillin and placebo refers to this specific study (see Fig. 7.1 on page 114). It is a real, numerical interval computed from the observed data in this specific study, as opposed to a pair of formulas to be used to compute intervals in studies not yet performed. The 95% credibility of this interval is this investigator's subjective probability that the difference Δ between the cure rates of these specific treatments, amoxicillin and placebo, is inside this specific interval, whereas 95% confidence is a claim that a particular approach will be correct in 95 studies out of 100. It is baffling that a claim about past and future performance in studies not yet even conceived of could ever be thought relevant to the interpretation of this particular study.

7.5.3 Confidence Intervals and Flat-Prior Credible Intervals

The attentive reader will have noticed that the confidence interval formula in Equation (7.8) is identical to the credible interval formula in Fig. 7.1 on page 114. Such defacto agreement between confidence and credible intervals happens often enough to be useful, but cannot be relied on in general. In particular, if the investigator does not use a flat, noninformative prior, then the confidence interval will not approximate the credible interval. In this text we will identify situations when confidence intervals can be treated as approximate credible intervals; in other cases, we strongly urge the reader not to treat confidence intervals as credible intervals and instead to use software designed to compute credible intervals.

7.5.4 Making Sense of p Values and Statistical Significance

Statistical Significance as a Decision Rule. In conventional statistical practice an observed difference is said to be *statistically significant* if the p value (probability of the data or more extreme data, assuming the null hypothesis) is 0.05 or less. This "test" is exactly equivalent to checking whether the 95% confidence interval includes the null hypothesis $\Delta = 0$. For example, the 95% confidence interval for the OME study is the same as the 95% credible interval, (0.17, 0.45). Since this interval does not contain $\Delta = 0$, the observed difference, $\hat{\Delta} = 0.3112$, is said to be *statistically significant*. This designation is an arbitrary verbal convention but is nevertheless a sort of holy grail for investigators using conventional statistical analyses. A "significant" difference is regarded as somehow confirmed and an insignificant difference is regarded as somehow disconfirmed.

It is true that there are situations in which it is necessary to make a go – no-go decision on the basis of available information; for example, should a new drug be approved for human use or not? However, the correct way to approach such decisions is to weigh the anticipated consequences of each decision in light of the prior information and information provided by the data and make the decision which has the more favorable expected consequences. An arbitrary, ad hoc, one-size-fits-all criterion such as statistical significance cannot possibly be expected to lead to the appropriate decision in every situation.

There is a separate, well-developed *theory of decisions*, which takes the posterior distribution of a relevant unknown quantity as one input, but also takes into account the economic and other consequences of making one decision instead of another. Conventional statistical practice makes a muddle of evaluating evidence and making decisions based on that evidence, and does neither very well. In particular, the idea of statistical significance as a decision rule is widely and deservedly criticized (see Parkhurst 1997). Bayesian statistical analysis disentangles evidence from decisions based on evidence. Evidence is entirely summarized by the posterior probability distribution, but that does not specify the appropriate decision.

This textbook is mostly about Statistics, the science of evidence; however, even statisticians cannot entirely avoid making decisions, and for that reason there is one chapter in which we venture into decision theory in order to decide upon the appropriate sample size for a study.

Reinterpreting p values. Like it or not, *p* values abound in the biomedical literature. Although some biomedical journals seem to informally discourage their use, it will be necessary to cope with them for the foreseeable future. A (fortunately dying) statistical practice found in older scientific papers is to report nothing but a *p* value – no confidence intervals, no descriptive statistics, just a two-sided *p* value. In certain highly constrained circumstances, the *p* value can be approximately converted into a posterior probability. Specifically, if the confidence interval approximates a credible interval (as it does for the difference of two-rates assuming a flat prior distribution), the two-sided *p* value is approximately equal to two times the smaller of the positive and negative tails of the posterior distribution of the difference Δ:

$$p \cong 2 \cdot \min\left(P\left(\Delta > 0|\text{data}\right), P\left(\Delta < 0|\text{data}\right)\right) \qquad (7.9)$$

It follows that the posterior probability that the true difference is in the same direction as the observed difference is $1 - p/2$:

$$\text{if } \hat{\Delta} > 0, \text{ then } P\left(\Delta > 0 \mid \text{data}\right) \cong 1 - p/2$$
$$\text{if } \hat{\Delta} < 0, \text{ then } P\left(\Delta < 0 \mid \text{data}\right) \cong 1 - p/2 \qquad (7.10)$$

For example, if a paper reports that the *p* value is 0.076 and there is some indication in the paper that the observed difference is positive, then P($\Delta > 0$ | Data) = $1 - 0.038 = 0.962$. In this hypothetical example, a statistically insignificant observed difference ("insignificant" because *p* > 0.05), which would conventionally be interpreted as evidence that the true difference is zero, actually ought to leave the investigator 96.2% convinced that the true difference is positive.

Although some useful evidence can sometimes be gleaned from a *p* value, there is more useful information in a credible interval and even more in a posterior distribution. Therefore there is no substitute for reporting the data in sufficient detail that the reader can compute his or her own posterior probabilities. In this case, that is not a heavy burden, since the relevant data are four numbers: the number of patients and the number of successes in the two treatment arms (amoxicillin and placebo).

7.6 QUANTIFICATION OF COMPARATIVE RISK

7.6.1 Prospective Studies

Michaud et al. (2001) reported an analysis of pancreatic cancer and coffee consumption. The data were obtained from two large studies, but we'll look at data from just one of them, the Nurses' Health Study (NHS). The NHS is a *prospective* (forward-looking) study, which means that one or more groups (*cohorts*) of participants were enrolled at roughly the same time (1976 and 1989 in the case of the NHS) and then followed for a number of years. Baseline data were obtained at the time of enrollment and at regular intervals (follow-ups) by various means, including questionnaires and samples. The following capsule history of the NHS was excerpted from the history page of NHS (2003):

> The Nurses' Health Study was established...in 1976...to investigate the potential long-term consequences of the use of oral contraceptives.... Registered nurses were selected to be followed prospectively...[B]ecause of their nursing education, they would be able to respond with a high degree of accuracy to brief, technically-worded questionnaires and would be motivated to participate in a long term study....
>
> Married registered nurses who were aged 30 to 55 in 1976...were enrolled in the cohort if they responded to our baseline questionnaire....
>
> Approximately 122,000 nurses out of the 170,000 mailed responded. Every two years cohort members receive a follow-up questionnaire with questions about diseases and health-related topics including smoking, hormone use and menopausal status.
>
> [I]n 1980, the first food frequency questionnaire was collected. Subsequent diet questionnaires were collected in 1984, 1986 and every four years since.... [Q]uestions related to quality-of-life were added in 1992 and repeated every four years. Because certain aspects of diet cannot be measured by questionnaire,... the nurses submitted 68,000 sets of toenail samples between the 1982 and 1984 questionnaires.... 33,000 blood samples were collected in 1989.... 'Response rates to our questionnaires are at 90% for each two-year cycle.
>
> A Second cohort was enrolled in 1989.

The advantages of an organized prospective study are, (1) the investigators can specify the particular type of individual to be recruited (e.g. married nurses), and (2) the investigators control what data are obtained and the way they are obtained. The alternative to a prospective study is a *retrospective* study in which data are obtained from participants' recollections of events and behaviors perhaps years in the past and/or from records maintained for other purposes.

Data extracted from Michaud's (2001) paper relating the possible risk factor (coffee consumption) to the adverse outcome (pancreatic cancer) are listed in Table 7.3.

The research question was whether the rate of pancreatic cancer is higher among all women who are heavy coffee drinkers (not just this particular sample of nurses). The model assumes that these nurses are representative random samples from two "boxes": Box N (women who do not drink coffee) and box C (women who drink four or more cups a day). The pancreatic cancer rates in the two boxes are denoted p_C and p_N. Box C is referred to as the *exposed* population — women who are exposed to the risk factor. Box N is the *unexposed* population. Pancreatic cancer is the *adverse outcome*, and being free of pancreatic cancer is the *favorable outcome*.

7.6.2 Relative Risk

In epidemiological studies like this one the comparison between the two rates is expressed as a ratio called the *relative risk* (RR), rather than as a difference (Δ). Relative risk is defined as the adverse outcome rate in the exposed population (coffee drinkers) divided by the adverse outcome rate in the unexposed population,

$$RR = \frac{p_{exposed}}{p_{unexposed}} = \frac{p_C}{p_N} \tag{7.11}$$

Relative risk is sometimes reported as *increased risk*, which is rate difference (Δ) expressed as a percentage of the denominator rate. For example, relative risk 1.15 indicates that the exposed population experienced 15% increased risk of the adverse outcome compared to the unexposed population:

$$\text{increased risk} = 100 \cdot (RR - 1) \tag{7.12}$$

When the two rates are equal, the difference Δ and the increased risk are both equal to zero, and the relative risk is equal to one. Consequently, if it were true that RR = 1, then the rate of pancreatic cancer would be the same for coffee

Table 7.3   Coffee-pancreatic cancer data from the NHS

Coffee consumption (at baseline)	n of nurses	Cases of pancreatic cancer	Relative freq. (per 100,000)
None	20,664	39	189
4+ Cups/Day	21,985	43	196

drinkers as for nondrinkers, and coffee drinking would not be considered a risk factor for pancreatic cancer. On the other hand, if it were true that the rate of pancreatic cancer was greater among coffee drinkers, then the relative risk would be greater than one. Consequently, if RR>1, then the cancer rate in the numerator p_C must exceed the rate in the denominator p_N, which means that coffee drinkers experience a higher rate of pancreatic cancer than nondrinkers. In that situation coffee drinking would be considered a risk factor for pancreatic cancer.

A risk factor is not necessarily a *causative* agent; it is simply a characteristic associated with a higher risk of the disease. Establishing that a risk factor likely to be a causative agent involves two components: Evidence for a biological pathway by which exposure to the risk factor promotes the adverse outcome, and the demonstration of a robust statistical association between the risk factor and the adverse outcome in a variety of studies that control for all other major factors known to influence the outcome.

Posterior Probabilities of Statements about RR. The primary research question in an epidemiological study of a risk factor such as coffee drinking is whether there is an increased risk in the exposed group. For example in Table 7.3 the observed relative risk of the adverse outcome (pancreatic cancer) is $196/186 = 1.054$, about a 5% increased risk associated with a four-cup-a-day coffee habit. However, these data refer to only 42 thousand women, and the question is what is the relative risk in the entire population. How strong is the evidence that there is an increased risk for the larger population of women? To answer that question we must compute the posterior probability, $P(RR > 1 \mid \text{data})$.

As usual, Bayes' rule will provide the answer, but there is a complication. Even for large samples the posterior distribution of RR remains skewed and not well approximated by a normal distribution. This phenomenon is illustrated in panel A of Fig. 7.2, where the exact posterior density function (the solid curve) is clearly skewed to the right and as a consequence the normal approximation (the dashed gray curve) is extremely poor. However, as panel B shows, the natural logarithm of the relative risk *is* approximately normally distributed, and the approximation is so accurate that it cannot be distinguished from the exact posterior distribution.

The procedure for computing the parameters μ and σ of the normal approximation to the posterior distribution of $\ln(RR)$ for the coffee-pancreatic cancer data is shown in Fig. 7.3. Once we have computed those parameter values, it is comparatively easy to compute the CI and relevant posterior probabilities. For example, in the coffee-pancreatic cancer study $\ln(RR)$ has an approximately normal distribution with $\mu = 0.0357$ and $\sigma = 0.2209$. Consequently the flat-prior 95% CI for the logarithm of the relative risk is approximately $\mu \pm 1.96 \cdot \sigma$; i.e., $0.0357 \pm 1.96 \times 0.2209$ or -0.397 to 0.469. Thus, there is 95% posterior probability that $\ln(RR)$ is somewhere in the interval $(-0.397, 0.469)$.

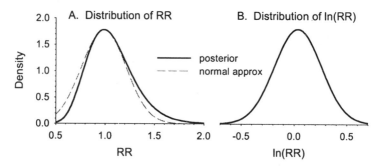

Figure 7.2 Posterior distributions of RR and ln(RR) for the coffee study. For ln(RR), the normal approximation is indistinguishable from the exact posterior distribution.

While relative risks are easy to interpret, their natural logarithms certainly are not and for that reason the CI is transformed back to the RR scale by taking antilogs (exp) of the endpoints. Thus the 95% CI for the relative risk is

	No coffee (N)	4+ cups/day (C)
Data		
No. of subjects (n)	20,664	21,985
No. of cancer cases (x)	39	43
Parameters of posterior distributions of rates:		
$\hat{p}$	0.001887	0.001956
sep	0.000302	0.000298
Approximate posterior distributions of ln(RR):		
$\widehat{RR} = \hat{p}_C/\hat{p}_N$	1.0363	
$\mu = \ln\left(\widehat{RR}\right)$	0.0357	
$\sigma = \sqrt{\left(sep_C/\hat{p}_C\right)^2 + \left(sep_N/\hat{p}_N\right)^2}$	0.2209	
Approximate CI's for ln(RR) and RR:		
95% CI for ln(RR): $\mu \pm 1.96 \cdot \sigma$	−0.3973	0.4687
95% CI for RR: $\exp(\mu \pm 1.96 \cdot \sigma)$	0.6721	1.5979

Figure 7.3 Normal approximation to the posterior distribution of ln(RR).

exp(−0.397) to exp(0.469) or (0.68, 1.60). In other words, the investigator is 95% certain that the rate of pancreatic cancer among four-cup-a-day coffee drinkers is somewhere between 32% below and 60% above the rate among nondrinkers.

Cis are useful in a generic sort of way, but do not answer the primary research question, "What is the probability that RR > 1?" To answer that question we must do a double translation, first to the log scale, then to the Z scale:

$$P\left(RR \leq b\right) = P\left(\ln\left(RR\right) \leq \ln\left(b\right)\right) = P\left(Z \leq \frac{\ln\left(b\right) - \mu}{\sigma}\right) \quad (7.13)$$

For example, $P(RR \leq 1) = P(\ln(RR) \leq 0)$ [because ln(1)=0]. Since ln(RR) is approximately normal with $\mu = 0.0357$ and $\sigma = 0.2209$, we can compute $P(\ln(RR) \leq 0)$ by translating 0 to the Z scale. The Z value corresponding to 0 on the ln(RR) scale is $Z = (0 - 0.0357)/0.2209 = -0.16$. The area to the left of − 0.16 in Table A.1 is 0.4359; consequently $P(RR \leq 1 \mid data) = 0.4359$ and $P(RR > 1 \mid data) = 1 - 0.4359 = 0.5641$, or about 56.4%. That answers the research question: there is about 56% posterior probability that coffee is a risk factor for pancreatic cancer. This is very weak evidence, since there is about an equal chance (44%) that coffee is a protective factor.

7.6.3   Interpreting Published Confidence Intervals

Relative risk is widely reported in biomedical studies. For example, a search for the phrase "relative risk" in Ovid Medline® (an online abstracting service that indexes information from about 3600 biomedical journals published worldwide) produced over 18000 articles, almost half of which (8700) contained confidence intervals. Among these, Jonsson et al. (2002), using data from a 23-year prospective study of 22,000 Swedish men, reported that the relative risk of death from cardiovascular disease (CVD) of obese men (BMI > 30) compared to normal-weight men (BMI 20–24.9) was 1.42 with 95% confidence interval (1.15, 1.75) (adjusted for differences in age, smoking habits, and other factors).

In conventional statistical reporting, a relative risk is said to be *statistically significant* if the 95% confidence interval does not include 1.00. The logic of this convention is that RR = 1.00 indicates equal risks in either group and RR > 1 corresponds to an increased risk in the exposed group. A relative risk is called "significant" if the confidence interval rules out the possibility that there is not an increased risk. Recall that statistical significance is sometimes expressed in terms of a p value, or significance level (see Section 7.5.4 on page 119). Older studies may report only a p value. If the reported p value is 0.05 or smaller, then the 95% confidence interval excludes RR = 1 and the relative risk is said to be statistically significant.

The FREQ procedure of SAS (SAS Institute 2003), the most widely used conventional statistical software package, computes confidence intervals for relative risks using the flat-prior CI formulas in Fig. 7.3 on page 124; consequently, *confidence intervals* for relative risks produced by that package (and probably most major statistical packages) can generally be interpreted as CI's and can be "reverse-engineered" to obtain the parameters of the normal approximation. (So-called "exact" or "nonparametric" confidence intervals should not be interpreted as approximate CIs.)

CIs for relative risks are interesting in a generic way, but to answer more relevant questions about the size of the risk we need the posterior distribution of RR. To illustrate this point, let us suppose that a 30% increased risk would be considered a public health problem. It follows that an important research question would be, "What is the posterior probability that RR > 1.3?" We will illustrate how to calculate this posterior probability for Jonsson's obesity-cardiovascular study described in the first paragraph of this section.

Jonsson reported that the estimated relative risk was $\widehat{RR} = 1.43$ and the 95% confidence interval was (1.15, 1.75), which, as we pointed out earlier, is interpretable as an approximate CI. Equations for converting an estimated RR and CI into the parameters μ and σ of the normal approximation of the posterior distribution of ln(RR) are

$$\mu = \ln\left(\widehat{RR}\right) = \ln(1.42) = 0.351$$

$$\sigma = \frac{\ln\left(\dfrac{UCL}{LCL}\right)}{2 \cdot 1.96} = \frac{\ln\left(\dfrac{1.75}{1.15}\right)}{3.92} = 0.107 \tag{7.14}$$

In Jonsson's study these turned out to be $\mu = 0.351$ and $\sigma = 0.107$.

Next we'll use Equation (7.13) on page 125 to compute the posterior probability that the true increased risk exceeds 30%, that is, $P(RR > 1.3 \mid \text{data})$. First translate 1.3 to the natural log scale: $P(\ln(RR) > \ln(1.3)) = P(\ln(RR) > 0.2624)$. Then use μ and σ to translate 0.2624 to the Z scale: $P(\ln(RR) > 0.2624) = P(Z > -0.82) = 1 - 0.2061 = 0.7939$. The conclusion is that there is 79% posterior probability that the relative risk is at least 1.3 (a 30% or greater increased risk).

7.6.4 Retrospective Studies and Odds Ratios

An epidemiological study of the connection between a possible risk factor and adverse outcome can be conducted *prospectively* or *retrospectively*. In a prospective study, people (for example nurses) are enrolled in the study and then are followed for some time period (perhaps decades, as in the Nurses' Health Study, or over generations, as in the Framingham study). Relative risks can be estimated from a prospective study. For example, the breast cancer rate for women who eat high-fat diets can be compared with the rate for women who eat low-fat diets.

In a *case–control study* (the most common kind of retrospective study) a sample of people (*cases*) with the adverse outcome (e.g., pancreatic cancer) are matched with otherwise similar people (*controls*) who did not experience the adverse outcome. Subjects in both groups are then interviewed (or their records are searched) to determine if they were exposed to the risk factor (e.g., heavy coffee drinking). A case–control study cannot be used to estimate rates and consequently cannot be used to estimate relative risk. To understand why this is true, look at the imaginary population in Table 7.4, in which the true relative risk is RR = 209/189 = 1.106, representing about an 11% increased risk of pancreatic cancer in the population of 11,023,000 women "exposed" to coffee, compared to the population of 9,017,000 "unexposed" women.

Imagine a case–control study, in which the investigator obtained a random sample of 400 pancreatic cancer cases (from hospital cancer units, for example) and another random sample of 400 people without pancreatic cancer (typically non cancer cases at the same hospitals). People in the second sample are called *controls*. If the samples perfectly represented the populations the data would be as as shown in Table 7.5. (Of course, real samples would deviate somewhat from perfection, but the point we're about to make is still true.)

Even though the samples of cases and controls are perfectly proportional to the population numbers, the relative frequencies of pancreatic cancer in the sample (Table 7.5) are grossly inflated compared to the rates in the population (Table 7.4). It should come as no surprise that the observed relative risk in the sample (1.05, or about a 5% increased risk) underestimates the true relative risk (1.106, or about an 11% increased risk): the increased risk is off by a factor of 2.

Table 7.4 The population (hypothetical).

Coffee consumption	Pancreatic cancer	No Pancreatic cancer	Cancer rate (per 100,000)
None	17,000	9,000,000	189
4+ cups/day	23,000	11,000,000	209

True relative risk = RR = 209/189 = 1.106.

Table 7.5 Samples of cases and controls (hypothetical).

Coffee consumption	Pancreatic cancer (cases)	No pancreatic cancer (controls)	Total number of subjects	Relative frequency (per 100,000)
None	170	180	350	48571
4+ cups/day	230	220	450	51111

Observed relative risk = $\widehat{RR}$ = 51111/48571 = 1.052.

This phenomenon is true in general — case–control studies estimate relative risk incorrectly. To get around this difficulty a different comparative risk measure, the *odds ratio*, must be used in case–control studies. The odds ratio is related to the adverse outcome rates in the exposed population (rate = p_E) and the unexposed population (rate = p_U) this way:

$$\text{odds ratio} = \frac{\text{odds of adverse outcome in exposed population}}{\text{odds of adverse outcome in unexposed population}}$$

$$= \frac{p_E / (1 - p_E)}{p_U / (1 - p_U)} \tag{7.15}$$

The odds ratio *can* be estimated from case-control data. For example, in the hypothetical population in Table 7.4 the odds of pancreatic cancer are $23{,}000/11{,}000{,}000 = 0.002091$ in the four-cup-a-day population and are 0.001889 in the coffee-avoiding population. Therefore the true odds ratio is $OR = 0.002091/0.001889 = 1.1069$. On the other hand, the observed odds ratio computed from the samples of cases and controls in Table 7.5 is the observed odds of pancreatic cancer for coffee drinkers ($230/220 = 1.0455$) divided by the odds of cancer for nondrinkers ($170/180 = 0.9444$); thus the estimated odds ratio is $1.0455/0.9444 = 1.1069$, exactly the same as the true odds ratio. The observed odds ratio is identical to the true odds ratio because in this artificial example the case and control samples happen to perfectly represent the case and control populations. For real samples the observed odds ratio will vary somewhat, but is a valid estimate of the population odds ratio. Here is the reason: Although a case–control study does not permit us to estimate either the rate or the odds of finding a cancer case among the coffee drinkers, it does permit us to estimate the odds of finding a coffee drinker among the cancer cases. In other words, the population ratio $23{,}000/17{,}000$ is validly estimated by the corresponding ratio $230/170$ in the sample of cases. Similarly, the population ratio $11{,}000{,}000/9{,}000{,}000$ is validly estimated by the ratio $220/180$ in the sample of controls. The ratio of these ratios is a sort of indirect odds ratio (IOR)

$$\text{IOR} = \frac{\text{odds of finding a coffee drinker in the cancer population}}{\text{odds of finding a coffee drinker in the noncancer population}} \tag{7.16}$$

that, we argued earlier, can be validly estimated from a case–control study because we have random samples from the cancer population (cases) and the non cancer population (controls).

However, what we really want to estimate is the standard odds ratio, first, because it is easier to interpret as a measure of comparative risk of the adverse outcome, and, second, because, as we'll learn shortly, the odds ratio (unlike the relative risk) can be a good estimate of the true relative risk when the adverse outcome is comparatively infrequent. It is easy to see that IOR is simply a rearrangement of OR and therefore the two are mathematically identical:

$$\begin{aligned} \text{IOR} \ &= \ \frac{23,000/17,000}{11,000,000/9,000,000} \\ &= \ \frac{23,000}{17,000} \cdot \frac{9,000,000}{11,000,000} \\ &= \ \frac{23,000/11,000,000}{17,000/9,000,000} \\ &= \ \text{OR} \end{aligned} \qquad (7.17)$$

In other words, the ratio of the odds on coffee drinking in the two outcome groups (cases vs. controls) is identical to the ratio of the odds on pancreatic cancer in the two exposure groups (exposed vs. unexposed).

Notice that the observed odds ratio (1.1069) is quite close to the true relative risk (1.106) whereas the observed relative risk (1.052) misses it rather badly. This is always true for case–control studies of comparatively rare adverse outcomes because in that case odds are extremely close to rates. For example, the rate of pancreatic cancer in the exposed population is 23,000/11,023,000, and the odds are 23,000/11,000,000. The denominators of these two fractions differ only in the fourth significant digit and so the decimal representations of the fractions will also differ only in the fourth significant digit. For this reason it is not unusual for the author of a review article of research on a particular risk factor to treat odds ratios from retrospective studies and relative risks from prospective studies as interchangeable.

Posterior Probabilities And Credible Intervals For Odds Ratios. Odds ratios are generally reported as the observed odds ratio and a confidence interval for the true odds ratio. For example, Zheng et al. (2002) reported a case–control study designed to identify possible risk factors for bladder cancer (see Fig. 7.4). The cases were a sample of Iowa residents with bladder cancer, and the controls were a matched sample of Iowa residents who did not have bladder cancer. The logic of the case–control study is that if, for example, exposure to solvents is a risk factor for bladder cancer, then we should find a higher proportion of people in solvent-using occupations among cases than among controls. For example, the authors reported that the estimated odds ratio for rubber and plastic workers was 3.1 with 95% confidence interval (1.2, 8.5).

One thing that is immediately apparent from this example is that case–control studies take less time and are less expensive than prospective studies. For that reason case–control studies are almost universally used for initial, exploratory studies of potential risk factors. If a series of well-controlled case–control studies consistently identify the same risk factor, then in vitro and/or

"Occupation and bladder cancer: a population-based, case–control study in Iowa."
Zheng, T., Cantor, K.P., Zhang, Y., Lynch, C.F.

While considerable efforts have been made to investigate the role of occupation
and industry in the risk of bladder cancer, many reported associations have not been
consistent, and strong evidence of increased risk is apparent for few occupational
groups. To further examine the issue, a large, population-based, case-control study
was conducted in the state of Iowa among both men and women. A total of 1452
incident bladder cancer cases and 2434 controls were included in the study.
Occupational history was collected from respondents for each job held for 5 years or
longer since age 16. Among men, excess risk was observed for industries including
plumbing, heating, and air conditioning (odds ratio [OR], = 2.2; 95% confidence
interval [CI], 1.0 to 5.0); rubber and plastic products (OR = 3.1; 95% CI, 1.2 to 8.5),
motor vehicle parts and supplies (OR = 4.5; 95% CI, 1.2 to 16.5), and occupations
including supervisors for transportation and material moving (OR = 6.5; 95% CI, 1.4
to 29.9), material-moving-equipment operators (OR = 1.9; 95% CI, 1.0 to 3.6),
automobile mechanics (OR = 1.6; 95% CI, 1.0 to 2.6), painters (OR = 2.7; 95% CI,
1.0 to 7.7), and metal- and plastic-working machine operators (OR = 2.0; 95% CI, 1.1
to 3.4). Among women, significant excess risk was observed for secondary school
teachers and record clerks. Housekeepers and butlers and workers in laundering and
dry cleaning were also at increased risk. In conclusion, these results suggest that
occupational exposures may play a significant role in the risk of bladder cancer.

Figure 7.4   Abstract of Zheng et al. (2002), © Lippincott Williams & Wilkins,
reprinted with permission.

animal experiments and/or a prospective study might be mounted to confirm the
risk factor.

7.6.5 Posterior Distribution of the Log Odds Ratio

The flat-prior posterior distribution of the natural logarithm of the odds ratio,
ln(OR), is approximately normal with parameters μ and σ computed as in
Fig. 7.5. The numbers of cases of bladder cancer (n_B) and controls (n_C) were taken
from Fig. 7.4, but the numbers of exposed individuals (auto mechanics) among
the bladder cancer cases (x_B) and among the controls (x_C) are hypothetical and
used here only to demonstrate the calculations. The observed odds ratio is 1.60
and the 95% credible interval is 1.003 to 2.552. The credible interval does not
include 1.00 so a conventional statistical report would conclude that a career as
an auto mechanic is a risk factor for bladder cancer.

	Cases (bladder cancer)	Controls
Data:		
Exposed: auto mechanics (x)	$x_B = 73$	$x_C = 78$
Total: (n)	$n_B = 1452$	$n_C = 2434$

Observed odds ratio:

$$\widehat{OR} = \frac{x_B / (n_B - x_B)}{x_C / (n_C - x_C)} \qquad \frac{35/1417}{37/2397} = 1.60$$

Parameters of the posterior distribution of $\ln(OR)$

$$\mu = \ln\left(\widehat{OR}\right) \qquad\qquad \ln(1.60) = 0.470$$

$$\sigma = \sqrt{\frac{1}{x_B} + \frac{1}{n_B - x_B} + \frac{1}{x_C} + \frac{1}{n_C - x_C}} \qquad \sqrt{\frac{1}{35} + \frac{1}{1417} + \frac{1}{37} + \frac{1}{1297}} = 0.238$$

95% Credible Interval for $\ln(OR)$ and OR:

$\ln(OR)$: $\mu \pm 1.96 \cdot \sigma$	$(0.003, 0.937)$
OR: $\exp(\mu \pm 1.96 \cdot \sigma)$	$(1.003, 2.552)$

Figure 7.5 Parameters of the normal approximation of the posterior distribution of $\ln(OR)$. The approximation is acceptable if there are at least 10 cases and 10 controls in each exposure category.

The posterior parameters of the distribution of $\ln(OR)$ can be used to compute relevant posterior probabilities via Equation (7.13) on page 125. For example,

$$
\begin{aligned}
P\left(OR > 1.0 | \text{data}\right) &= P\left(\ln(OR) > 0 | \text{data}\right) = P\left(Z > \frac{0 - \mu}{\sigma}\right) \\
&= P\left(Z > \frac{0 - 0.470}{0.238}\right) = P(Z > -1.97) \\
&= 1 - 0.024 = 0.976
\end{aligned}
$$

So the investigator would be about 98% convinced that auto mechanics in general (not just the 151 in the sample) have an increased rate of bladder cancer. Of course, the calculation is correct only if the auxiliary and hidden assumptions are all true; among these are the assumptions that the samples are representative of (indistinguishable from, *exchangeable* with) bladder cancer cases and non bladder cancer cases in general, and that ascertainment of occupational history was not biased by knowledge of the subject's cancer status.

Suppose the data are to be used to inform a policy decision, perhaps a state mandate that large repair shops must provide adequate ventilation for their mechanics. In that case the policy makers would have to set some threshold of increased risk that would trigger action on their part. For the sake of illustration, suppose that overwhelming evidence of a 15% or greater increased risk would trigger legislative attention. The relevant posterior probability in this case would be the posterior probability that the odds ratio exceeds 1.15

$$P(OR > 1.15 | \text{data}) \;=\; P\big(\ln(OR) > \ln(1.15)\big) \;=\; P\big(\ln(OR) > 0.140\big)$$
$$=\; P\left(Z > \frac{0.140 - \mu}{\sigma}\right) \quad=\; P\left(Z > \frac{0.140 - 0.470}{0.238}\right)$$
$$=\; P(Z > -1.39) \qquad\qquad=\; 0.92$$

This is strong, but not overwhelming, evidence of an increased risk, and the response might be to fund further studies of the issue with the intent of taking legislative action if the risk is verified.

Working with published estimates and confidence intervals.   The methods presented in Section 7.6.3 on page 125 apply to odds ratios as well as relative risks. For example, let us compute the posterior probability of a 50% increased risk of bladder cancer for painters. In other words $P(OR > 1.5 \mid \text{data})$. Zheng et al. reported an observed odds ratio of 2.7 with 95% confidence interval 1.0 to 7.7 (see Fig. 7.4 on page 130). The parameters of the posterior distribution of ln(OR), computed using Equation (7.14) on page 126, are

$$\mu \;=\; \ln\big(\widehat{OR}\big) \qquad=\; \ln(2.7) \qquad=\; 0.99$$
$$\sigma \;=\; \ln\left(\frac{UCL/LCL}{2 \cdot 1.96}\right) \;=\; \ln\left(\frac{7.7/1.0}{3.92}\right) \;=\; 0.52$$

Therefore the required posterior probability is

$$P(OR > 1.5 | \text{data}) \;=\; P\big(\ln(OR) > \ln(1.5)\big) \;=\; P\big(\ln(OR) > 0.405\big)$$
$$=\; P\left(Z > \frac{0.405 - \mu}{\sigma}\right) \quad=\; P\left(Z > \frac{0.405 - 0.99}{0.52}\right)$$
$$=\; P(Z > -1.125) \qquad=\; 0.87$$

In other words, there is 87% posterior probability that there is at least a 50% increased risk of bladder cancer in the population of painters.

7.7 COMPUTER ANALYSIS OF RATES AND PROPORTIONS

7.7.1 WinBUGS

WinBUGS is a versatile and flexible computer package for computing posterior distributions. Instructions for computing posterior distributions, CIs, and tail areas for single Bernoulli rates and differences, relative risks, and odds ratios of two rates are in Appendix B.

7.7.2 The SAS System

The SAS System is a widely used computer package for conventional (frequentist) statistical analysis. It has several procedures for analyzing rates. Here we show how to use PROC FREQ to compute confidence intervals for odds ratios.

Data from a case-control study are shown in this table:

Table 7.6   Aggregated case-control data.

		Bladder cancer	
		Yes	No
Solvent exposure	Yes	65	40
	No	55	65

Data in this form are said to be *aggregated*, that is, the 225 people who were the source of these data have been cross-classified by exposure status and outcome. The original raw or *disaggregated* data constituted a data table with 225 rows (one row per person). The original table no doubt had additional information on each person, such as age and smoking history. A fragment of the original data table is shown below; E and U stand for exposed and unexposed and C and N for cancer and not cancer.

Table 7.7   Disaggregated data underlying Table 7.6.

Initials	Age	Cigs/Day	Solvent	Bladder Cancer
JBL	54	20	E	C
RGM	61	10	U	C
GTO	48	0	U	N
$\cdots$ 220 rows not shown $\cdots$				
NAT	73	0	E	N
WTS	59	30	E	C

```
DATA SASexample;
  INPUT Solvent $ Cancer $ freq;
CARDS;
Exposed    Cancer    65
Exposed    Not       40
Unexposed  Cancer    55
Unexposed  Not       65
;;;;

PROC FREQ DATA=SASexample;
  WEIGHT freq;
  TABLES Solvent*Cancer / CL RISKDIFF RELRISK RDIFF;
RUN;
```

Figure 7.6   Analyzing aggregated frequencies with PROC FREQ.

The procedure for analyzing aggregated data is shown in Fig. 7.6; the first part of the output is the table of exposure status by outcome, or *crosstabulation*, shown below. Notice that the rows (solvent) are arranged in alphabetical order as are the columns (cancer). The arrangement of the table is determined by the order of the variables in the TABLES statement — the first variable (solvent) defines the rows and the second (cancer) defines the columns. Each of the four *cells* of the table contains the raw frequency count for that group of subjects as well as the relative frequency expressed as a percentage of the grand total, of the row total, and of the column total. Thus, 65 solvent-exposed individuals had bladder cancer, which represents 28.89% of the entire sample, 61.9% of the row total, and 54.17% of the column total, that is, 61.9% of the exposed group and 45.83% of the unexposed group had bladder cancer:

```
               Table of Solvent by Cancer

          Solvent    Cancer

          Frequency|
          Percent
          Row Pct
          Col Pct  |Cancer  |Not     | Total

          Exposed  |     65 |     40 |    105
                   |  28.89 |  17.78 |  46.67
                   |  61.90 |  38.10 |
                   |  54.17 |  38.10 |

          Unexpose |     55 |     65 |    120
                   |  24.44 |  28.89 |  53.33
                   |  45.83 |  54.17 |
                   |  45.83 |  61.90 |

          Total         120      105      225
                      53.33    46.67   100.00
```

The next part of the output consists in two statistical tables. The first of these assumes that the adverse outcome is in column 1 of the crosstabulation. The Row 1 risk, 0.6190, is the observed rate of the adverse outcome (cancer) in the group tabulated in row 1 (the exposed group). ASE is the standard error of that rate, and the columns labeled "95% Confidence Limits" are interpretable as an approximate 95% CI for a flat prior distribution. Similarly, the Row 2 risk is the observed cancer rate in the unexposed group, and the Risk and ASE in the row labeled "Difference" are $\hat{\Delta}$ and sed, that is, that the posterior distribution of the difference between the cancer rates of exposed and unexposed individuals is approximately normal with $\mu = 0.1607$, $\sigma = 0.0657$, and 95% CI (0.0320, 0.2895):

```
            Statistics for Table of Solvent by Cancer

                       Column 1 Risk Estimates

                                  (Asymptotic) 95%      (Exact) 95%
                Risk      ASE     Confidence Limits    Confidence Limits

Row 1          0.6190    0.0474    0.5262   0.7119     0.5191   0.7121
Row 2          0.4583    0.0455    0.3692   0.5475     0.3671   0.5517
Total          0.5333    0.0333    0.4681   0.5985     0.4659   0.5999

Difference     0.1607    0.0657    0.0320   0.2895
```

Examine the crosstabulation carefully to verify that the first column is indeed the adverse outcome. If this is not the case, then the statistical table entitled "Column 2 Risk Estimates" (not shown here) should be used. If the first row of the crosstabulation is not the exposed group, then the difference will be reported as the rate in the unexposed group minus the rate in the exposed group.

The relative risk and odds ratio are reported in the next table. If the upper right cell of the crosstabulation is the exposed group and adverse outcome, then the row labeled "Odds Ratio" is the odds ratio as defined in this chapter. The estimated odds ratio $\widehat{OR}$ is listed in the "Value" column and the 95% Confidence Limits can be interpreted as an approximate 95% CI, assuming flat prior distributions on the two rates, that is the estimated odds ratio is 1.92 with 95% CI (1.13, 3.27). This interpretation is correct if the exposed group and adverse outcome are either in the upper left or lower right cell of the crosstabulation, otherwise the odds ratio computed by SAS is the reciprocal of the usual odds ratio.

The relative risk estimate and approximate credible interval is found in one or the other of the "Cohort" rows, depending on the configuration of the crosstabulation. If the upper left cell of the crosstabulation is the exposed group and adverse outcome, then the Col1 Risk is the usual relative risk. That is the case here, so the relative risk estimate is 1.35 with 95% CI (1.06, 1.73). On the other hand, if exposed group and adverse outcome is in the upper right cell of the crosstabulation, then the Col2 Risk is the usual relative risk. If the exposed group is in the second row of the crosstabulation the reciprocal of the usual relative risk is reported. Recall that the relative risk estimate is invalid for a Case-Control study.

Estimates of the Relative Risk (Row1/Row2)

Type of Study	Value	95% Confidence Limits	
Case-Control (Odds Ratio)	1.9205	1.1270	3.2725
Cohort (Col1 Risk)	1.3506	1.0565	1.7267
Cohort (Col2 Risk)	0.7033	0.5241	0.9438

Analyzing disaggregated data requires some modifications to the data and procedure steps. This SAS program reads the disaggregated data file in Table 7.7 and produces output similar to the above:

```
DATA SASExample;
    INPUT Initials $ Age CigsDay Solvent Cancer;
DATALINES ;
JBL     54 20 E  C
RGM     61 10 U  C
GTO     48  0 U  N
 ·  ·  · 220 rows not shown · · ·
NAT     73  0 E  N
WTS     59 30 E  C
RUN;
PROC FREQ DATA=SASExample;
    TABLES Solvent*Cancer / CL RISKDIFF RELRISK RDIFF;
RUN;
```

Figure 7.7 Analyzing disaggregated data with PROC FREQ.

7.8 EXERCISES

7.1 Random pullovers of 481 male and 138 female drivers found that 77 males and 16 females had detectable amounts of alcohol by a Breathalyzer test. Compute μ and σ, and sketch the approximate posterior distribution of the difference of the drunk driving rates $\Delta = p_M - p_F$. Compute the approximate 95% confidence interval for the difference. Compute the approximate posterior probability that the difference is positive. Compute the approximate posterior probability that the difference is bigger than 0.05 (5 percentage points).

7.2 A hypothetical study in a driving simulator compared 66 drivers wearing monofocal vs. 43 wearing multifocal lens implants. The table shows (somewhat simplified) data on the numbers of subjects (x) who failed to correctly recognize a "Truck Crossing" sign. Compute μ and σ of the approximate posterior distribution of the difference between the recognition rates $\Delta = p_{Mult} - p_{Mon}$. Obtain an approximate 95% confidence interval for the difference and say if it is statistically significant. Compute the approximate posterior probability that Δ exceeds 0.10 (10 percentage points).

	Monofocal	Multifocal
n	66	43
x	10	15

7.3 In a lawsuit for sex discrimination, the statistical expert for the plaintiff, a female employee, stated that over the previous 5 years the rate of promotions for females was lower than the rate for males and that the p-value for the difference, $\Delta = p_M - p_F$, was 0.034. What can you say about $P(\Delta > 0 \mid \text{data})$?

7.4 Compute μ and σ of the posterior distribution of $\ln(RR)$ using the data in Table 7.1. Compute the posterior probability that $RR \leq 0.9$.

7.5 Jonnson et al. (2002) also reported that the relative risk of cardiovascular death for underweight (BMI<20) compared to normal weight men was 1.07 with confidence interval 0.78 to 1.46.
(a) Is this 7% increased risk "statistically significant" in the sense that the 95% CI excludes 1.00?
(b) Compute the approximate posterior probability that $RR > 1$.

7.6 Obtain a 95% CI for the odds ratio using the data in Exercise 7.2.

7.7   Obtain μ and σ for the posterior distribution of ln(OR) for rubber and plastic workers in Zheng's study (Fig. 7.4 on page 130), and compute the approximate posterior probability that OR > 1.15.

7.8   Two hundred sixty dieters volunteered for a study of a fat-reducing dietary supplement. They were randomly assigned to receive a placebo or the supplement. One side effect is reduction in iron, a key component of hemoglobin. Here are the data on percentages of subjects with lower iron after 10 weeks of treatment.

	Placebo	Supplement
n	130	130
% with reduced iron	17%	28%

   (a) Describe the approximate posterior distribution of the log relative risk of reduced iron.
   (b) Compute the approximate posterior probability that the relative risk exceeds 1.
   (c) Compute the approximate posterior probability that the relative risk exceeds 1.5.

7.9   A fictional study investigated whether oral contraceptive use increased the rate of heart attacks. The fictional data are in the table below:

	Used oral contraceptives	Never used oral contraceptives
n of subjects	57	167
No. having heart attacks	23	35

   (a) Describe the posterior distribution of the log odds ratio.
   (b) Obtain a 95% CI for the odds ratio.
   (c) Compute the posterior probability that the odds ratio exceeds 1.0.
   (d) Would a conventional statistical report a statistically significantly increased risk? Explain why or why not.

7.10   Analyze the data in Exercise 7.7 using SAS PROC FREQ. Use oral contraceptives as the exposed group and heart attack as the adverse outcome. You will have to calculate the number of individuals experience the favorable outcome (no heart attack).

7.11   Analyze the data in Exercise 7.7 using WinBUGS. See Appendix Section B.4 for instructions.

7.9 REFERENCES

Carver, R.P., "The Case Against Statistical Significance Testing," *Harvard Educational Review* Vol. 48 (1978), No. 3, pp. 378–399.

Fisher, R.A., *The Design of Experiments*. Edinburgh; Oliver and Boyd, 1935.

Goodman, S.N., "Toward Evidence-Based Medical Statistics. 1: The *p* Value Fallacy," *Annals of Internal Medicine*, Vol. 130 (1999), pp. 995–1005.

Jonsson, S., Hedblad, B., Engstrom, G., Nilsson, P., Berglund ,G., and Janzon, L., "Influence of Obesity on Cardiovascular Risk. Twenty-Three-year Follow-up of 22,025 Men from an Urban Swedish Population," *International Journal of Obesity & Related Metabolic Disorders*, Vol. 26, No. 8 (Aug. 2002), pp. 1046–1053.

Michaud, D.S., Giovannucci, E., Willett, W.C., Colditz, G.A., and Fuchs, C.S., "Coffee and Alcohol Consumption and the Risk of Pancreatic Cancer in Two Prospective United States Cohorts," *Cancer Epidemiology, Biomarkers & Prevention*, Vol. 10 (May 2001), pp. 429–437.

Nurses Health Study, "History," http://www.channing.harvard.edu/nhs/, 2003.

Parkhurst, David F., "Commentaries on Significance Testing," www.indiana.edu/~stigtsts/, 1997.

SAS Institute, "Statistical computation: The FREQ procedure," *SAS OnlineDoc, Version 9*, http://v9doc.sas.com/sasdoc/, 2003.

van Balen, F. A., de Melker, R. A., Touw-Otten, F. W., "Double-Blind Randomized Trial of Co-Amoxiclav versus Placebo for Persistent Otitis Media with Effusion in General Practice," Lancet, Vol. 348, No. 9020 (1996), pp. 713-716.

Zheng, T. Cantor, K.P. Zhang, Y., Lynch, C.F., "Occupation and Bladder Cancer: A Population-based, Case-control Study in Iowa," *Journal of Occupational & Environmental Medicine*, Vol. 44, No. 7 (Jul. 2002), pp. 685-691.

8

Inference on Means

The previous chapter dealt with analyzing the success rate of a binary outcome such as survival vs. death or cancer vs. no cancer. In this chapter we begin to analyze the mean (or the median) value of a numerical outcome, such as the systolic blood pressure of hypertensive patients treated with a calcium channel blocker or the concentration of 16-androstene steroids in boar fat. For example, column T of Table 4.5 on page 60 lists consonant recognition scores of $n=20$ profoundly deaf adults implanted with a third generation cochlear implant. The mean of these 20 observations is $\bar{x} = 60.95$. What does this tell us about how this implant would function if implanted in the larger population of profoundly deaf adults (estimated to be about 400,000)? Bayes' rule, of course, gives the answer; but the calculation is not as simple as it was for analyzing the success rate of a Bernoulli process.

8.1 MODELS FOR MEASUREMENT DATA

The population of profoundly deaf patients implanted with third generation cochlear implants is modeled as a *box of numbers*. Imagine a box of plastic poker chips with a number engraved on each; each chip represents one profoundly deaf person implanted with a third generation implant, and the number engraved on that chip is that person's consonant recognition score, a number between 0 and 100. The box contains up to 400,000 chips, and the 20 observations in column T of Table 4.5 on page 60 can be thought of as a sample of $n=20$ chips from the box. The question is, what can this rather small sample tell us about the contents of the box? As usual, Bayes' rule provides the answer.

The application of Bayes' rule requires us to make a list of possible models for the contents of the box, assign each model a prior probability, compute the likelihood of obtaining the observed data given each model, and insert all these numbers into a Bayes' rule spreadsheet. The difficulty with this recipe is that the number of possible models is huge (there are 400,000 chips in the box, and each chip could be engraved with any number between 0 and 100, for a total of $101^{400,000}$ possible models. Clearly, listing all *possible* models for a box of numbers is hopelessly complicated, and we must devise some way to shorten the list. What is often done is to assume that the numbers engraved on the chips in the box have an approximately normal distribution as depicted in Fig. 8.1. If the

141

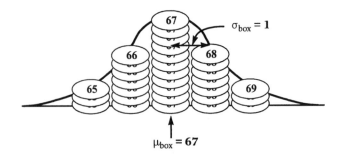

Figure 8.1   An approximately normal box of numbers.

box is approximately normal then there are only two unknown quantities: the parameters μ and σ of the normal distribution of the numbers in the box. We will use the symbols μ_{box} and σ_{box} to stand for these unknown quantities.

For the box representing the hypothetical population of 400,000 profoundly deaf adults wearing third generation cochlear implants, μ_{box} is the average of all the speech recognition scores in the box. It is the population counterpart of the sample average $\bar{x}$. Similarly, σ_{box} is the standard deviation of all the speech recognition scores in the box and is the population counterpart of the sample standard deviation s [Equation (4.2) on page 58]. Fig. 8.1 is a box model with an approximately normal distribution; however, while this box contains only 34 chips, the box of all profoundly deaf adults wearing third generation implants contains hundreds of thousands of chips. The question is, what can a sample of fewer than a hundred chips tell us about the box average, μ_{box}?

For flat priors and moderately large samples from approximately normal boxes there is an acceptable normal approximation to the posterior distribution of the box mean μ_{box}. As usual, the normal approximation consists of equations for the parameters μ and σ of the posterior distribution. The posterior distribution of μ_{box} is approximately normal with parameters μ and σ as shown in Fig. 8.2. The parameter μ of the posterior distribution is the sample mean $\bar{x}$, and its parameter σ is the *standard error of the mean*, or sem. The term "standard error" is well established in conventional statistical practice. We will use it to denote the parameter σ of a posterior distribution, thus sep is the standard error of a proportion or rate, sed is the standard error of a difference, and now sem is the standard error of a mean.

The normal approximation in Fig. 8.2 is acceptable if (1) n is larger than about 55, (2) the contents of the box have a reasonably normal distribution, and (3) the investigator uses flat priors for μ_{box} and $\ln(\sigma_{box})$ to represent a state of near-ignorance about the contents of the box prior to observing the data. If the contents of the box are not approximately normally distributed, then n must be larger to produce an accurate normal approximation; in that case it can be useful

Data:	
n	20
$\bar{x}$	60.95
s	22.55
Posterior distribution of μ_{box}:	
$\mu = \bar{x}$	60.95
$\sigma = \text{sem} = \dfrac{s}{\sqrt{n}}$	5.04
95% Credible interval:	
$\bar{x} \pm 1.96 \cdot \text{sem}$	50.1 to 70.8

Figure 8.2 Approximate posterior distribution of μ_{box}.

to search for a reexpression of the data to make the contents of the box more normal (see Section 8.4). If, instead of using flat priors, the investigator incorporates relevant prior information into the analysis, the approximation requires some modification. The next section shows how to modify the approximation for small n.

8.2 THE *t* FAMILY OF DISTRIBUTIONS

When the sample size n is small, the posterior distribution of μ_{box} is symmetric and bell-shaped, but it is nevertheless not normal. This is illustrated in Fig. 8.3, which shows the posterior distribution based on a sample of $n=7$ drawn from a perfectly normal box. The data are $\bar{x} = 61$, and $s = 12.25$. The posterior distribution, a member of the t family of distributions, is shown as a solid line. The normal approximation (dashed line) is too concentrated — too low in the tails and two high in the center.

The t distribution is actually a family of bell-shaped curves. In addition to the familiar center and spread parameters, μ and σ, there is a third, shape-determining parameter called the *degrees of freedom* (df) and denoted by the Greek letter nu (ν). The symbol $t(\nu)$ stands for a quantity having a standard t distribution with ν df. Fig. 8.4 shows t distributions with 3, 9, and 21 df; these are *standard* distributions in the sense that $\mu=0$ and $\sigma=1$. The normal distribution is shown as the limiting case of the t family as the number of degrees of freedom ν approaches infinity.

Later we will use expressions like $P(t(6) \leq 1.2)$ to represent the area under the standard $t(6)$ density curve to the left of 1.2 on the horizontal axis. Be alert

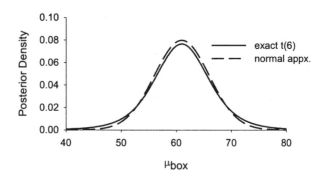

Figure 8.3   Exact t(6) posterior distribution of μ_{box} and the normal approximation. The normal approximation concentrates probability too close to the center.

not to interpret the expression $t(6)$ as "t times six"; the only role of the number of degrees of freedom (6) in that expression is to indicate which row in Table A.2 to use.

The t approximation to the posterior distribution of μ_{box} was first proposed by W. S. Gossett (1876–1937), writing under the nom de plume "Student." He was a research chemist for the Guinness Brewery, but had the problem of analyzing agricultural experiments with small sample sizes. He discovered that the normal approximation does not produce valid confidence intervals for small samples because the "1.96σ" rule is not accurate. It was he who discovered that the t family of distributions produced correct confidence intervals for μ_{box}. Student's t

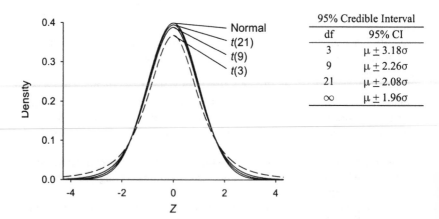

Figure 8.4   Some members of the t family and their 95% credible intervals.

approximation of the posterior distribution of the mean of a normal box, μ_{box} is given by

$$
\begin{aligned}
\mu &= \bar{x} \\
\sigma &= \text{sem} \\
\nu &= n\text{-}1
\end{aligned}
\tag{8.1}
$$

assuming flat priors for μ_{box} and $\ln(\sigma_{box})$.

The *standard* $t(\nu)$ distribution has parameters $\mu = 0$ and $\sigma = 1$. If we had a table similar to Table A.2 for each member of the t family, it would then be possible to compute posterior probabilities such as $P(\mu_{box} > 28 \mid data)$; however, such a table would require 100 pages for just the first 50 members of the t family. Instead, we have provided a *quantile* table in Table A.2. There is one row in the quantile table for each of the first 50 or so members of the t family. Table 8.1 reproduces a single row of Table A.2 — the row corresponding to the $t(6)$ distribution. This row would be used to compute posterior probabilities and credible intervals based on a sample of size n = 7.

The column heads of Table 8.1 list a few left tail areas, and the body of the table lists the t values (quantiles) that cut off those tail areas. The symbol for the qth quantile of the $t(\nu)$ distribution is $t_q(\nu)$. For example, the 25th percentile (the 0.25th quantile) of the $t(6)$ distribution is $t_{0.25}(6) = -0.72$. The 97.5th percentile, $t_{0.975}(6) = 2.45$, is particularly important because it determines the width of the 95% credible interval (CI):

$$
95\% \text{ CI: } \mu \pm t_{0.975}(\nu) \cdot \sigma
\tag{8.2}
$$

[We reiterate the warning not to interpret (ν) as a multiplier — its only function is to identify which row of Table A.2 to use.]

We now know that the normal CI for the mean consonant recognition score in Fig. 8.2 on page 143 can be improved. The df parameter is $\nu = n-1 = 19$, and the 97.5th percentile of $t(19)$ is $t_{0.975}(19) = 2.09$ (see row 19 of Table A.2); consequently, the 95% CI for μ_{box} is $\bar{x} \pm 2.09 \cdot$ sem, that is, $60.95 \pm 2.09 \times 5.04$, or 50.4 to 71.5, which is slightly wider than the normal credible interval, because the normal 97.5th percentile 1.96 was replaced by the corresponding $t(19)$ percentile 2.09.

Table 8.1   Fragment of Table A.2: quantiles of the standard t distribution.

	Left tail area											
ν	0.005	0.01	0.025	0.05	0.1	0.25	0.75	0.9	0.95	0.975	0.99	0.995
6	-3.71	-3.14	-2.45	-1.94	-1.44	-0.72	0.72	1.44	1.94	2.45	3.14	3.71

Computing Left Tail Areas.   Table A.2 permits somewhat crude calculation of tail areas. For example, suppose that we want to compute the posterior probability that third generation implants give the average patient at least 50% word understanding, that is, $\mu_{box} > 50$. We know from Equation (8.1) that the posterior distribution of μ_{box} is t(19) with $\mu = 60.95$ and $\sigma = 5.04$. The procedure for computing the tail area is the same as for a normal distribution: first use the Z transformation [Equation (6.3)] to covert the value 50 into a standard t(19) value, and then look up the tail area in Table A.2

$$
\begin{aligned}
P\left(\mu_{box} > 50 | \text{data}\right) &= P\left(t(19) > \frac{50-\mu}{\sigma}\right) \\
&= P\left(t(19) > \frac{50-60.95}{5.04}\right) \\
&= P\left(t(19) > -2.17\right)
\end{aligned}
$$

So far so good; however, we run into the problem that we do not have a detailed left-tail table for the t(19) distribution. The best we can do with Table A.2 is bracket the tail area: In row 19 we read that $P(t(19) \leq -2.54)$ = 0.01 and that $P(t(19) \leq -2.09) = 0.025$. Since -2.17 is between -2.54 and -2.09, the corresponding left tail area must be between 0.01 and 0.025, and the right tail area we seek must be between 0.975 and 0.99.

The precise left tail area can be computed with the spreadsheet function, TDIST. The spreadsheet formula for $P(t(\nu) \leq b)$ is

$$=\text{IF}(b<0,\text{TDIST}(-b,\text{nu},1),1-\text{TDIST}(b,\text{nu},1)) \qquad (8.3)$$

where b and nu are cell addresses, such as A1 and A2, of the cells containing the numerical values of b and ν Inserting $\nu = 19$ and $b = -2.17$ into expression (8.3) gives an left tail area of 0.0214 (which is indeed between 0.01 and 0.025) and therefore an upper tail area of 0.9786. A small spreadsheet, TTailArea.xls is available for downloading from the textbook website; Fig. 8.5 shows how to set up the spreadsheet to compute the tail area to the left and right of -2.17.

	A	B	C
1	Tail areas for t(df)		
2		(enter) df =	19
3		(enter) b =	-2.17
4	Left Tail	P(t(df) ≤ b) =	0.0214
5	Right Tail	P(t(df) > b) =	0.9786

Figure 8.5   Computing tail areas with tTailArea.xls.

8.3 COMPARING TWO MEANS

Table 4.5 on page 60 lists consonant recognition scores for 42 patients wearing two generations of cochlear implants. Third generation (T) devices incorporate changes suggested by experience with second generation (S) devices and are expected to provide improved speech recognition for the average patient. We will model these data as samples from two boxes of numbers. Box T contains consonant recognition scores that would be obtained if all eligible profoundly deaf patients were implanted with third generation implants. Box S contains scores that would be obtained if all eligible individuals had been implanted with second generation implants.

The data in Table 4.5 can be thought of as a sample of $n = 22$ chips from box S and a sample of $n = 20$ chips from box T. The population means (box means) are denoted μ_T and μ_S. The box standard deviations, denoted σ_T and σ_S, measure individual differences in speech recognition among deaf individuals wearing third and second generation devices, respectively. The research question is whether third generation devices provide more speech understanding to the average patient. Once again, we will use the symbol Δ to stand for the difference $\mu_T - \mu_S$. The research project sought to answer the question, "How strong is the evidence that $\mu_T > \mu_S$?" which is the same as asking, "How strong is the evidence that $\Delta > 0$?" To answer the question we must compute the posterior probability $P(\Delta > 0 \mid \text{data})$.

Expressing the research question in terms of Δ also allows us to answer questions about the size of the difference. For example, speech and hearing experts might consider anything less than a 15–point improvement to be clinically unimportant. Thus an important secondary question is, "How strong is the evidence that $\Delta > 15$?" To answer this we must compute the posterior probability $P(\Delta > 15 \mid \text{data})$. The next subsection shows how to compute posterior probabilities for Δ using a $t(\nu)$ approximation.

8.3.1 Sampling from Approximately Normal Boxes

Recall the discussion of auxiliary hypotheses in Section 7.4. It made the point that posterior probability rests on the truth of sometimes unstated auxiliary hypotheses; in the present case, one auxiliary hypothesis is that the contents of the two boxes are approximately normally distributed. Under that umbrella, the posterior distribution of the difference Δ can be approximated by a member of the t family as shown in Fig. 8.6, where we see that the posterior distribution of the mean difference (Δ) between the two generations of cochlear implants has an approximate $t(38)$ distribution with parameters $\mu = 24.8$ and $\sigma = 6.5$.

Armed with the parameters of the posterior t distribution, we can compute a 95% credible interval as shown in Fig. 8.6, and, with the help of the spreadsheet

Data:		
	S	T
n	22	20
$\bar{x}$	36.18	60.95
s	19.23	22.55
Posterior parameters of each box mean:		
$\mu = \bar{x}$	36.18	60.95
$\sigma = \text{sem}$	4.10	5.04
$\nu = \text{df} = n\text{-}1$	21	19
Posterior parameters of the difference:		
$\mu = \hat{\Delta} = \bar{x}_T - \bar{x}_S$	24.77	
$\sigma = \text{sed} = \sqrt{\text{sem}_T^2 + \text{sem}_S^2}$	6.50	
$\nu = \dfrac{\text{sed}^4}{\dfrac{\text{sem}_T^4}{\nu_T} + \dfrac{\text{sem}_S^4}{\nu_S}}$	38	
95% Credible Interval: $\hat{\Delta} \pm 2.02 \cdot \text{sed}$	(11.6, 37.9)	

Figure 8.6 t approximation of the posterior distribution of Δ

tTailArea.xls we can also compute the posterior probabilities of statements about the difference Δ. For example, the posterior probabilities that third generation devices produce 15 percentage points more speech recognition than second generation devices: is

$$P(\Delta > 15|\text{data}) \;=\; P\left(t(38) > \frac{15-\mu}{\sigma}\right) \;=\; P\left(t(38) > \frac{15-24.8}{6.5}\right)$$
$$=\; P(t(38) > -1.51) \;=\; 0.930$$

(The right tail area 0.930 was computed with the spreadsheet tTailArea.xls.)
    Computing interval and half-interval probabilities for a t distribution is essentially the same as for the normal distribution; the endpoint(s) of the interval are converted to Z values and the tail area to the left of each Z value is obtained from a table. The difference here is that the t tail areas are obtained from a spreadsheet rather than a printed table. The novel part of the t approximation is computing the approximate degrees of freedom using a formula originally proposed by Frank Satterthwaite:

$$\nu = \frac{\text{sed}^4}{\dfrac{\text{sem}_T^4}{\nu_T} + \dfrac{\text{sem}_S^4}{\nu_S}} \tag{8.4}$$

8.3.2   Interpreting Conventional Statistical Reports

Most conventional statistical software uses Satterthwaite's approximation to compute the confidence interval for Δ. Such confidence intervals can therefore be interpreted as flat-prior credible intervals and, if necessary, can be reverse engineered to deduce the parameters μ and σ of the posterior distribution of Δ.

In the older literature, confidence intervals may not be reported. In such cases a p value and sometimes a t statistic (t value) is reported:

$$t = \frac{\hat{\Delta}}{\text{sed}} = \frac{\bar{x}_T - \bar{x}_S}{\text{sed}} = \frac{24.77}{6.40} = 3.81 \tag{8.5}$$

We discussed the p values earlier in Section 7.5.4; the t value or t statistic is the basis of the famous t test of statistical significance: if the absolute value of the t statistic exceeds the 0.975th quantile (97.5th percentile) of the $t(\nu)$ distribution, then the observed difference $\hat{\Delta}$ is said to be *significantly significant*:

$$\hat{\Delta} \text{ is statistically significant if } t_{0.975}(\nu) \le \frac{|\hat{\Delta}|}{\text{sed}} \tag{8.6}$$

In the case of the cochlear implant data the t statistic equals 3.81 and there are 38 degrees of freedom. In row 38 of Table A.2 on page 302 we find that the 97.5th percentile of the $t(38)$ distribution is 2.02. Since the percentile is less than the observed t value ($2.02 < 3.81$), the difference would be declared statistically significant in a conventional statistical analysis. However, we made the case earlier that significance testing in general is not useful either as a way to summarize what the investigator has learned from the data or as a way to make a decision based on the data. Nevertheless, in the older literature the results of a significance test may be all that is reported.

If a confidence interval is reported, then the t test is superfluous since it will declare the difference significant if and only if the null hypothesis $\Delta=0$ is not inside the confidence interval.

The proof is fairly simple: The confidence interval is

$$\hat{\Delta} \pm t_{0.975}(\nu) \cdot \text{sed}$$

If the confidence interval excludes the point 0, then either 0 is less than the lower endpoint of the interval or is greater than the upper endpoint. Suppose that 0 less than the lower endpoint of the confidence interval:

$$0 < \hat{\Delta} - t_{0.975}(\nu) \cdot \text{sed}$$

Transfer the negative term to the left side of the equation,

$$t_{0.975}(\nu) \cdot \text{sed} < \hat{\Delta}$$

and divide through by sed,

$$t_{0.975}(\nu) \leq \frac{\hat{\Delta}}{\text{sed}}$$

Since $t_{0.975}$ is positive, so is $\hat{\Delta}$; consequently, $\hat{\Delta} = |\hat{\Delta}|$ and therefore,

$$t_{0.975}(\nu) \leq \frac{|\hat{\Delta}|}{\text{sed}}$$

which is what we set out to prove. The other possibility is that 0 is greater than the upper endpoint of the confidence interval. A similar line of reasoning implies

$$t_{0.975}(\nu) \leq \frac{-\hat{\Delta}}{\text{sed}}$$

But in this case $-\hat{\Delta}$ must be positive; consequently $-\hat{\Delta} = |\hat{\Delta}|$ and

$$t_{0.975}(\nu) \leq \frac{|\hat{\Delta}|}{\text{sed}}$$

which was what we set out to prove. The conclusion is that the t-test [Equation (8.6)] is equivalent to checking whether the confidence interval excludes the point $\Delta = 0$.

Some older research papers report p values only. This practice has been widely criticized and should be avoided in favor of a standard of statistical reporting that is sufficiently detailed that the reader can compute his or her own posterior probabilities. We earlier explained how to extract some meaning from the reported p value for a difference of rates [see Equation (7.11)], and those methods apply here as well; a p value less than or equal to 0.05 implies that the 95% confidence interval does not contain $\Delta = 0$. Furthermore, if (as is usually the case) the confidence interval was computed using Satterthwaite's approximation (see Fig. 8.5), then it can be interpreted as a flat-prior CI and $1 - p/2$ can be interpreted as the posterior probability that the true difference Δ has the same sign (positive or negative) as the observed difference $\hat{\Delta}$. In particular if $p \leq 0.05$, there is a better than 97.5% chance that the sign of $\hat{\Delta}$ is correct.

Some caution is required in interpreting confidence intervals as approximate CI's and half p values as posterior probabilities. The reader is safest when the author explicitly states that "unequal variance" or "Satterthwaite" formulas were used. Any confidence interval described as "nonparametric" or "exact" probably cannot be interpreted as a CI.

8.3.3 Interpreting p Values

The frequent and (in the opinion of many statisticians) inappropriate use of p values in conventional statistics forces us to devote more time to the topic than it deserves. We agree in advance that the topic is puzzling; the only defense we can offer for presenting it here is to give the reader some strategies for interpreting p values encountered in research reports.

As we understand the concept, the p value is the probability of obtaining the observed difference $\hat{\Delta}$ or value of $\hat{\Delta}$ that is further from 0 in a hypothetical study like the one being analyzed. The probability is computed assuming that the true difference is zero: $\Delta=0$ (the so-called null hypothesis). We commented at length on this concept in Section 7.5.4 and will not repeat that discussion here. We will simply begin by stating that the formula for computing the p value is:

$$p = 2 \cdot P\left(t(\nu) \geq \frac{|\hat{\Delta}|}{\text{sed}} \right) \tag{8.7}$$

For example, in the case of the cochlear-implant data analyzed in Fig. 8.6 the p value is $2 \cdot P(t(38) \geq 3.81)$. This number can be approximated using row 38–39 of Table A.2 or can be computed with the spreadsheet ttailArea.xls (see Fig. 8.5). The approximate value is between 0.01 and 0.00 and the precise value is 0.0002. We stated earlier that a p value can be re-interpreted as a posterior probability, and we now propose to prove it.

The primary research question in a comparative study of two drugs, surgical interventions, medical devices, etc., is whether there is strong evidence that one is better than the other; in other words, a highly relevant calculation is the probability that one drug (intervention, device, etc.) is better than the other, that is,

$$P(\Delta > 0 | \text{data}) = P\left(t(\nu) > \frac{0 - \mu}{\sigma} \right) = P\left(t(\nu) > -\frac{\hat{\Delta}}{\text{sed}} \right) \tag{8.8}$$

Let us convert Equation (8.8) into the posterior probability that the observed difference, $\hat{\Delta}$, has the correct sign. Suppose first that $\hat{\Delta}$ is negative (in which case $-\hat{\Delta} = |\hat{\Delta}|$). The probability that the sign is correct is the probability that the true difference, Δ, is also negative:

$$
\begin{aligned}
P\left(\Delta < 0|\text{data}\right) &= 1 - P\left(\Delta \geq 0|\text{data}\right) = 1 - P\left(t(\nu) \geq \frac{0-\mu}{\sigma}\right) \\
&= 1 - P\left(t(\nu) \geq -\frac{\hat{\Delta}}{\text{sed}}\right) = 1 - P\left(t(\nu) \geq \frac{|\hat{\Delta}|}{\text{sed}}\right) = 1 - \frac{p}{2}
\end{aligned}
$$

The last equality is by virtue of Equation (8.7).

By a similar argument, if $\hat{\Delta}$ is positive the probability that it has the correct sign is $P(\Delta > 0|\text{data}) = 1 - p/2$. Thus if the investigator has no relevant prior information about the contents of the two boxes other than that they are normally distributed, then the posterior probability that the observed difference has the correct sign is

$$
\begin{aligned}
P\left(\Delta > 0|\text{data}\right) &\cong 1 - \frac{p}{2} \quad \text{if} \quad \hat{\Delta} > 0 \\
P\left(\Delta < 0|\text{data}\right) &\cong 1 - \frac{p}{2} \quad \text{if} \quad \hat{\Delta} < 0
\end{aligned}
\tag{8.9}
$$

In conventional statistics, the difference is said to be statistically significant if the p value is less than or equal to 0.05 ($p \leq 0.05$); consequently, if the difference is significant, then there is at least 97.5% posterior probability that the sign of the observed difference is correct. Fig. 8.7 demonstrates that the p value criterion is equivalent to the confidence interval excluding zero.

There are two possibilities to consider: Either the confidence interval is entirely to the right of 0 or it is entirely to the left. If the confidence interval is entirely to the right of zero, as illustrated in Fig. 8.7, then the area to the left of zero (black) must be less than or equal to the area to the left of the confidence interval which, by definition is 2.5%. According to Equation (8.9) the black area is $p/2$ and since it is clearly smaller than the area to the left of the confidence interval, we conclude that $p/2 \leq 0.025$, hence $p \leq 0.05$, which is the p value criterion of statistical significance. The other case is a mirror image of Fig. 8.7. Since the 95% confidence interval, as usually computed approximates a flat-prior CI it is possible to say that the p value would be less than 0.05 and the difference would be considered statistically significant if the flat prior CI excludes the null hypothesis $\Delta = 0$.

We reiterate that this connection between the p value, which is not a Bayesian concept, and the 95% credible interval (CI), which is a Bayesian concept, is an approximation and is valid only if the prior distribution is flat, or *noninformative*. If the investigator has prior information that suggests that some values of Δ are more probable than others, then the conventional confidence

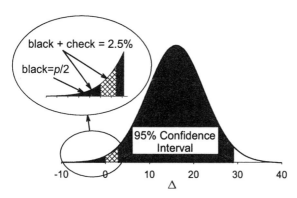

Figure 8.7   If the 95% confidence interval excludes zero then the p value is less that 0.05

interval does not approximate his or her credible interval and conventional p values could not be re-interpreted as posterior probabilities for that person. We do not advocate routinely interpreting $p/2$ as a posterior probability. The principal point we wished to make is that conventional practice; i.e. declaring a difference to be *insignificantly different from zero* when the p value exceeds 0.05, is risky, since a p value slightly above 0.05 is actually fairly strong evidence that the difference is *not* zero and is very likely to be in the direction indicated by the estimated difference. For example suppose that an observed difference is positive and the associated p value is 0.08. The difference would, by conventional statistical practice, be declared statistically insignificant, yet there would be approximately 96% posterior probability that the true difference was positive.

8.4   MAKING DATA NORMAL BY TRANSFORMATIONS

8.4.1   A Single Nonnormal Sample

In order to apply the methods of this chapter to data from nonnormal boxes the data must first be transformed to approximate normality. For example, the data below are from an imaginary driving simulator study involving $n = 48$ drivers:

Driving simulator data (% time in the wrong lane)
0.03, 0.03, 0.03, 0.03, 0.03, 0.08, 0.08, 0.08, 0.08, 0.14, 0.14, 0.14, 0.19, 0.19, 0.19, 0.19, 0.19, 0.19, 0.19, 0.25, 0.25, 0.31, 0.31, 0.31, 0.31, 0.36, 0.36, 0.36, 0.36, 0.42, 0.42, 0.47, 0.47, 0.47, 0.58, 0.64, 0.69, 0.75, 0.81, 0.92, 1.25, 1.36, 1.75, 1.97, 2.86, 2.97, 4.36, 5.53

The data values are the percentages of time that the driver placed any part of the vehicle in the wrong lane during a 30-minute simulated drive on a rural two-lane highway. Descriptive statistics are reported in the "Raw" column of Table 8.2. The data are extremely skewed, with mean 0.71 and median 0.31. The quartiles are Q1 = 0.165 and Q3 = 0.665. Consequently the quartile range is QR = Q3 − Q1 = 0.500. Any observation more the three OR's above Q3 is an extreme outlier, that is, any observation larger than 2.16 (see "*Outliers*" on page 56). In this case there are four extreme outliers indicating that the data almost certainly come from a nonnormal population.

This extreme nonnormality means that it is not appropriate to analyze these data using the methods of Section 8.3, because those methods were predicated on the auxiliary hypothesis that the observations were approximately normal. The most common way to deal with grossly non-normal data is to search for a transformation (perhaps the logarithm, reciprocal, or square root) to make the data appear more normal. Perfect normality is not necessary, since the methods of Section 8.3 are quite robust to moderate amounts of nonnormality. In practice it is enough that the transformed data be roughly symmetric (mean $\cong$ median) and that there be no extreme outliers.

There are systematic ways to search for a good transformation but it is often possible to find one by trial and error. For the simulated driving simulator data the log transformation $y = \ln(x)$ is as good as any at removing the outliers and making the mean and median more nearly equal (after transformation the mean is −1.16 and the median is −1.18. The transformed data are simply the natural logs of the raw data, for example the largest observation, 5.53 transforms to $\ln(5.53) = 1.71$ and the smallest transforms to $\ln(0.03) = -3.58$. Stemplots of raw and transformed data are in the middle panel of Fig. 8.8.

Table 8.2   Descriptive Statistics for Raw and Transformed Data.

	Raw: x	Transformed: $y = \ln(x)$
n	48	48
	Moments	
Mean	0.71	-1.16
Standard deviation	1.11	1.30
	Quantiles	
Min	0.03	-3.51
Q1	0.17	-1.77
Median	0.31	-1.17
Q3	0.67	-0.41
Max	5.53	1.71

The purpose of collecting these data was to discover the median time spent in the wrong lane for all drivers, not just the 48 drivers in the sample. There is no standard symbol for the median, so we'll just invent one,

$$\tilde{\mu}_{box} = \text{median of a box of numbers} \qquad (8.10)$$

The derivation of the posterior distribution of the box median is diagrammed in Fig. 8.8; the reader will probably find it helpful to consult the figure as we work through the derivation.

The useful thing about the median or any other quantile is that it transforms correctly: medians transform into medians, quartiles into quartiles, and so on, provided the transformation is *monotone* (always increasing or always decreasing), like the log transformation we're using in this example. For example, in Table 8.2 the raw median is 0.31, the transform of the median is $\ln(0.31) = -1.17$, and the median of the transformed data is -1.17. All quantiles transform correctly, apart from interpolation error when a sample quantile is the average of two adjacent values:

$$\text{Transformed quantiles equal quantiles of transformed data} \qquad (8.11)$$

For example, the first quartile of the raw data is the average of the 12th and 13th observations, $Q1 = (0.14 + 0.19)/2 = 0.165$. The first quartile of the transforms is $(\ln(0.14) + \ln(0.19))/2 = -1.81$. The transformed first quartile is $\ln(0.165) = -1.80$, which, apart from interpolation error, is equal to the first quartile of the transformed data. For large samples and for populations the interpolation error is negligible.

Moments, on the other hand, do not transform correctly. For example, the mean of the raw data is 0.71 and the transformed mean is $\ln(0.71) = -0.34$. However, the mean of the transformed data values is completely different, -1.16:

$$\text{Transformed moments are not equal to moments of transformed data} \qquad (8.12)$$

The final link in the derivation of the posterior distribution of the box median is provided by the symmetry of the normal distribution. If a distribution is symmetric, then the mean and median are exactly equal. Thus the mean and median of the *transformed* box will be exactly equal if the transformation has completely normalized the data. At this point we need another symbol,

$$_{T}\mu_{box} = \text{mean of the box of transformed numbers} \qquad (8.13)$$

Now we have the key relationship: the mean and median of the transformed box are equal, and the median of the untransformed box transforms correctly; consequently, for the log transformation,

$$\ln\left(\tilde{\mu}_{\text{box}}\right) = {}_T\mu_{\text{box}} \qquad (8.14)$$

and most usefully, exponentiating each side of the equation, we get

$$\tilde{\mu}_{\text{box}} = \exp\left({}_T\mu_{\text{box}}\right) \qquad (8.15)$$

Equation (8.15) is extremely useful because we know how to compute the posterior distribution of the mean of the transformed box ${}_T\mu_{BOX}$, because the contents of the box were rendered nearly normal by the transformation. It is safe to analyze such data using the methods of this chapter; specifically, we deduce from the t approximation in Equation (8.1) on page 145 that the posterior distribution of the mean of the transformed box is approximately $t(47)$ with $\mu = -1.16$ and $\sigma = \text{sem} = 1.30/\sqrt{48} = 0.187$. The 97.5th percentile of the $t(47)$ distribution is 2.01; thus, an approximate 95% credible interval for ${}_T\mu_{BOX}$ is $-1.16 \pm 2.01 \times 0.187$ or $(-1.54, -0.78)$. In other words, there is 95% posterior probability that

$$-1.54 \leq {}_T\mu_{\text{box}} \leq -0.78 \qquad (8.16)$$

If we exponentiate each member of Equation (8.16), the middle member is converted into $\tilde{\mu}_{BOX}$ (the median of the untransformed box), and the result is an approximate 95% credible interval for the box median

$$0.21 \leq \tilde{\mu}_{\text{box}} \leq 0.46 \qquad (8.17)$$

A WinBUGS program for this analysis is presented in Fig. 8.9. For details of running WinBUGS see Appendix B. The program can be modified to handle any monotone normalizing transformation; for example, if the square root successfully normalizes the data then two changes are needed: the log must be replaced with the square root:

```
y[i] <- log(x[i])   is replaced by   y[i] <- sqrt(x[i])
```

and the antilog (exp) must be replaced with the anti square root (second power):

```
mu.tilde <- exp(t.mu)   is replaced by   mu.tilde <- pow(t.mu,2)
```

Note that WinBUGS reported almost exactly the same confidence interval for $\tilde{\mu}_{BOX}$ as the approximate hand computation culminating in Equation (8.17); thus it appears that the hand computation method is reliable for moderate to large samples.

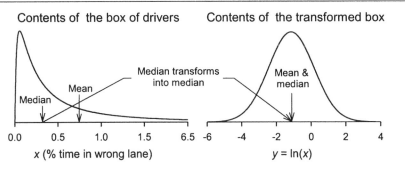

Contents of the box of drivers   Contents of the transformed box

1. ↓ Sample ↓

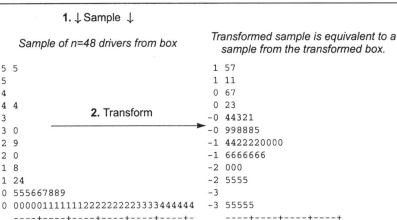

Sample of n=48 drivers from box

Transformed sample is equivalent to a sample from the transformed box.

3. ↓ Bayes' rule ↓

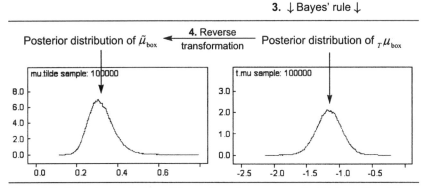

Figure 8.8 Posterior distribution of the median of a non-normally distributed box. The posterior distribution graphs are output from the WinBUGS program in Fig. 8.9 (mu.tilde = $\tilde{\mu}_{box}$ and t.mu = $_T\mu_{box}$).

```
MODEL DriveSim {
      # noninformative prior distribution of mu and tau
      # of the transformed box
            t.mu ~ dnorm(0,.001)
            t.tau ~ dgamma(.001,.001)
            t.sigma <- pow(t.tau,-2)

      # transform the data
            for (i in 1:n) {
                  y[i] <- log(x[i])
            }

      # likelihood of the data
            for (i in 1:n) {
                  y[i] ~ dnorm(t.mu,t.tau)
            }

      # back transform to median of original box
            mu.tilde <- exp(t.mu)
}

DATA list(n=48, x=c(0.03,0.03,0.03,0.03,0.03,0.08,0.08,
0.08,0.08,0.14,0.14,0.14,0.19,0.19,0.19,0.19,0.19,0.19,
0.19,0.25,0.25,0.31,0.31,0.31,0.31,0.36,0.36,0.36,0.36,
0.42,0.42,0.47,0.47,0.47,0.58,0.64,0.69,0.75,0.81,0.92,
1.25,1.36,1.75,1.97,2.86,2.97,4.36,5.53))

INITIAL VALUES list(t.mu=0,t.tau=1)
```

Output

node	mean[a]	sd[b]	MC error	2.5%[c]	median	97.5%	sample
mu.tilde	0.32	0.06195	1.941E-4	0.2157	0.3141	0.4576	100000
t.mu	-1.158	0.1916	6.024E-4	-1.534	-1.158	-0.7818	100000

a. μ of the posterior distribution.
b. σ of the posterior distribution (standard error).
c. 2.5% and 97.5% are endpoints of the 95% credible interval.

Figure 8.9   WinBUGS analysis of a sample from one nonnormal box.

8.4.2   Comparing the Medians of Two Nonnormal Boxes

The data in the upper panels of Fig. 8.10 are from a fictional study of whether a therapeutic drug has the side effect of impairing driving performance. The study involved 96 drivers evenly divided into a group of 48 receiving a placebo and a group of 48 receiving the drug under investigation. The measure of driving performance is the *crossing count*, number of times that each driver crossed the centerline of the highway into the wrong lane during a 30-minute simulated drive. The data in Fig. 8.10 are not real data from any study, but are statistically similar to data observed in several driving simulator studies (Weiler et al. 2000). The raw data are extremely nonnormal — the distributions (shown as stemplots) are extremely skewed, with most of the subjects crossing the centerline fewer than 10 times but with four extreme outliers in the drug data and two in the placebo data, indicating extreme nonnormality and causing the means to be about twice the as large as the medians (see *Outliers*, page 56). Therefore it is necessary to identify a normalizing transformation.

8.4.3   Box–Cox Power Transformations

The statisticians G.E.P. Box and D. R. Cox (1964) suggested that a power transformation could be used to approximately normalize the distribution of positive-valued observations:

$$y = \frac{(x+c)^{\lambda} - 1}{\lambda} \quad \text{if } \lambda \neq 0$$

$$y = \ln(y+c) \quad \text{if } \lambda = 0 \tag{8.18}$$

Here x stands for a raw, untransformed observation and y stands for its transformed value; the symbols λ and c are constants selected by the analyst. The *power* λ, typically between -1 and 1, is chosen to reduce the skewness of x and make the distribution of y as normal as possible. When $\lambda = 0$ and $c = 0$, the transformation is $y = \ln(x)$. If the variable x is equal to zero for some subjects, as it is in Fig. 8.10, then the usual practice is to add a positive *offset*, c, before applying the transformation, to avoid taking the log of zero or raising zero to a negative power. In Fig. 8.10 the offset is $c = 0.5$ and the power is $\lambda = 0$ so that the transformation is $y = \ln(x + 0.5)$.

   Box and Cox (1966) developed an automated method for identifying the best choice of the power λ; however, the offset c has to be discovered by trial and error. For counted data (number of lane crossings for example) the usual choice of c is 0.5 and the best choice of λ is often somewhere between 0.5 (the square root) and 0 (the logarithm). An implementation of the Box–Cox procedure (BoxCox.sas) as

	Placebo	Drug
		9 9
		9
		8
		8
		7 8
		7
		6
		6
		5
	5 0	5 13
Raw data (x)	4	4
	4	4
	3	3 5
	3	3 1
	2 ⑦	2
	2	2 24
	1 57	1 6
	1 00124	1 01234
	0 55556667788	0 5555666677888
	0 000000000011111112222222223334	0 00000111222333333344
	----+----+----+----+----+--	----+----+----+----
	Multiply Stem by 10	Multiply Stem by 10
n	48	48
Q1 Med Q3	1.0 2.5 7.0	2.5 5.0 11.5
$\bar{x}$	5.62	12.29
s	8.51	20.06
		4 6
		4 04
	3 9	3 69
	3 ③	3 124
	2 5779	2 5678
Transformed	2 0011444	2 0011144
data (y)	1 57777999	1 5577779999
	1 333	1 3333333
	0 99999999	0 999
	0 4444444	0 4444
	-0	-0
	-0 777777777	-0 77777
	----+----+-	---+----+--
Q1 Med Q3	0.41 1.09 2.01	1.08 1.71 2.48
$\bar{y}$	1.15	1.73
s	1.20	1.31

Figure 8.10 Transforming nonnormal data. The transformation $y = \ln(x+0.5)$ eliminates outliers and makes the transformed data more normal. Circles and arrow: monotone transformations are order-preserving and quantile-preserving.

a SAS *macro* (i.e. a user-written procedure) is available for downloading from the textbook website, see Section 8.7 for details. Box–Cox transformations are generally unsuccessful for bounded variables such as percentages and proportions, for which special transformations have been developed. The power transformation in Fig. 8.10 was found by the Box–Cox method; however, it is often possible to find an effective transformation by trial and error.

A transformation that works well for the raw crossing count data in Fig. 8.10 is $y = \ln(x + 0.5)$, that is, the Box–Cox transformation with $\lambda = 0$ and $c = 0.5$. This transformation makes the data much more normal in the sense that after transformation the placebo mean is approximately equal to the placebo median and the drug mean is approximately equal to the drug median and there are no outliers, as can be seen in the bottom panels of Fig. 8.10. It is important to understand that the same transformation must be used on both the placebo and the drug samples.

8.4.4 Posterior Distribution of the Difference of two Medians

This analysis involves two boxes of numbers, called box T (treatment) and box P (placebo). Each box contains the crossing count values for the very large population of drivers who would have been eligible for the simulator study. Box T contains the crossing-count values that would have been observed from these drivers under the treatment condition, and Box P contains the values that would have been observed under the placebo condition. The unknown parameters we are interested in are the median crossing counts in the two boxes,

$$\tilde{\mu}_{\text{trt}} = \text{median of box T} \qquad (8.19)$$
$$\tilde{\mu}_{\text{ctr}} = \text{median of box C}$$

What we are primarily interested in is whether the treatment group's driving performance is impaired by the drug, consequently the goal of the analysis is to compute the posterior distribution of the difference between the medians,

$$\tilde{\Delta} = \tilde{\mu}_{\text{trt}} - \tilde{\mu}_{\text{ctr}} \qquad (8.20)$$

(Despite the tilde notation, the difference of two medians is not a median.)

We learned in Section 8.4.1 how to compute the posterior distribution of a single box median, and the extension to two box medians is not difficult, however, there is one disappointment: there is no approximation that can be computed by hand. The only feasible method is to use Bayesian software such as the WinBUGS program listed Fig. 8.11.

```
MODEL CrossingCounts {
    # flat prior distribution of mu and tau
    # in two transformed boxes
        t.mu.trt ~ dnorm(0,.001)
        t.tau.trt ~ dgamma(.001,.001)
        t.mu.ctr ~ dnorm(0,.001)
        t.tau.ctr ~ dgamma(.001,.001)

    # transform the data and specify
    # the likelihood in two samples
        for (i in 1:n.trt) {
            y.trt[i] <- log(x.trt[i]+c)
            y.trt[i] ~ dnorm(t.mu.trt,t.tau.trt)
        }
        for (i in 1:n.ctr) {
            y.ctr[i] <- log(x.ctr[i]+c)
            y.ctr[i] ~ dnorm(t.mu.ctr,t.tau.ctr)
        }

    # back transform to medians of original boxes
    # and compute the difference of the medians
        mu.tilde.trt <- exp(t.mu.trt)-c
        mu.tilde.ctr <- exp(t.mu.ctr)-c
        delta.tilde <- mu.tilde.trt - mu.tilde.ctr
}

DATA list(c=0.5,n.trt=48,n.ctr=48,
x.trt=c(0,0,0,0,0,1,1,1,1,2,2,2,3,3,3,3,3,3,3,4,4,5,5,5,5,6,
6,6,6,7,7,8,8,8,10,11,12,13,14,16,22,24,31,35,51,53,78,99),
x.ctr=c(0,0,0,0,0,0,0,0,0,1,1,1,1,1,1,1,2,2,2,2,2,2,2,2,3,3,
3,4,5,5,5,5,6,6,6,7,7,8,8,10,10,11,12,14,15,17,27,50))

INITIAL VALUES
list(t.mu.trt=0,t.tau.trt=1,t.mu.ctr=0,t.tau.ctr=1)
```

Figure 8.11   WinBUGS program for comparing two nonnormal boxes.

The output in Table 8.3 indicates that the 95% posterior CI of the difference of the box medians, $\tilde{\Delta}$ or "delta.tilde," is 0.27 to 5.24. This shows that there is at least 97.5% posterior probability that the difference is positive (see Fig. 8.7 on page 153), which is strong evidence that the drug is somewhat impairing; however, the impairment could be negligible. Therefore, although the CI indicates fairly strong evidence that there is some impairment, a more focused response to the research question is needed. For example, the drug treatment seems to have

Table 8.3   WinBUGS Output: Posterior Moments and Quantiles.

node	mean	sd	MC error	2.5%	median	97.5%	sample
delta.tilde	2.548	1.258	0.004243	0.2738	2.479	5.235	100000
mu.tilde.ctr	2.699	0.5728	0.00176	1.723	2.648	3.971	100000
mu.tilde.trt	5.247	1.121	0.003837	3.365	5.144	7.746	100000

the effect of doubling the median crossing count. An interesting question, therefore, is what is the probability that the impression is correct, that is,

$$P\left(\frac{\tilde{\mu}_{trt}}{\tilde{\mu}_{ctr}} \geq 2 \Big| data\right) \tag{8.21}$$

To compute this probability and also the probability that the difference is positive, insert the following lines before the final closing brace "}" in the WinBUGS program in Fig. 8.11:

```
# compute the ratio and selected posterior probabilities
    ratio <- mu.tilde.trt/mu.tilde.ctr
# "treatment doubles the median"
        PRatgt2 <- step(ratio-2.0)
# "the difference is bigger than 0"
        PDifgt0 <- step(delta.tilde-0.0)
```

(For an explanation of how to use the step function to compute posterior probabilities see Appendix Section B.4.)

The output produced by these lines is in Table 8.4. It shows that the posterior probability that the difference is positive is 98.6% (the calculation is accurate to an MC error of ±0.04). Furthermore, there is 46.2% probability (accurate to ±0.15) that the drug doubles the median crossing count. This calculation used 100,000 simulated observations from the posterior distribution; for more accuracy it would be necessary to run more simulations (see Appendix B for an explanation of the accuracy of Monte-Carlo computation).

Table 8.4   WinBUGS output: Posterior Probabilities of Sentences of Interest.

node	mean	sd	MC error	2.5%	median	97.5%	sample
PDifgt0	0.9863[a]	0.1163[b]	3.654E−4	1.0	1.0	1.0	100000
PRatgt2	0.4617[c]	0.4985	0.001581	0.0	0.0	1.0	100000
ratio	2.032	0.626	0.001922	1.077	1.942	3.497	100000

a. Posterior probability that the difference of medians is positive.
b. Output in strikeout type is irrelevant and should be ignored.
c. Posterior probability that the ratio of medians exceeds 2.0.

The normalizing transformation used in this example was $y = \ln(x+c)$, and its reverse, or inverse, transformation was $x = \exp(y) - c$, with c assigned the value 0.5 in the DATA list. The WinBUGS program can be adapted to any monotone normalizing transformation by changing the lines containing the transformation and reverse transformation. To use a Box-Cox power transformation other than the logarithm, the normalizing transformation for the data in the treatment group is

```
y.trt[i] <- (pow(x.trt[i]+c,lambda)-1)/lambda
```
(8.22)

and the reverse transformation is

```
mu.tilde.trt <- pow(t.mu.trt*lambda+1,1/lambda)-c
```
(8.23)

with similar expressions for the control group. Values of both lambda and c must be assigned in the DATA list.

8.5 ANALYZING DESIGNED EXPERIMENTS

8.5.1 Randomization, Control, and Blinding

Chapters 7 and 8 deal with making comparisons between groups of subjects; examples include coffee drinkers vs. abstainers, patients wearing one type of cochlear implant vs. another, and people driving after taking a particular drug vs. a placebo. The first example was an *observational study*, and the other two are *experiments*. In an experiment the investigator controls which subjects receive a particular treatment (or are exposed to a particular condition), whereas in an observational study whether or not the subject was exposed to a possible risk factor is the result of many unknown factors, including the subjects' own choices.

Experimental studies are considered to be the gold standard for identifying effective medical interventions and elucidating cause-and-effect relationships. On the other hand, when the evidence for a supposed cause-and-effect relationship, such as the connection between smoking and lung cancer in humans, is derived entirely from observational studies, it is always possible to propose alternative explanations for the observed relationship. For example, it has been argued there is evidence that the intensity of smoking has a genetic component. Therefore, it is argued, if smoking is partly genetic, then it might be linked with a genetic predisposition to cancer. In other words, perhaps genes cause smoking and also cause cancer. If that were the explanation, then inducing people to stop smoking would not prevent the cancer that is their genetic destiny.

Ideally, an experiment is *randomized, placebo-controlled*, and *double-blind*. We'll explain these concepts with a hypothetical example: Four hundred subjects were recruited to test the efficacy of an antihistamine for reducing allergy symptoms.

The subjects were randomly assigned to drug or placebo and sat for 5 hours in a pollen chamber exposed to 3000 grains of pollen per cubic meter of air. Subjects recorded their symptoms on a paper-and-pencil scanner form every half hour, and the average number of symptoms was calculated for each subject.

Double blinding means that neither the subject nor the investigators is aware of who received the placebo and who received the drug. In practice, this is accomplished by having the drug and placebo doses packaged by an independent group unconnected with the study sponsor or with the investigator. The packaging group would be responsible for preparing 200 doses of placebo and 200 of the drug, perhaps in individual pill bottles. The doses would be labeled "Subject 1," "Subject 2," etc., in random order and delivered to the study site. At this point only the packaging group would know whether subject 1 received drug or placebo, although for patient safety in case of an adverse reaction the identity of the dose is under a tamper-evident peel-off label on the pill bottle. The bottles are returned to the packaging group, and data from subjects corresponding to peeled-off labels are either not used in the analysis or recorded as treatment failures. After the experiment is completed and the data have been certified and locked, the packaging group *unblinds* the investigator by revealing the contents of the pill bottles.

In complex experiments it may be necessary to use multiple placebos (called dummies). For example, in a comparison of drug A (delivered in a gelatin capsule) and drug B (delivered as a tablet) vs. no drug, each subject would receive a gelatin capsule and a tablet, one or the other (or both) of which would be inactive. Such an experiment would be described as randomized, placebo-controlled, double-blind, double-dummy.

Randomization is the responsibility of the packaging group. The mechanics of randomizing five doses of drug and five of placebo are illustrated in Table 8.5. The method consists in listing the required doses (five drug, five placebo) in column A, entering random numbers in column B and sorting column A in order of column B. The packaging group delivers the bottles to the study site with the identity of the contents (drug or placebo) concealed by a tamper-evident peel-off label.

8.5.2 Varieties of Experimental Designs

Crossover Designs. In a *crossover design* each subject experiences all of the treatments. For example, in a fictional driving simulator study 10 subjects were recruited to participate in a drug study. The purpose of the study was to determine if a drug (D) produced more driving impairment than a placebo (P). Each subject had three driving sessions (periods), one week apart, at the same time on the same day of the week. The first drive was for practice; in the second period each subject received placebo or drug at random, and in the third period the subject received the other treatment (see Table 8.6). Each subject receives

Table 8.5   Randomization using computer-generated random numbers.

Unsorted		Sorted by Random Numbers		
Contents of bottle	Random numbers	Contents of bottle (concealed under label)	Random numbers	Label on bottle
D	0.5941	D	0.0580	Subj 01
D	0.9866	P	0.1485	Subj 02
D	0.0580	D	0.3833	Subj 03
D	0.3833	P	0.4463	Subj 04
D	0.7920	P	0.5436	Subj 05
P	0.8313	D	0.5941	Subj 06
P	0.5436	D	0.7920	Subj 07
P	0.1485	P	0.8313	Subj 08
P	0.9535	P	0.9535	Subj 09
P	0.4463	D	0.9866	Subj 10

both placebo and drug at different time periods and thus each is his or her own control.

As shown in Table 8.7, five subjects received the drug in period 1 and received placebo in period 2; and another five received placebo in period 1 and drug in period 2. An independent packaging group prepared two doses for each subject, one dose for period 1 and one for period 2, and randomly assigned subjects to *order* (drug first or placebo first). The randomization is *balanced* in the sense that equal numbers of subjects receive drug first or placebo first. This way of conducting the study would be described as a *randomized, placebo-controlled, two-arm, two-period crossover design.*

In our fictional study, poor driving performance was measured by steering instability, a measure (in meters) of how far the driver weaves left and right of his or her habitual position in the lane. Drivers with high instability often cross over into the oncoming lane. Panel A of Table 8.7 displays the steering instability data as they were collected at the study site. At that point, the investigator was still *blinded* (in ignorance of treatment-placebo assignment). Panel B shows how the data were assigned to treatment or control after the investigator was *unblinded.*

Table 8.6   Structure of a crossover experiment.

	Practice phase	Experimental phase	
Period:	0	1	2
Medication:	No medication	Drug or placebo	Placebo or drug

Table 8.7   Blinded (A) and unblinded (B) steering instability data.

	A. Blinded Data			B. Unblinded Data		
Subj	Order	Period 1	Period 2	Drug	Placebo	Difference
1	DP	0.45	0.47	0.45	0.47	−0.02
2	DP	0.47	0.49	0.47	0.49	−0.02
3	PD	0.48	0.51	0.51	0.48	0.03
4	DP	0.48	0.47	0.48	0.47	0.01
5	DP	0.51	0.47	0.51	0.47	0.04
6	PD	0.45	0.50	0.50	0.45	0.05
7	PD	0.48	0.48	0.48	0.48	0.00
8	DP	0.60	0.48	0.60	0.48	0.12
9	PD	0.44	0.52	0.52	0.44	0.08
10	PD	0.48	0.50	0.50	0.48	0.02
				n		10
		Data analysis: differences		$\bar{d}$		0.0310
		of paired observations are		s		0.0441
		analyzed as a single sample.		sem		0.0139
				ν		9
				95% CI		-0.0044 to 0.0624

For example, after unblinding, the investigator knew that the period 1 observation for subject 1 was obtained under the drug condition.

These data are analyzed using the single-box analysis in Fig. 8.2, despite the fact that there are two treatments. We are able to treat this as a single-box analysis by working with individual differences, because the data from each subject are *paired observations*. The model for this experiment can be thought of as a box of plastic chips (like Fig. 8.1). The chips in the box represent the population of people for whom the drug would be appropriate. However, each chip bears three numbers — the steering instability score the person would receive under the drug condition, the steering instability score the person would receive under the placebo condition, and the difference (d) between the two. The research question is whether the mean value of the difference is positive, meaning that the average person would drive more erratically in the drug condition than in the placebo condition; consequently, although the experiment reveals all three numbers for each subject, it is the differences (d) that are analyzed.

The box mean of the differences is denoted Δ_{box}. The research question is whether $\Delta_{box} > 0$. To answer it we need the posterior distribution, which is computed at the bottom of Table 8.7. The calculation indicates that the posterior distribution of Δ_{box} is approximately $t(9)$ with parameters $\mu = \bar{d} = 0.310$ and $\sigma = $ sem $ = 0.0139$. The 97.5th percentile of the standard t distribution with $\nu = 9$

degrees of freedom is 2.26 (see Table A.2 on page 302), so the credible interval is $0.310 \pm 2.26 \times 0.139$ or -0.0044 to 0.0624.

If the differences are nonnormal, it may be necessary to transform the observations before computing differences. Back-transformation of the posterior distribution to the original measurement scale using WinBUGS is possible, but requires statistical techniques (mixed general linear models) that are beyond the scope of this book.

Completely Randomized Designs. Table 8.8 is an example of a *completely randomized design.* In this experimental design each subject experiences one, and

Table 8.8   Analysis of data from a completely randomized experiment.

A. Data			
Placebo arm		Drug arm	
Subject	Symptom count	Subject	Symptom count
3	21.8	1	10.3
5	22.6	2	9.5
7	18.7	4	11
10	19.4	6	13.6
12	19.9	8	11.9
13	20.2	9	10.6
15	20.5	11	12.8
16	18.8	14	10.2
18	20.1	17	11.6
20	19.3	19	11.6
B. Statistics			
n_P	10	n_D	10
ν_P	9	ν_D	9
$\overline{x}_P$	19.21	$\overline{x}_D$	11.19
s_P	1.25	s_D	1.25
$sem_P = s_P/\sqrt{n_P}$	0.395	$sem_D = s_D/\sqrt{n_D}$	0.395

C. Parameters of the normal approximation of the posterior distribution of Δ

$\mu = \hat{\Delta} = \overline{x}_D - \overline{x}_P$	8.020
$\sigma = \text{sed} = \sqrt{\text{sem}_P^2 + \text{sem}_D^2}$	0.559
$\nu = \dfrac{\text{sed}^4}{\dfrac{\text{sem}_P^4}{\nu_P} + \dfrac{\text{sem}_D^4}{\nu_D}}$	18

only one, randomly assigned treatment. In this fictional experiment, 20 subjects were recruited to test the efficacy of an antihistamine for reducing allergy symptoms. The subjects were randomly assigned to drug or placebo. Subjects sat for 5 hours in a pollen chamber exposed to 3000 grains of pollen per cubic meter of air and recorded their symptoms on a scanner form every half hour. The data in Table 8.8 are the average numbers of symptoms reported by each subject.

Data from a completely randomized design are analyzed using the "two-box" analysis presented in Section 8.3.1. The analysis in Table 8.8 indicates that the posterior distribution of the difference of the mean symptom counts is approximately $t(18)$ with $\mu = 8.020$ and $\sigma = 0.559$, and with the number of degrees of freedom $\nu = 18$ (in round numbers) calculated using Satterthwaite's formula [Equation (8.4) on page 149]. The 97.5^{th} percentile of the t(18) distribution is 2.12 (see Table A.2 on page 302); consequently the 95% confidence interval for the difference Δ is $8.020 \pm 2.12 \times 0.559$ or 6.83 to 9.21. The parameters μ and σ can be used to compute posterior probabilities as in Section 8.3.1

If data from a completely randomized experiment are nonnormal, then it may be necessary to find a normalizing transformation and back-transform the posterior distribution to the original scale as explained in Section 8.4.4.

8.6   COMPARING MEANS AND ANALYZING EXPERIMENTS IN SAS

SAS has several procedures for comparing means and analyzing experiments. For comparing two means, PROC TTEST computes confidence intervals but it does not appear to apply the Satterthwaite approximation to the confidence interval, and therefore cannot be recommended for general use. The MIXED procedure can be instructed to apply Satterthwaite's approximation and is recommended for general use for the purpose of comparing means, in problems from those as simple as comparing two means to those as complex as comparing treatment means in a crossover design.

8.6.1   Comparing Two or More Means

The basic syntax for comparing two or more means is

```
PROC MIXED DATA=lib_name.file_name;
   CLASS categorical_var;
   MODEL response_var = categorical_var / DDFM=SATTERTH;
   REPEATED / GROUP = categorical_var;
   LSMEANS categorical_var / CL DIFF;
RUN;
```

Here *categorical_var* defines the groups to be compared (gender, for example), and *response_var* is the variable being compared (reaction time, for example).

Fig. 8.12 shows how to use PROC MIXED to compute a 95% confidence interval for the difference between male and female reaction times in Table 4.1. The confidence interval can be interpreted as an approximate 95% posterior CI for an analyst with a noninformative (flat) prior distribution. The posterior distribution of the difference is approximately $t(29)$ with $\mu = 0.050$, and $\sigma = 0.225$. The t value is defined in Equation (8.5), and the p value is defined in Equation (8.7) can be re-interpreted as a posterior tail area using Equation (8.9). In this case $P(\Delta > 0 \mid \text{data}) \approx 1 - p/2 = 0.983$. It is left as an exercise to the reader to verify that hand calculation as illustrated in Fig. 8.6 produces the same results.

```
DATA ReacTime;
    INPUT Semester $ Initials $ Gender $ Age Distance Time;
DATALINES;
Fall    AEI    M   24  11.0   0.153
Fall    SUN    F   22  28.0   0.241
Fall    SLA    M   23  9.00   0.139
... 35 lines not shown ...
Spring GTN     M   22  8.70   0.133
Spring IRD     F   21  24.0   0.221
Spring BHB     F   21  20.0   0.201
RUN;

PROC MIXED DATA=ReacTime;
    CLASS Gender;
    MODEL Time=Gender / DDFM=SATTERTH;
    REPEATED / GROUP=Gender;
    LSMEANS Gender / DIFF CL;
RUN;
```

OUTPUT (edited for clarity):

Least Squares Means

Effect	Gender	Estimate	Standard Error	DF	Lower	Upper
Gender	F	0.2389	0.02000	19	0.1970	0.2807
Gender	M	0.1889	0.01036	20	0.1672	0.2105
Key:		$\bar{x}$	sem	ν	95% CI	

Differences of Least Squares Means

Gender−Gender		Estimate	Standard Error	DF	t-Val	Pr>|t|	Lower	Upper
F	M	0.04999	0.02252	28.6	2.22	0.0345	0.003897	0.09609
Key:		$\hat{\Delta}$	sed	ν	t value	p value	95% CI	

Figure 8.12   SAS PROC MIXED calculation of the 95% confidence interval for the difference of mean male and female reaction times in Table 4.1.

8.6.2   Analyzing Completely Randomized Experiments

In a hypothetical study three types of implantable medical device are available (A, B, and G3). Types A and B are currently approved; type G3 is experimental. The response variable is a measure of how much lost functionality is restored to the patient wearing the device (0 = none, 100 = complete restoration). The data are available in a Microsoft Excel® table (downloadable as DeviceStudy.xls). The first column (DevType) is a categorical variable identifying the type of device worn by the subject, and the second column (FunRes) is the response variable:

Table 8.9   Contents of DeviceStudy.xls

DevType	FunRes
G3	87
G3	78
— 17 rows not shown —	
G3	81
B	40
B	65
— 17 rows not shown —	
B	26
A	59
A	61
— 17 rows not shown —	
A	44

The SAS program in Fig. 8.13 first imports the data table and then compares the three device types. The p values in the table of differences indicate that there is approximately 99.9% posterior probability $(1 - 0.0019/2)$ that the mean of patients implanted with device B is smaller than that for G3 (assuming a noninformative prior distribution), and about 99.8% posterior probability $(1 - 0.0044/2)$ that device A provides less benefit than G3. The credible intervals for each of these differences excludes $\Delta = 0$; however, the credible interval for the difference between device types A and B does include $\Delta_{AvsB} = 0$. The posterior probability that device A provides more benefit than device B is approximately 66% $(1 - 0.6876/2)$, however the investigator judged that the difference would have to be at least 10 points to be clinically important. The posterior probability of that is

$$P\left(\Delta_{A-B} > 10 | data\right) = P\left(t(38) > \frac{10 - \mu}{\sigma}\right) = P\left(t(38) > \frac{10 - 2.15}{5.3057}\right)$$
$$= P\left(t(38) > 1.48\right) = 0.0736$$

```
PROC IMPORT OUT=DeviceStudy
    FILE="C:\TEMP\DeviceStudy.xls"
    DBMS=EXCEL REPLACE;
RUN;
PROC MIXED DATA=DeviceStudy;
    CLASS DevType;
    MODEL FunRes=DevType / DDFM=SATTERTH;
    REPEATED / GROUP=Type;
    LSMEANS Type / DIFF CL;
RUN;
```

OUTPUT (edited for clarity)

Least Squares Means

Effect	DevType	Estimate	Standard Error	DF	Lower	Upper
DevType	A	57.70	3.6383	19	50.0849	65.3151
DevType	B	55.55	3.8617	19	47.4673	63.6327
DevType	G3	73.95	3.9500	19	65.6826	82.2174
	Key:	$\bar{x}$	sem	ν	95% CI	

Differences of Least Squares Means

DevType—DevType		Estimate	Standard Error	DF	t-Val	Pr>|t|	Lower	Upper
A	B	2.15	5.3057	37.9	0.41	0.6876	-8.5921	12.8921
A	G3	-16.25	5.3703	37.7	-3.03	0.0044	-27.1240	-5.3760
B	G3	-18.40	5.5241	38	-3.33	0.0019	-29.5831	-7.2169
	Key:	$\hat{\Delta}$	sed	ν	t value	p value	95% CI	

Figure 8.13 Analysis of a completely randomized experiment comparing three medical devices. Three differences between device types are reported. Device G3 is clearly superior to either A or B.

because the posterior distribution of each difference is approximately $t(\nu)$ with $\mu = \hat{\Delta}$ and σ=sed.

Thus, there is no evidence of a clinically important difference between device types A and B.

Notice that differences are computed in alphabetical order, for example B minus G3, which is negative indicating that the average benefit for patients implanted with device G3 was larger than that for those implanted with B.

8.6.3   Analyzing Crossover Experiments

Crossover experiments are somewhat trickier to analyze. The problem is that there are two confounding variables that might have to be statistically removed from the treatment comparisons: time and carryover. Therefore, this analysis is covered in the next chapter.

8.7   SUPPLEMENT: THE BOX-COX PROCEDURE

A *macro* is a user-written SAS procedure. A SAS macro called BoxCox.sas is available for download from the textbook website. The SAS program to read the reaction time data (Table 4.1 on page 48) is listed in Fig. 8.14 and the output is in Fig. 8.15. The BoxCox.sas macro must be downloaded from the textbook website as a plain text file and saved on the user's hard drive or on a floppy disk. The program in Fig. 8.14 assumes that the program was downloaded into the directory C:\temp\. See Appendix C for instructions on running SAS programs.

The syntax of the macro is,

%BoxCox(var_name + offset, data_file_name, cat_var, range);

The percent sign and semicolon are mandatory, and the arguments are

- var_name: the variable in need of transformation
- offset: (optional) an offset c may be added to the variable
- data_file_name: the name of the SAS data file containing the data to be analyzed
- cat_var: the name of the categorical variable that defines the groups to be compared
- range: the range of lambda values to be checked

For example in Fig. 8.14 the variable to be transformed is distance, the file name is ReacTime, the groups to be compared are defined by gender, and the analyst chose to use no offset and to try λ values between -1 and 1 in steps of 0.25. Thus the *macro call* is,

%BoxCox(distance, ReacTime,gender, -1 to 1 by 0.25);

The relevant part of the output is the low resolution graph shown in Fig. 8.15. The best choice of λ is at the low point of the graph ($\lambda = 0.25$, the fourth root); consequently a good transformation to try is

$$y = \frac{x^{0.25} - 1}{0.25}$$

```
DATA REACTIME;
    INPUTsemester $ initials $ gender $
    age distance time;
DATALINES;
FallAEI    M      24     11.0   0.153
FallSUN    F      22     28.0   0.241
FallSLA    M      23      9.0   0.139
    ...  34 lines not shown  ...
SpringGTN  M      22      8.7   0.133
SpringIRD  F      21     24.0   0.221
SpringBHB  F      21     20.0   0.201
;;;;

%INCLUDE "C:\TEMP\BOXCOX.SAS";
%BOXCOX(REACTIME,DISTANCE,GENDER,-1 TO 1 BY .25);
```

Figure 8.14 SAS program to identify a good Box-Cox transformation for the distance variable. The included program, BoxCox.sas, must be downloaded from the textbook website.

The analyses presented in this section treat the power λ as if it were a known quantity; however, in a proper Bayesian analysis it would be treated as another unknown quantity with a prior distribution. The result would be to somewhat increase the dispersion of the posterior distribution of the box median. Put another way, the analyses presented in this section slightly understate the widths of credible intervals for box medians and differences of box medians.

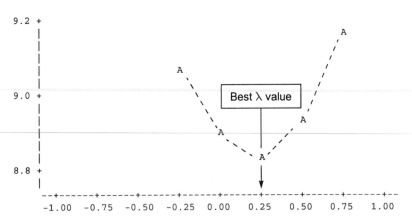

Figure 8.15 Output of BoxCox.sas. Dotted lines were added here to clarify the graph. The best λ value, 0.25, is at the lowest point of the graph.

8.8 EXERCISES

8.1 An economic survey of a sample of 225 US wage earners showed an average of $\bar{x}=\$23.50$ was spent per week dining out. The standard deviation of the sample was reported to be $s=\$12.00$. Obtain a 95% credible interval for the mean of all wage earners.

8.2 Dehydroepiandrosterone (DHEA) is produced by the adrenals. There is weak evidence that it might strengthen the immune system and slow down the effects of aging. That was the motivation for the study summarized in the table below. The study involved collecting blood samples from 30 practitioners of transcendental meditation and 51 controls who did not practice it. Notice that the data are reported as $\bar{x} \pm$ sem. [Source: Glaser, et al. 1992.]

Obtain the three parameters of the approximate t posterior distribution of Δ. Compute the approximate 95% credible interval for Δ and compute $P(\Delta > 0 \mid \text{data})$.

Subjects were not randomly assigned to the two arms of this experiment so it is not exactly clear what "boxes" are being sampled here. Explain why the difference between the means might not necessarily be due to Transcendental Meditation.

Serum DHEA-S (mean $\pm$ sem) for women aged 45-49

		Non-TM sample		TM sample
Age group	n	DHEA-S level (μg/dl)	n	DHEA-S level (μg/dl)
45–49	51	88 ± 12	30	117 ± 11

8.3 The table below reports the results of a randomized experiment comparing patients wearing a nicotine patch or a placebo patch. The response variable is the number of cigarettes smoked per day. Note that the data are reported as mean $\pm s$, not sem. Obtain the normal approximation of the posterior distribution of Δ defined as $\mu_P - \mu_A$. Both n's exceed 55 so you do not have to compute ν. Compute the approximate 95% credible interval. Compute the normal approximation of $P(\Delta > 0 \mid \text{data})$.

Cigarettes per day with and without active nicotine patch (mean $\pm s$).

Active ($n=119$)	Placebo ($n=119$)
28.8 ± 9.4	30.6 ± 9.8

8.4   Here are some t and p values for differences Δ. Assuming that the observed difference is positive, convert these into approximate posterior probabilities of the form $P(\Delta > 0 \mid \text{Data}) = ???$. In which cases would a conventional statistical report pronounce the difference significant? Why?
   (a) $t = 2.01, \nu = 16$
   (b) $t = 3.01, \nu = 27$
   (c) $p = 0.073$
   (d) $p = 0.0001$

8.5   An investigator measured fecal coliform counts at 25 sites upstream and 18 sites downstream of the outlet of a sewage treatment facility. Her statistics are shown in the table below. Do you see evidence of skewness and extreme outliers? What is the evidence? What additional data would you need to compute the posterior distribution of the difference?

Fecal Coliform Statistics

	Upstream	Downstream
n	25	18
$\overline{x}$	40	500
s	90	800
Q1	0	40
Median	5	150
Q3	10	1000
Max	350	4500

8.6   Two hundred forty-four male alcoholics suffering from secondary hypertension participated in a study to determine the efficacy of a new antihypertensive agent. The men were assigned at random to either the control group or the treatment group. Men in the control group received a placebo. Statistics for arterial pressure at 30 days post treatment for the subjects who completed the study are shown in the table below:

	Placebo	Treatment
n	100	144
Mean ($\overline{x}$)	140.3	102
Standard deviation (s)	15.0	7.2

   (a)   Describe the approximate posterior distribution of the difference (P minus T) between the means of the two populations.
   (b)   Compute an approximate 95% CI for the difference.
   (c)   Compute the approximate posterior probability that the difference is positive.

8.7 The data in the table below are the survival times in days of closely and poorly matched skin grafts. Each of the eleven burn patients in the study received both types of graft on different parts of his or her body. [Source: Holt et al. 1974]. In some of the subjects (marked with a +) graft failure did not occur; however, for the purposes of this assignment, treat those numbers as survival times.

Compute $\bar{x}$, s, and sem for the differences (close – poor). Compute the approximate 95% credible interval for the difference. Compute the posterior probability that the mean difference is greater than 0.

Skin graft survival time data

Patient	Close match	Poor match
1	37	29
2	19	13
3	57+	15
4	93	26
5	16	11
6	22	17
7	20	26
8	18	21
9	63	43
10	29	15
11	60+	40

8.8 Modify Fig. 8.11 to compare median male and female distances using the data in Table 4.1 on page 48. The samples have skewed distributions with at least one outlier, so a transformation is needed. The fourth root transformation does an excellent job of normalizing the data. To use it, replace the log transformation with the fourth root transformation [equations (8.22) and (8.23) with lambda = 0.25, c = 0], and use the following DATA statement:

```
DATA list(lambda=0.25,c=0,n.fem=20, n.mal=21, x.fem=c(28,
29, 38, 60, 40, 71, 31, 26, 26, 113, 14, 10, 15, 23.5, 30,
19, 2, 15, 24, 20), x.mal=c(11, 9, 6, 32, 23, 29, 24, 25,
18, 26, 11, 42, 11, 18, 13, 15, 19.5, 20, 21.5, 7, 8.7))
```

(a) Compute the posterior 95% CI for the difference of male and female median distances (female – male).

(b) Use the WinBUGS step( ) function to compute the posterior probability that the difference is positive, and the posterior probabilities that the ratio of medians exceeds 1.25 and that it exceeds 1.50.

8.9   REFERENCES

Box, G. E. P. and Cox, D. R., "An Analysis of Transformations," *Journal of the Royal Statistical Society*, Vol. 26 (1964), pp. 211–243; discussion 244–252.

Glaser, J.L., Brind, J.L., Vogelman, J.H., Eisner, M.J., Dillbeck, M.C., Wallace, R.K., Chopra, D., Orentreich, N., "Elevated serum dehydroepiandrosterone sulfate levels in practitioners of the Transcendental Meditation (TM) and TM-Sidhi programs," *Journal of Behavioral Medicine*, Vol 15, No. 4 (Aug 1992), pp. 327-41.

Holt, J.D. and Prentice, R.L., "Survival Analysis in Twin Studies and Matched Pairs Experiments," Biometrika, Vol. 61 (1974), pp. 17-30.

Weiler, John, M., Bloomfield, John R., Woodworth, George G.,Grant, Angela R., Layton, Teresa A., Brown, Timothy L., McKenzie, David R., Baker, Thomas W., and Watson, Ginger S., "Effects of Fexofenadine, Diphenhydramine, and Alcohol on Driving Performance, A Randomized, Placebo Controlled Trial in the Iowa Driving Simulator," *Annals of Internal Medicine*, Vol. 132 (2000), pp. 354-363.

9

Linear Models and Statistical Adjustment

9.1 ETHICAL TREATMENT OF HUMAN SUBJECTS

Randomized, placebo-controlled, blinded experiments are considered the gold standard for identifying cause-and-effect relationships and for reliably assessing the efficacy of drugs and other medical interventions, in some cases, however, experiments are ethically objectionable or are too expensive to use for preliminary, exploratory investigations. An experiment on the effect of smoking on human health would require randomly selecting nonsmokers and inducing them to take up the habit, which would be ethically impermissible. Experiments to determine the most cost-effective way to sterilize humans or to study the effects of untreated syphilis are historical facts but are now morally repugnant.

Since October 1, 2000, the National Institutes of Health has required all key personnel who are conducting human subjects research and receiving NIH funding to certify that they have received education on the protection of human subjects; a web-based tutorial is available at http://cme.nci.nih.gov/.

In 1979, the National Commission for the Protection of Human Subjects of Biomedical and Behavioral Research wrote their report, *Ethical Principles and Guidelines for the Protection of Human Subjects of Research*, commonly called the Belmont Report. The three fundamental ethical principles that guide the ethical conduct of research involving human participants are:

- ***Respect for Persons,*** Individuals should be treated as autonomous agents
- ***Beneficence,*** Human participants should be treated in an ethical manner not only by respecting their decision and protecting them from harm, but also by making efforts to secure their well-being.
- ***Justice,*** Unless there is clear justification, research should not involve persons from groups that are unlikely to benefit from subsequent applications of the research.

9.2 THE NEED FOR STATISTICAL ADJUSTMENT

These ethical and moral considerations make it important and necessary to make effective use of nonexperimental, observational data and to consider

179

nonrandomized experimental designs (see pages 3–44 of Kadane 1996). In this chapter we introduce the concept of statistical adjustment as a way of interpreting evidence provided by a nonexperimental study. In a typical observational study — for example an investigation of whether industrial solvents are a risk factor for bladder cancer or whether diesel exhaust is a risk factor for asthma — two or more groups of subjects (exposed and unexposed) are compared with respect to some response variable (bladder cancer, FEV_1, etc.).

The problem with this sort of face value comparison is that the two groups may differ in other ways in addition to the factor being investigated. For example, the exposed group may be older or may have a higher proportion of heavy smokers. In that case the exposure is said to be *confounded* with smoking. As used by statisticians, the term means to be mixed up so that the components are hard to separate. It is true that the solvent-exposed group was observed to have a higher rate of bladder cancer but it is not clear that the higher rate is due to solvent exposure. It is possible that workers in the exposed group are also heavier smokers than nonexposed workers. If so, then it is impossible to say if their increased rate of bladder cancer is the result of solvent exposure or the result of heavy smoking. The investigator is unable say whether the solvent or the smoking is responsible for the increased rate of cancer.

The purpose of this and the next chapter is to introduce methods for statistically removing (adjusting) the influence of one or more confounding factors. These methods involve making a statistical model of the combined influences of the various factors (solvent exposure, age, drinking, smoking, etc.) in order to isolate the specific influence of the risk factor (solvent exposure). The most commonly used tool for this purpose is multiple regression.

9.3   REGRESSION

Regression is representation of the mean value of a response variable as a function of one or more explanatory variables. A regression can be reported as a graph, a table, or an equation. For example, Fig. 9.1 refers to 24 prelingually profoundly deaf children who received a cochlear implant and were tested immediately after connection and then at 12-month intervals. The *response variable* plotted on the vertical axis (y) is the word recognition score, and the two *explanatory variables* (x) are age at connection of the implant and number of years of experience with the implant. One explanatory variable (months since connection) is plotted on the horizontal axis, and the other (age at connection) is indicated in the legend and represented by the shape of the plotting symbols.

Generic symbols for response and explanatory variables are y and x; in this case, y stands for a child's word recognition score, and x is actually a vector of two numbers (age, experience).

Each point on the graph is an average word recognition score (symbolized $\bar{y}_x$) for children with a given combination of age and experience. For example, $\bar{y}_{(2,36)}$ represents average speech understanding score of children who

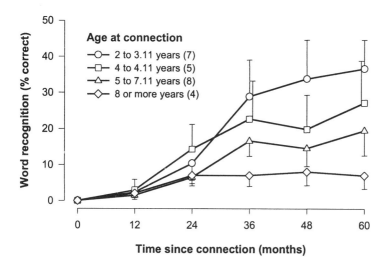

Figure 9.1 Average word recognition as a function of age at connection and years of experience with an implant (mean ± sem). Source: Tyler et al. 2000, © Annals of Otology, Rhinology & Laryngology.

were implanted at ages 2–3 and tested 36 months later. The "whiskers" above or below each plotted average are called *error bars* and generally represent the standard error of the mean (sem). The number of children in each age category is shown in parentheses in the figure legend.

9.3.1   Linear Regression

Linear regression is a technique for approximating a regression graph with one or more straight lines as shown in Fig. 9.2. The equations for the four straight lines are

$$
\begin{aligned}
\hat{y}_{(2,M)} &= 0.71 \times M \quad \text{for ages 2 to 3.11} \\
\hat{y}_{(4,M)} &= 0.46 \times M \quad \text{for ages 4 to 4.11} \\
\hat{y}_{(5,M)} &= 0.35 \times M \quad \text{for ages 5 to 7.11} \\
\hat{y}_{(8,M)} &= 0.13 \times M \quad \text{for ages 8 and up}
\end{aligned}
\tag{9.1}
$$

where M is the number of months since connection and the first subscript refers to age at connection. The symbol $\hat{y}_{(A,M)}$ stands for the linear regression approximation of the average $\bar{y}_{(A,M)}$. For example, the approximate average word

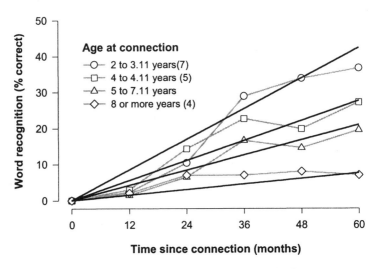

Figure 9.2   Linear regression approximation of the relationship between age, experience, and word recognition.

recognition score at 36 months for children implanted at age 2 or 3 years is $\hat{y}_{(2,36)} = 0.71 \times 36 = 25.6$.

Linear regression is widely used in the analysis of observational data and can be an excellent approximation; however, linearity is not guaranteed, and the linear regression in Fig. 9.2, for example, is not a particularly good approximation to the observed data at 12 or 24 months for the youngest children. The linear regression for these children (the steepest straight line) is *misspecified*, meaning that it fails to capture the shape of the relationship between months of experience and word recognition, which, for children who were implanted at age 2 or 3, features slow initial growth, at first lagging behind the next older group, followed by such rapid growth that this group ultimately reaches higher levels of speech understanding than any other age group. This example shows that it is important to remember that linear regression is not necessarily a good approximation. There is a very large literature on methods for detecting various aspects of model misspecification; the topic is covered in most courses and textbooks on linear regression.

Simple (Bivariate) Linear Regression.   In simple, bivariate linear regression there is one explanatory variable and therefore one straight line. For example, Fig. 9.3 shows the linear regression approximation of the relationship between speech recognition (y) in adult cochlear implant recipients and log duration of profound deafness (x) before receiving the implant. The straight line

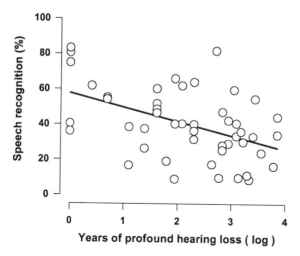

Figure 9.3 Speech recognition as a function of log-duration of profound deafness in adult cochlear implant patients. Source: Iowa Cochlear Implant Project, unpublished data. Supported in part by research grant 5 P50 DC00242 from the NIH, and RR 00059 from the General Clinical Research Center Program.

approximates the average value of y at a given value of x. The equation for this particular straight line is $\hat{y}_x = 68.31 - 0.97 \cdot x$ so, for example the linear regression approximation of the average speech recognition score for people implanted after 20 years of profound deafness is $68.31 - 0.97 \times 20 = 48.9$.

9.3.2 Computing the Slope and Intercept

The bivariate regression equation can be written in slope–intercept form,

$$\hat{y}_x = a + b \cdot x \qquad (9.2)$$

or as a proportionality,

$$\hat{y}_x = \bar{y} + b \cdot (x - \bar{x}) \qquad (9.3)$$

The formulas for the slope and intercept involve first and second moments:

$$\text{Slope:} \quad b \;=\; \frac{s_{xy}}{s_x^2} \;=\; r\frac{s_y}{s_x} \qquad (9.4)$$

$$\text{Intercept:} \quad a \;=\; \bar{y} - b \cdot \bar{x} \qquad (9.5)$$

Obs	Data		Deviations		Cross-products		
i	x	y	$(x-\bar{x})$	$(y-\bar{y})$	$(x-\bar{x})^2$	$(x-\bar{x})\times(y-\bar{y})$	$(y-\bar{y})^2$
1	x_1	y_1	$(x_1-\bar{x})$	$(y_1-\bar{y})$	$(x_1-\bar{x})^2$	$(x_1-\bar{x})\times(y_1-\bar{y})$	$(y_1-\bar{y})^2$
2	x_2	y_2	$(x_2-\bar{x})$	$(y_2-\bar{y})$	$(x_2-\bar{x})^2$	$(x_2-\bar{x})\times(y_2-\bar{y})$	$(y_2-\bar{y})^2$
$\vdots$	$\vdots$	$\vdots$	$\vdots$	$\vdots$	$\vdots$	$\vdots$	$\vdots$
n	x_n	y_n	$(x_n-\bar{x})$	$(y_n-\bar{y})$	$(x_n-\bar{x})^2$	$(x_n-\bar{x})\times(y_n-\bar{y})$	$(y_n-\bar{y})^2$
Sums $\Rightarrow$	$\sum x_i$	$\sum y_i$	$0\checkmark$	$0\checkmark$	SS_x	SP_{xy}	SS_y
$\left(\dfrac{Sum}{n}\right)\Rightarrow$	$\bar{x}$	$\bar{y}$	$\left(\dfrac{Sum}{n-1}\right)\Rightarrow$		s_x^2	s_{xy}	s_y^2
			$\sqrt{\left(\dfrac{Sum}{n-1}\right)}\Rightarrow$		s_x		s_y

Figure 9.4 Computation of means, standard deviations, and covariance. Steps are: compute means, then deviations, then cross products, then second moments.

The *first moments*, $\bar{y}$ and $\bar{x}$, are simply the average values of y (speech recognition) and x (duration of profound deafness). *Second moments*, s_x^2, s_y^2, and, s_{xy}, are averages of squares or cross-products of x and y expressed as deviations from $\bar{x}$ and $\bar{y}$. The average squared deviations, s_x^2 and s_y^2, are called the *variance of* x and the *variance of* y, respectively, and the average product of deviations, s_{xy}, is called the *covariance between* x *and* y. An efficient layout for computing these statistics is presented in Fig. 9.4.

The covariance, s_{xy}, being the average product of percentage times duration, is expressed in percent-years, analogous to torque (foot-pounds) or electrical energy (watt-hours). The *Pearson correlation coefficient* r is a unit-free version of the covariance,

$$r = \frac{s_{xy}}{\sqrt{s_x^2 \cdot s_y^2}} = \frac{s_{xy}}{s_x \cdot s_y} \tag{9.6}$$

The correlation coefficient is a measure of how closely the graph of y vs. x adheres to a straight line. The squared correlation coefficient, generally written R^2 rather than r^2, is claimed by some to be a measure of "goodness of fit," in the sense that if $R^2 = 1$, then the data points lie on a straight line and the closer R^2 is to 0, the more scattered the points are around a straight line.

In Fig. 9.5, we have inserted the speech recognition and duration of deafness values underlying Fig. 9.3 into a layout like Fig. 9.4 in to obtain first and second moments. Inserting those values in Equations (9.4), (9.5), and (9.6), we compute:

slope: $b = \dfrac{s_{xy}}{s_x^2} = \dfrac{-10.42}{1.298} = -8.03$

intercept: $a = \bar{y} - b \cdot \bar{x} = 40.4 - (-8.03) \times 2.17 = 57.8$ (9.7)

correlation: $r = \dfrac{s_{xy}}{s_x \cdot s_y} = \dfrac{-10.42}{1.139 \times 19.77} = -0.463$

Therefore the equation for the straight line in Fig. 9.3 is

$$\hat{y}_x = 57.8 - 8.03 \cdot x$$

9.3.3   Residuals and Residual Variation

Residuals are prediction errors. For example, the second patient in Fig. 9.5 was deaf for 8 years, so his or her x value is ln(8), or 2.08. This patient achieved 61.96% speech recognition, but the *predicted* mean speech recognition score for patients who had been deaf for 8.0 years is

$$\hat{y}_{2.08} = 57.8 - 8.03 \times 2.08 = 41.1   \tag{9.8}$$

The number "2.08" in the subscript is the value of the x value for subject 2, which is denoted by x_2.

Obs	Data		Deviations		Cross-products		
i	x	y	$(x-\bar{x})$	$(y-\bar{y})$	$(x-\bar{x})^2$	$(x-\bar{x})\cdot(y-\bar{y})$	$(y-\bar{y})^2$
1	1.39	26.3	−0.78	−14.1	0.614	11.05	198.8
2	2.08	62.0	−0.09	21.6	0.008	−1.96	466.6
3	0.00	40.4	−2.17	0.0	4.709	−0.00	0.0
	42 of the 48 subjects (4 through 45) are not shown						
46	0.41	61.7	−1.76	21.3	3.114	−37.58	453.7
47	3.22	30.0	1.05	−10.4	1.100	−10.91	108.2
48	1.79	19.0	−0.37	−21.4	0.143	8.09	458.0
Sums	103.97	1937.3	0.00	0.00	60.99	−489.59	18374.6
Sum/48	$\bar{x}$:2.17	$\bar{y}$:40.4			Sum/47 s_x^2: 1.298	s_{xy}: −10.42	s_y^2: 391.0
					Root s_x: 1.139		s_y: 19.77

Figure 9.5   First and second moments of speech recognition and log duration of deafness.

Similarly, the x value for the ith subject is x_i, and consequently the predicted y value for the ith subject is

$$\hat{y}_{x_i} = 57.8 - 8.03 \cdot x_i$$

Keeping track of subscripts on subscripts is not fun, so we will use the symbol $\hat{y}_i$ to represent the predicted value for the ith subject with the understanding that it actually depends on the x value for that subject. Thus, the number we calculated in Equation (9.8) is the predicted value for the second subject, $\hat{y}_2 = 41.1$, and his or her observed value is $y_2 = 62.0$.

The *residual* (e) for the second subject is the difference between his or her observed and predicted values

$$\hat{e}_2 = y_2 - \hat{y}_2 = 62.0 - 41.1 = 20.9$$

and the residual for the ith subject is

$$\hat{e}_i = y_i - \hat{y}_i \tag{9.9}$$

Every subject has a residual; some are positive, some are negative, but they sum to 0. If a residual is exactly equal to 0, then that subject's y value was predicted without error. If all the residuals are equal to 0 then all subjects were perfectly predicted and the plotted data will lie on a straight line. Thus the tightness of the prediction is measured by the *scatter* of the residuals — not by the average of the residuals but by their variation. The scatter of the residuals and hence the scatter of data points around the regression line is called the *residual standard deviation*

$$s_{y|x} = \sqrt{\frac{\sum \hat{e}_i^2}{n-2}} \tag{9.10}$$

The residual standard deviation can be computed directly from the residuals or using the algebraic relationship

$$s_{y|x} = s_y \cdot \sqrt{1-r^2} \cdot \sqrt{\frac{n-1}{n-2}} \cong s_y \cdot \sqrt{1-r^2} \tag{9.11}$$

For the data analyzed in Fig. 9.5 we have $n = 48$, $s_y = 19.77$, and $r = -0.463$ [from Equation (9.7)], so the residual standard error is

$$s_{y|x} = 19.77 \times \sqrt{(1 - (-0.463)^2)} \times \sqrt{47/46} = 17.71 \tag{9.12}$$

The word *residual* means "left over," and the residual standard deviation has that interpretation. The standard deviation $s_y = 17.71$ summarizes all of the variation of y. Part of that variation is "explained" by differences in duration of deafness and part of it (the residual scatter around the regression line) is unexplained. Thus the residual standard summarizes the "left-over," unexplained variation in y. If s_y is the total standard deviation and $s_{y|x}$ is the unexplained standard deviation, then $s_y - s_{y|x}$ must be the explained part of the standard deviation; consequently the amount of variation explained by x is

$$s_y - s_{y|x} \cong s_y - s_y \cdot \sqrt{1 - r^2} = s_y \cdot \left(1 - \sqrt{1 - r^2}\right)$$

and the proportion of the standard deviation of the response variable y that is explained by subjects having different values of the explanatory variable x is

$$\frac{s_y - s_{y|x}}{s_y} \cong 1 - \sqrt{1 - r^2} \tag{9.13}$$

For example, the correlation coefficient for the data in Fig. 9.3 was r = 0.463, thus the proportion of the standard deviation "explained" by duration of deafness is

$$1 - \sqrt{1 - 0.463^2} \cong 0.11 \tag{9.14}$$

In other words knowing how long a person had been deaf prior to implantation would reduce our uncertainty about his/her post-implantation speech recognition score by only about 11%.

We have gone into this topic at some length because R^2 is often described as the proportion of variance accounted for by the explanatory variable(s). That statement is literally correct, since variance is squared standard deviation; however it is also quite misleading, because people are inclined to interpret the word "variance" as it is used in ordinary speech to mean "deviation from some standard." It is deviation from prediction that is meaningful, not squared deviation. For the adult cochlear implant data $R^2 = 0.463^2 = 0.19$, or 19%; however, according to Equation (9.14), knowledge of a patient's duration of deafness reduces the standard deviation of our forecast of his speech understanding after implantation by only 11%, not 19%.

This disparity is even more striking with large values of the correlation coefficient. For example, a correlation of r = 0.99 is said, conventionally, to mean that the explanatory variable accounts for $0.99^2 = 0.9801$, or about 98% of the "total variance" of y, presumably leaving only 2% unexplained. In fact, the explanatory variable accounts for only $1 - \sqrt{1 - 0.99^2} = 0.8589$, or about 86%, of the total standard deviation of y, leaving about 14% unexplained.

9.3.4   Posterior Predictive Distribution

One application of the regression equation $\hat{y}_x = a + b \cdot x$ is to forecast the value of the response variable from advance knowledge of the explanatory variable(s). For example, the equation could be used to forecast how much speech understanding a cochlear implant candidate can expect to achieve with an implant. Let us use the symbol y_{new} for the as yet unobserved speech understanding score of an implant candidate, and the symbol x_{new} for the number of years he or she has been deaf. Prior to implantation, y_{new} is an unknown quantity, and therefore what is known about it can be described by a probability distribution, called the *posterior predictive distribution*. This distribution is posterior to observing the 46 patients who have already received implants and posterior to learning the value of x_{new} for the new patient, but prior to observing his or her speech recognition score (y_{new}). If the analyst had no prior information about the slope, intercept, and residual standard deviation, if the residuals are approximately normal, and if the regression is approximately linear, then the posterior predictive distribution is approximately $t(n-2)$ with parameters μ and σ given by

$$\mu = \hat{y}_{x_{new}}, \text{ and } \sigma = s_{y|x} \cdot \sqrt{1 + \frac{1}{n} + \frac{\left(x_{new} - \bar{x}\right)^2}{SS_x}} \qquad (9.15)$$

For example, suppose that the implant candidate has been profoundly deaf for 5 years ($x_{new} = 5$). The posterior predictive distribution of his speech understanding score after implantation is approximately $t(44)$ with parameters

$$\begin{aligned}
\mu &= 68.29 - 0.970 \times 5 &= 63.44 \\
\sigma &= 19.66 \times \sqrt{1 + \frac{1}{46} + \frac{\left(5 - 22.8\right)^2}{3757.24}} &= 20.68
\end{aligned} \qquad (9.16)$$

From this we can compute, for example, the probability that the candidate will achieve 50% or better speech understanding, $P(y_{new} > 50 \mid \text{data})$, by converting 50 to a Z value and looking up the tail area. The Z value is $(50-\mu)/\sigma$ $= (50 - 63.44)/20.67 = -0.65$, and the lower tail area from tTailArea.xls is about 0.26; consequently the upper tail area is $P(y_{new} > 50 \mid \text{data}) \cong 0.74$. The surgeon could advise her patient that he has about a three chances in four, or 3 to 1 odds, of achieving at least 50% speech recognition after receiving the implant.

9.4   STATISTICAL ADJUSTMENT VIA MULTIPLE REGRESSION

Multiple regression involves two or more explanatory variables. Fig. 9.1 on page 181 is an example in which the explanatory variables are age at implantation and number of months since implantation. One of the most

Table 9.1 Salaries, experience, and gender of 35 employees.

Females				Males	
Experience (years)	Salary ($)	Experience (years)	Salary ($)	Experience (years)	Salary ($)
6	38,456	7	41,206	0	30,500
4	37,038	7	41,410	3	36,044
1	32,702	6	40,286	0	30,470
6	40,504	3	35,640	4	37,740
6	39,190	1	31,404	3	34,788
3	35,262	7	40,880	2	34,168
0	28,968	3	35,212	0	30,536
1	31,474	5	38,330	0	30,050
3	35,612	7	41,226	2	34,578
2	34,022	3	35,292	5	39,010
6	39,456	1	31,120		
4	35,802	5	37,944		
3	34,946				

important applications of multiple regression is to disentangle the influences of confounded explanatory variables. For example, Fig. 9.1 makes it clear how speech understanding is influenced both by experience and by age at implantation. The purpose of this section is to show how to use multiple regression to isolate the individual influences of several explanatory variables.

The example we'll use to introduce multiple linear regression and statistical adjustment involves the fictional data in Table 9.1. The data are the salaries of 25 female and 10 male employees. The males noticed that their average salary, $33,790, was about $3,000 lower than the average female salary, $36,536. On its face, this looks like a case of gender discrimination; however, the story is not so simple. It turns out that the average female has 4 years of experience and the average male has only about 2 years of experience; therefore, we are comparing the average salary of more experienced females with the average salary of less experienced males. It stands to reason that more experienced employees should be paid more, regardless of gender, so there may be no salary discrimination when experience is taken into account. In order to make a valid comparison between male and female salaries it is necessary to use multiple regression to disentangle the influences of experience and gender.

A graph of the raw data (Fig. 9.6) makes it clear that salary increases with experience in a highly predictable way, but it is also clear that in almost every case males are paid more than females with the same number of years of experience. To see this, notice that of three new hires, the woman has the lowest salary, likewise for the three employees with two years of experience and for the employees with four or five years of experience. There is only one instance in which a male earned less than a female with the same amount of experience. The graph makes clear how all but one male could equal or out-earn equally experienced females, yet the males have a smaller average salary: no male has more that five years of experience.

A *multiple regression* involves computing regression lines for males and females separately. For convenience let's use S to stand for salary and E to stand for years of experience. Using those symbols, the equations for the male and female regression lines in Fig. 9.7 are

$$\text{Females:} \quad \hat{S} = 30,194 + 1,584 \cdot E$$
$$\text{Males:} \quad \hat{S} = 30,567 + 1,748 \cdot E$$

(9.17)

The female (male) equation was computed by inserting the female (male) salary data for x and y in Fig. 9.4 on page 184.

Comparing the raw average salaries of females and males was invalid because the two groups did not have the same amount of job experience; therefore, in order to make a valid salary comparison it is necessary somehow to compare males and females with the same amount of experience. The standard way of doing this is to work out the average amount of experience without regard to gender (which turns out to be 3.4 years) and then to compare the *predicted* salary of a male having 3.4 years of experience with that of a female having 3.4 years of experience.

These predicted salaries, called *adjusted means* (or *least squares means*), are indicated on the graph of the two regression lines in Fig. 9.7. The numerical values of the adjusted means and their difference are in Table 9.2 with 95% CI's. The table shows that after adjusting for years of experience, the average female employee earned about $827 *less* than a male with the same number of years of

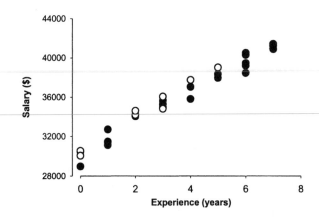

Figure 9.6 Employee's salaries and years of experience.

Table 9.2   Raw and adjusted mean salaries for male and female employees.

Gender	Raw mean[a]	Adjusted mean[b]	95% Credible interval
Female	$36,535	$35,584	$35,341 to $35,827
Male	$33,788	$36,411	$35,923 to $36,898
Difference	$2,746	−$826.59	−$1,371.62 to −$281.57

a. Raw means were computed with SAS PROC MEANS.
b. Adjusted means, mean difference, and credible intervals were computed with SAS PROC MIXED.

experience (95% CI $282 to $1,372). The conclusion is that, on average, men actually earn more than equally experienced women — just the reverse of the initial impression based on unadjusted salaries.

9.5   DESIGN VARIABLES

9.5.1   Combining Equations via Boolean Variables

The generic form of a multiple regression equation expresses the predicted value of the response variable y as a a linear combination of *design variables* (the x's) weighted by *regression coefficients* (the β's):

$$\hat{y}_x = \beta_1 \cdot x_1 + \beta_2 \cdot x_2 + \cdots + \beta_k \cdot x_k \qquad (9.18)$$

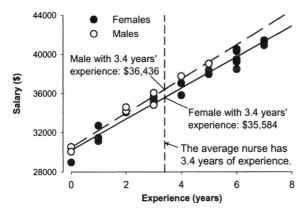

Figure 9.7   Regression lines and adjusted means for males and females.

The important thing about Equation (9.18) is that it expresses a *multiple* regression in a *single* equation despite the fact that multiple regression models, such as Fig. 9.2 and Equation (9.1), and Fig. 9.7 and Equation (9.17), feature more than one straight line. How can one equation produce many straight lines? To answer that question, let's revisit the salary equations pictured in Fig. 9.2:

$$\text{Females:} \quad \hat{S} = 30,194 + 1,584 \cdot E$$
$$\text{Males:} \quad \hat{S} = 30,567 + 1,748 \cdot E \tag{9.19}$$

The trick to combining these into a single equation is to use *Boolean variables*. A Boolean variable is a variable that indicates whether a sentence is true or false. In this case the two Boolean variables we need are the *gender indicators*,

$$F \quad = \quad \begin{cases} 1 & \text{if gender } = \text{ female} \\ 0 & \text{if gender } \neq \text{ female} \end{cases}$$
$$M \quad = \quad \begin{cases} 1 & \text{if gender } = \text{ male} \\ 0 & \text{if gender } \neq \text{ male} \end{cases} \tag{9.20}$$

Now use the Boolean variables to combine the two equations in (9.19):

$$\hat{S} = (30194 + 1584 \cdot E) \cdot F + (30567 + 1748 \cdot E) \cdot M \tag{9.21}$$

The trick works. For example if the subject is a female with E years of experience then the Boolean variables are $F = 1$, and $M = 0$ so that

$$\hat{S} \quad = \quad (30194 + 1584 \cdot E) \cdot 1 + (30567 + 1748 \cdot E) \cdot 0 \tag{9.22}$$
$$= \quad 30194 + 1584 \cdot E$$

which is the female salary equation. Clearly the role of the Boolean variables is to select the appropriate equation for males or females.

Go back to Equation (9.21) and multiply through by the Boolean variables to get the *multiple regression model*

$$\hat{S} \quad = \quad (30194 + 1584 \cdot E) \cdot F + (30567 + 1748 \cdot E) \cdot M$$
$$\hat{S} \quad = \quad 30194 \cdot F + 1584 \cdot (E \cdot F) + 30567 \cdot M + 1748 \cdot (E \cdot M) \tag{9.23}$$
$$\downarrow \qquad \downarrow \downarrow \qquad \downarrow \downarrow \qquad \downarrow \downarrow \qquad \downarrow \downarrow$$
$$\hat{y} \quad = \quad \beta_1 \cdot x_1 \ + \ \beta_2 \cdot x_2 \ + \ \beta_3 \cdot x_3 \ + \ \beta_4 \cdot x_4$$

The bottom line in Equation (9.23) simply supplies generic symbols for the components of the line above it. Thus y is salary; the female intercept, 30,194, is called β_1; the female slope, 1,584, is called β_2; etc. Likewise, the Boolean variable

F is called x_1, and the product E·F is called x_2; etc. The β's are called *regression coefficients*, and the x's are called *design variables*. Thus, the predicted salary for a given person is the sum of products of regression coefficients times design variables. Table 9.3 may help clarify the nature of the design variables.

The layout in Table 9.3, with a regression coefficient written at the bottom of each column of design variable values, makes it particularly easy to compute the predicted salary for each person; for example, the predicted salary for the second woman is the sum of the design variable values in her row of the data matrix times the regression coefficients,

$$\begin{aligned}
\hat{y}_2 &= \beta_1 \cdot 1 + \beta_2 \cdot 4 + \beta_3 \cdot 0 + \beta_4 \cdot 0 \\
&= 30194 \times 1 + 1584 \times 4 + 30567 \times 0 + 1748 \times 0 \\
&= 36530
\end{aligned}$$

9.5.2 Reference Categories

To illustrate the concept of a reference category, suppose that employees are classified by training rather than gender. The categories are H (high school or less), B (BS or BA), and G (a graduate degree). The Boolean variables for the three categories of training are called H, B and G; for example, if the employee has a BS and one year of graduate school, but no graduate degree then $H=0$, $B=1$, and $G=0$. For every employee, the three Boolean variables must sum to 1, because that employee has to be in one of the three categories, thus $H + B + G = 1$. Any one of the Boolean variables can be computed from the other two. For example,

Table 9.3   Data and Design Variables for first 3 and last 3 Employees

Response variable	Explanatory variables		Boolean variables		Design variables			
Salary	Gender	Experience	Male	Female	F	E·F	M	E·M
y	G	E	F	M	x_1	x_2	x_3	x_4
38,456	F	6	1	0	1	6	0	0
37,038	F	4	1	0	1	4	0	0
32,702	F	1	1	0	1	1	0	0
30,050	M	0	0	1	0	0	1	0
34,578	M	2	0	1	0	0	1	2
39,010	M	5	0	1	0	0	1	5
		Regression coefficients:			30194	1584	30567	1748

the Boolean variable B is $B = 1 - H - G$. The category chosen to be redundant and therefore computed from the other two is called the *reference category.*
Suppose that the regression equations for the three categories of employees are,

Training	Predicted salary
High school or less	$\hat{S} = a_H + b \cdot E$
Bachelor's degree	$\hat{S} = a_B + b \cdot E$
Graduate degree	$\hat{S} = a_G + b \cdot E$

Here E is the employee's number of years of experience; a_H, a_B, and a_G are the intercepts of the three straight lines; and b stands for the slope, which in this example is assumed to be the same in all three groups. The three straight lines can be expressed as a single regression model by combining them via the Boolean variables with the redundant Boolean variable B replaced by the expression $1 - H - G$ The regression model that combines all three equations is.

$$\hat{S} = (a_H + b \cdot E) \cdot H + (a_B + b \cdot E) \cdot (1 - H - B) + (a_G + b \cdot E) \cdot G \qquad (9.24)$$

To express this in the form of a multiple regression equation, perform the multiplications and collect similar terms to get

$$\hat{S} = a_B \cdot 1 + \quad b \cdot E + (a_H - a_B) \cdot H + (a_G - a_B) \cdot G$$
$$\quad \downarrow \downarrow \quad \downarrow \downarrow \qquad\qquad \downarrow \downarrow \qquad\qquad \downarrow \downarrow \qquad\qquad (9.25)$$
$$\hat{y} = \beta_1 \cdot x_1 + \beta_2 \cdot x_2 \quad\quad + \beta_3 \cdot x_3 \quad\quad + \beta_4 \cdot x_4$$

The most interesting thing about the multiple regression model in Equation (9.25) is the interpretation of the regression coefficients: β_1 is the intercept for the reference category (B), β_2 is the slope b, which is common to the three categories, β_3 is the *difference* between the intercept for category H and for the reference category, and β_4 is the difference between the intercept for category G and the reference category. The conclusion is that the regression coefficient of the Boolean indicator variable of an *included* category (H or G) is the difference between that category's intercept and that of the reference category (B).
The purpose of this section was to prepare us to interpret published regression coefficients. The last step in our preparation is to demonstrate that β_3 and β_4, the regression coefficients for the Boolean indicator variables H and G, are in fact *adjusted mean differences* and therefore are comparisons of categories, free of the effects of the confounding variables in the regression model. To demonstrate this final point, suppose for the sake of illustration that the average number of years of experience for all employees is 3.9 (the exact value is

immaterial because it is going to cancel out); then the adjusted means for the three categories of employees are the predicted salaries of employees with 3.9 years of experience:

Training	Predicted salary
High school or less	$\hat{S} = a_H + b \cdot 3.9$
Bachelor's degree	$\hat{S} = a_B + b \cdot 3.9$
Graduate degree	$\hat{S} = a_G + b \cdot 3.9$

Consequently, using Equation (9.25), the adjusted mean differences are,

Contrast	Adjusted Mean Difference
H vs B:	$a_H - a_B = \beta_3$
G vs B:	$a_G - a_B = \beta_4$

We conclude that

If each continuous explanatory variables has the same slope in every group, then the regression coefficient for a category is the adjusted mean difference between that category and the reference category.

9.6 INTERPRETING PUBLISHED MULTIPLE REGRESSIONS

Adjusted means are generally not reported in papers, although that would be an extremely useful practice. Instead, what is typically reported is a list of *regression coefficients* along with their standard errors and their *t* values or *p* values. For example, Table 9.4, is a fairly typical published report of a multiple regression. It reports the results of fitting a multiple regression equation to the salaries of 563 employees of a state agency (Finkelstein and Levin 2001). The response variable *y* is salary, and the explanatory variables are the number of years of service and three categorical variables: ethnicity (white or nonwhite), education (Bachelor's or Master's), and job title (Chairman, Agent, Associate, or Assistant).

The employee's tenure, or number of years with this employer, is a continuous explanatory variable, which in this regression model has only one slope. The slope can be interpreted as a salary increment for each additional year of tenure regardless of ethnicity, job title, or education.

The reference categories in the regression model are the omitted categories: the Bachelor's degree, the job title Assistant, and nonwhite ethnicity. Consequently the regression coefficients for the included categories are adjusted mean differences relative to the reference category for that variable. For example, all other things being equal, the average Agent earned an estimated $2404.44 more than the average Assistant (the reference category), the average Associate earned $918.82 more than the average Assistant, and employees with Master's degrees earned on average $898.55 more than employees with Bachelor's degrees, all other things being equal.

The purpose of this analysis was to check for evidence of racial discrimination in salary levels. The regression coefficient for the Boolean variable "White" is $394.80, which we know is the adjusted mean difference between the salaries of white and black employees. Thus, all other things being equal, the average white employee earned $394.80 more than more than his or her nonwhite counterpart with the same job title, level of education and years of tenure. The table reports estimates and standard errors of regression coefficients. The estimate and standard error are interpreted as the parameters μ and σ of the posterior distribution of the true regression coefficient; thus, for example, an approximate 95% credible interval for the adjusted mean difference between the salaries of white and black employees is $394.80 \pm 1.96 \times \$137.67$, or $125 to $665.

A potential defect of this analysis is what appears on the face of it to be a misspecified regression model; it seems unlikely that the salary increment for an additional year of tenure would the same for all categories of employees. Surely department chairpersons receive larger raises than assistants. It seems more realistic therefore to use a model with different slopes in different categories of employees. In that case the regression coefficients of Boolean variables would no longer interpretable as adjusted mean differences, and it would be necessary to use statistical software to compute adjusted mean differences; instructions for doing this are in Section 9.7.

Table 9.4   Multiple regression analysis of the salaries of 563 employees.[a]

Explanatory variable	Estimate of regression coefficient	Standard error
Constant	$9291.51	n/a
Tenure (years)	$59.06/year	8.47
Degree: Master's	$898.55	140.36
Title: Associate	$918.82	174.41
Agent	$2404.44	170.58
Chairman	$5221.19	232.28
Ethnicity: White	$394.80	137.67

a. Source: Evidence introduced in Bazemore v. Friday, 478 U.S. 385 (1986), reproduced in Finkelstein and Levin (2001), Table 13.6.1, Step 6, page 390, ©Springer-Verlag.

Table 9.5   Data layout for salary study.

Salary ($)	Tenure (years)	Education	Job Title	ethnicity
y	x	BS	Assistant	NW
y	x	BS	Assistant	W
y	x	BS	Associate	NW
y	x	BS	Associate	W
y	x	BS	Agent	NW
y	x	BS	Agent	W
y	x	BS	Chairman	NW
y	x	BS	Chairman	W
y	x	MS	Assistant	NW
y	x	MS	Assistant	W
y	x	MS	Associate	NW
y	x	MS	Associate	W
y	x	MS	Agent	NW
y	x	MS	Agent	W
y	x	MS	Chairman	NW
y	x	MS	Chairman	W

Design variables and model specification for the salary data.   The underlying data consisted of salary and tenure for employees in 16 categories as depicted in Table 9.5. Here only one row is shown for each of the 16 categories of employees; however, the full data table has 563 rows of data with many employees in each category.

The estimated regression equation reported in Table 9.4 has seven regression coefficients (counting the constant). The design variables for each of the sixteen categories of employees are shown in Table 9.6. (In the full data table there were 563 rows of data, one row for each employee.) Design variable x_1 is the constant; x_2, the number of years of tenure, is a continuous variable, and x_3 through x_7 are Boolean variables for categories. For example, variable x_5 indicates whether or not the employee has the title Agent (1 = yes, 0 = no). The estimated regression coefficients were listed in Table 9.4; consequently the estimated regression equation is (in round numbers),

$$\hat{y}_x = 9292 \cdot x_1 + 59 \cdot x_2 + 899 \cdot x_3 + 919 \cdot x_4 + 2404 \cdot x_5 + 5221 \cdot x_6 + 395 \cdot x_7 \quad (9.26)$$

The response variable wears two hats because it is an estimate of a predicted value. To illustrate how the equation works, consider nonwhite agents with an MS and 5 years of tenure. The design variables, from Table 9.6, are (1,5,1,0,1,0,0). Consequently, the estimated value of the mean salary for this group of employees is:

$$\hat{y}_x = 9292 \cdot 1 + 59 \cdot 5 + 899 \cdot 1 + 919 \cdot 0 + 2404 \cdot 1 + 5221 \cdot 0 + 395 \cdot 0$$
$$= 12890 \tag{9.27}$$

To investigate the influence of ethnicity, we compute the estimated mean salary of the corresponding group of white employees:

$$\hat{y}_x = 9292 \cdot 1 + 59 \cdot 5 + 899 \cdot 1 + 919 \cdot 0 + 2404 \cdot 1 + 5221 \cdot 0 + 395 \cdot 1 \tag{9.28}$$
$$= 13285$$

The difference between these estimates is the adjusted mean difference between white agents and black agents having both an MS and 5 years of tenure,

$$\hat{\Delta}_{ADJ} = 13285 - 12890 = 395 \tag{9.29}$$

which indicates that white agents with a MS and 5 years of tenure earn an estimated \$395 per year more than nonwhite agents with the same education and tenure. As expected, the adjusted mean difference is the regression coefficient for the category of white employees.

Table 9.6   Design Variables

Explanatory Variables				Design Variables						
				Int	Yrs	MS	Assc	Agnt	Chmn	White
Yrs[a]	Ed	Title	Ethnicity	x_1	x_2	x_3	x_4	x_5	x_6	x_7
v	BS	Assistant	NW	1	v	0	0	0	0	0
v	BS	Assistant	W	1	v	0	0	0	0	1
v	BS	Associate	NW	1	v	0	1	0	0	0
v	BS	Associate	W	1	v	0	1	0	0	1
v	BS	Agent	NW	1	v	0	0	1	0	0
v	BS	Agent	W	1	v	0	0	1	0	1
v	BS	Chairman	NW	1	v	0	0	0	1	0
v	BS	Chairman	W	1	v	1	0	0	1	1
v	MS	Assistant	NW	1	v	1	0	0	0	0
v	MS	Assistant	W	1	v	1	0	0	0	1
v	MS	Associate	NW	1	v	1	1	0	0	0
v	MS	Associate	W	1	v	1	1	0	0	1
v	MS	Agent	NW	1	v	1	0	1	0	0
v	MS	Agent	W	1	v	1	0	1	0	1
v	MS	Chairman	NW	1	v	1	0	0	1	0
v	MS	Chairman	W	1	v	1	0	0	1	1

a.v stands for the observed value of years of tenure for each employee

9.7   COMPUTING ADJUSTED MEANS AND DIFFERENCES IN SAS

SAS procedures creates design variables through the interaction of the CLASS statement and the MODEL statement. The fictional salary data in Table 9.1 will be used to illustrate these concepts. The data table, Salaries.xls, is available for downloading from the textbook website and can be imported into a SAS data table via PROC IMPORT (see Appendix subsection C.3.2), or opened and pasted into a DATA step:

```
DATA SALARIES;
    INPUT Gen $ Yrs Sal;
DATALINES;
F   1    31404
F   3    35262
(21 data lines not shown)
F   7    41226
F   7    41410
M   0    30470
M   2    34168
(6 data lines not shown)
M   4    37740
M   5    39010
;;;;
```

The variables in the SAS data table SALARIES are:

Variable name	Description	Values
Gen	Gender	M or F
Yrs	Experience (years)	Number of years of experience
Sal	Salary	Salary in dollars

The regression model with different intercepts for males and females but the same slope for both genders is specified as follows:

```
PROC MIXED DATA=SALARIES;
CLASS Gen;
MODEL Sal = Gen Yrs / NOINT Solution;
LSMEANS Gen / Diff CL;
RUN;
```

This program constructs the design variables, computes the regression coefficients associated with each design variable, and calculates adjusted mean salaries and the adjusted mean difference.

Design variables are determined by the CLASS statement, the *effects* listed after the equal sign in MODEL statement, and the presence or absence of the NOINT option. To display the design variables, run these optional commands:

```
PROC GLMMOD DATA=SALARIES;
    CLASS Gen;
    MODEL Sal = Gen Yrs / NOINT;
RUN;
```

The output includes a description of the design variables:

Parameter Definitions

Column Number	Name of Associated Effect	CLASS Variable Values Gen	Comments
1	Gen	F	X1 = (gen="F")
2	Gen	M	X2 = (gen="M")
3	Yrs		X3 = yrs

and a list of the numerical values of the design variables:

Design Points

Observation Number	Sal	Column Number 1	2	3	
1	31404	1	0	1	
2	35262	1	0	3	
−21 lines not shown−					
24	41226	1	0	7	↑
25	41410	1	0	7	*Females*
26	30470	0	1	0	*Males*
27	34168	0	1	2	↓
−6 lines not shown−					
34	37740	0	1	4	
35	39010	0	1	5	

Declaring gen (gender) to be a CLASS, or categorical, variable instructed the software to create the Boolean variables X1 (female gender) and X2 (male gender), which are listed as items 1 and 2 under the heading "Column number" in the design variable table. The NOINT option told the software not to create a constant design variable; instead, the constant is split between the two genders. The continuous explanatory variable yrs (years of experience) is the third design variable X3. The output from the MIXED procedure includes estimates and standard errors of regression coefficients (requested by the SOLUTIONS statement):

Solution for Fixed Effects						
Effect	Gen	Estimate	Standard Error	DF	t Value	Pr > |t|
Gen	F	30060	223.12	32	134.73	<.0001
Gen	M	30713	205.54	32	149.42	<.0001
Yrs		1618.70	47.5311	32	34.06	<.0001

The estimated female intercept is $30,060 with standard error $223. Assuming a non-informative prior, these are the parameters μ and σ of the posterior distribution of the true female intercept. The shape of the distribution is approximately $t(32)$. Similarly the posterior distribution of the male intercept is approximately $t(32)$ with parameters $30,713 and $206, and the posterior distribution of the slope is approximately $t(32)$ with parameters $\mu=47.43$ and $\sigma=32$ ($/year). The columns labeled "t Value" and "Pr > $|t|$" are the t value and p value. Recall that the quantity $1-p/2$ can be interpreted as the probability that the estimate of an unknown quantity has the correct sign. In this case those probabilities are all a least 0.9995; however, that is hardly useful, since we already know that starting salaries (β_1 and β_2) and raises (β_3) are positive.

The LSMEANS command requests that the software report the adjusted means (least squares means) for males and females, and the DIFF and CL options request that the adjusted mean difference be reported with a 95% credible interval:

Least Squares Means									
Effect	Gen	Estimate	Standard Error	DF	t Value	Pr > |t|	Alpha	Lower	Upper
Gen	F	35564	120.21	32	295.86	<.0001	0.05	35319	35809
Gen	M	36216	197.93	32	182.98	<.0001	0.05	35813	36620

Differences of Least Squares Means									
Effect	Gen_Gen	Estimate	Standard Error	DF	t Value	Pr > |t|	Alpha	Lower	Upper
Gen	F–M	652.40	240.19	32	-2.72	0.0106	0.05	-1141.7	-163.2

As before, the p values for the adjusted means are not particularly interesting; however the p value for the difference, 0.0106, is interpretable as the probability of discrimination: $P(\Delta_{adj} < 0 \mid data) \approx 1 - p/2 = 0.9947$. The quantity labeled "Alpha" is 0.05 by default, which requests that 95% confidence intervals (approximate 95% CI's) be reported. If, for example, 90% CI's are desired, THE required LSMEANS statement is

```
LSMEANS Gen / Diff CL ALPHA=.10;
```

A regression model with different intercepts and slopes for males and females is specified as follows:

```
PROC MIXED DATA=SALARIES;
   CLASS Gen;
   MODEL Sal = Gen Yrs*Gen / NOINT SOLUTION;
   LSMEANS Gen / Diff CL;
RUN;
```

The model statement can also be written

```
MODEL Sal = Gen Yrs(Gen) / NOINT SOLUTION;
```

Design variables are determined by the CLASS statement, the effects after the equal sign in the MODEL statement, and the presence or absence of the NOINT option. To display the design variables, run these optional commands:

```
PROC GLMMOD DATA=SALARIES;
   CLASS Gen;
   MODEL Sal = Gen Yrs / NOINT;
RUN;
```

Note that the CLASS, MODEL, and NOINT statements must be exactly the same as in PROC MIXED (but leave out the SOLUTION request). The output from the GLMMOD procedure includes a description of the design variables:

		Parameter Definitions	
Column Number	Name of Associated Effect	CLASS Variable Values Gen	Comments
1	Gen	F	$X1 = (gen="F")$
2	Gen	M	$X2 = (gen="M")$
3	Yrs*Gen	F	$X3 = yrs \cdot (gen="F")$
4	Yrs*Gen	M	$X4 = yrs \cdot (gen="M")$

indicating that there are four design variables — two Boolean variables indicating gender and two variables that are products of a continuous and a

Boolean variable.   Variable X3 could be called "female years of experience" and X4 "male years of experience."
The procedure also outputs a list of numerical values of the four design variables:

	Design Points					
Observation			Column Number			
Number	Sal	1	2	3	4	
1	31404	1	0	1	0	
2	35262	1	0	3	0	
−21 lines not shown−						
24	41226	1	0	7	0	↑
25	41410	1	0	7	0	*Females*
26	30470	0	1	0	0	*Males*
27	34168	0	1	0	2	↓
−6 lines not shown−						
34	37740	0	1	0	4	
35	39010	0	1	0	5	

notice that the constant design variable (a column of 1's) has been split between the genders, and, in addition, the variable yrs (years of experience) has been split between the genders.
As a general rule a symbolic product involving a categorical variable and a continuous variable, yrs*gen for example, is *expanded* into as many design variables as there are levels of the categorical variable — in this case X3 and X4.
The output produced by the MIXED procedure with separate intercepts and slopes is

Solution for Fixed Effects						
			Standard			
Effect	Gen	Estimate	Error	DF	t Value	Pr > |t|
Gen	F	30194	239.49	31	126.07	<.0001
Gen	M	30467	267.86	31	113.74	<.0001
Yrs*Gen	F	1585.37	52.5121	31	30.19	<.0001
Yrs*Gen	M	1748.17	103.48	31	16.89	<.0001

The terminology used in output generated by SAS software requires some explanation:

- An "estimate" generally is a *maximum likelihood estimate* (MLE) or *a restricted maximum likelihood (REML)* estimate. For a flat or nearly flat prior it is the most probable value of the unknown quantity, and it sometimes (but not always) approximates the parameter μ of the posterior distribution of the unknown quantity. That interpretation is valid for regression coefficients, adjusted means, and adjusted mean differences.
- A "standard error" generally (but not always) is an approximation to the parameter σ of the posterior distribution. That interpretation is valid for regression coefficients, adjusted means, and adjusted mean differences.
- An "Effect" is an unknown quantity — the true regression coefficient of a design variable. Thus the "Gen F Effect" is the true female intercept. Likewise, the "Yrs*Gen F Effect" is the true female slope.
- "Least Squares Means" are adjusted means and "Differences of Least Squares Means" are adjusted mean differences.
- The words "Lower" and "Upper" denote the endpoints of a confidence interval. For regression coefficients, adjusted means, and differences, they can be interpreted as the endpoints of a credible intervals. By default 95% confidence intervals are reported, but other levels of confidence can be requested by means of the ALPHA option described earlier.

The estimated salary model is

$$\widehat{Salary} = 30194 \cdot (Gender = "F") + 1585 \cdot Experience \cdot (Gender = "F")$$
$$+30467 \cdot (Gender = "M") + 1748 \cdot Experience \cdot (Gender = "M")$$

However, the analysis was not conducted to develop a predictive model but to determine the influence of gender on salary controlling for experience. That question is answered by the adjusted mean difference estimate produced by the LSMEANS statement:

			Least Squares Means						
Effect	Gen	Estimate	Standard Error	DF	t Value	Pr > |t|	Alpha	Lower	Upper
Gen	F	35584	119.28	31	298.31	<.0001	0.1	35382	35786
Gen	M	36411	239.13	31	152.26	<.0001	0.1	36005	36816

			Differences of Least Squares Means						
Effect	Gen_Gen	Estimate	Standard Error	DF	t Value	Pr > |t|	Alpha	Lower	Upper
Gen	F—M	-826.59	267.23	31	-3.09	0.0042	0.1	-1279.69	-373.49

In this case, the adjusted mean difference (expressed as female mean minus male mean) is −$827, with standard error $267 and 95% confidence interval, or approximate 95% CI, −$1372 to −$282.

9.8 ANALYSIS OF CROSSOVER AND RELATED DESIGNS

In a crossover design each subject receives each of the treatments in different time periods. For example, Section 8.6.3 describes a crossover experiment to compare a drug (D) with a placebo (P). Each subject receives drug or placebo in the first period, and then placebo or drug in the second period, thus serving as his or her own control. The problem with crossover studies is that the placebo may have a different effect depending on whether it is administered before or after the drug, and the same is true of the drug. In other words, the order in which the treatments are administered is a confounding variable and its influence must be removed in order to validly compare the treatments. It should come as no surprise that a form of multiple regression is used to compute the adjusted mean difference between the drug and placebo. The data to be analyzed are in Table 8.7 A; however, in that table there are two responses (period 1 and period 2) per row. The first step is to rearrange the data table with one response per row:

```
DATA Crossover;
    INPUT ord $ sub per trt $ y;
DATALINES;
DP  1 1 D 0.45
DP  1 2 P 0.47
DP  2 1 D 0.47
DP  2 2 P 0.49
PD  3 1 P 0.48
PD  3 2 D 0.51
DP  4 1 D 0.48
DP  4 2 P 0.47
DP  5 1 D 0.51
DP  5 2 P 0.47
PD  6 1 P 0.45
PD  6 2 D 0.50
PD  7 1 P 0.48
PD  7 2 D 0.48
DP  8 1 D 0.60
DP  8 2 P 0.48
PD  9 1 P 0.44
PD  9 2 D 0.52
PD 10 1 P 0.48
PD 10 2 D 0.50
;;;;
```

For example the first subject received the drug in period 1 and the placebo in period 2 (order = DP) and the third subject experienced the reverse order. The variables in the data table are order (ord), a subject identification number (sub), the period (per), the treatment administered in that period (trt), and the response variable (y). For example, subject 1's responses, which occupy two rows of the data table, were 0.45 under the drug condition in period 1 and 0.47 under the placebo condition in period 2.

The analysis using PROC MIXED involves two novelties, first the treatment effect is influenced by order, and second, some observations came from the same subject and some from different subjects:

```
PROC MIXED DATA=CROSSOVER;
    CLASS ord sub per trt;
    MODEL y=trt ord trt*ord / NOINT SOLUTION;
    REPEATED per / SUBJECTS=sub TYPE=UN;
    LSMEANS trt / DIFF CL;
RUN;
```

The CLASS, MODEL and LSMEANS statements should by now be familiar. The inclusion of trt*ord in the MODEL statement specifies that the order of administration modifies the effects of the treatments. The REPEATED statement indicates that observations with the same subject number (sub) are repeated observations of the same person, and that the order in which the observations were made is specified by the period variable (per). TYPE=UN (unstructured) is the *covariance structure* and specifies that the variability of the observations can be different in the different periods.

The relevant part of the voluminous output of the procedure is the table of least squares means (adjusted means) and differences:

			Least Squares Means						
Effect	trt	Estimate	Standard Error	DF	t Value	Pr > |t|	Alpha	Lower	Upper
trt	D	0.502	0.01019	9	49.28	<.0001	0.05	0.4790	0.5250
trt	P	0.471	0.01019	9	46.24	<.0001	0.05	0.4480	0.4940

			Differences of Least Squares Means						
Effect	trt_trt	Estimate	Standard Error	DF	t Value	Pr > |t|	Alpha	Lower	Upper
trt	D-P	0.031	0.01468	9	2.11	0.0639	0.05	-0.00221	0.06421

The posterior distribution of the order-adjusted mean difference between the drug and placebo is approximately $t(9)$ with $\mu=0.031$ (Estimate) and $\sigma=0.0147$ (Standard Error). The 95% CI is approximately -0.002 to 0.064. The p value 0.639 is reported under the heading "Pr>|t|"; consequently, the posterior probability that the difference is positive is $P(\Delta_{adj} > 0 \mid \text{data}) \approx 1-p/2 = 0.968$. However, the more relevant question is whether the difference is clinically important. If, for example, the investigator judges that a true difference less than 0.025 would be clinically unimportant, then the relevant probability is,

$$P\left(\Delta_{adj} > 0.025 \middle| data\right) \approx P\left(t(9) > \frac{0.025 - 0.031}{0.0147}\right)$$
$$= P\left(t(9) > -0.408\right)$$
$$= 0.654$$

That is, there are about 2 to 1 odds that the drug has a clinically important effect relative to placebo. This is rather weak evidence either of clinical importance or unimportance. A study such as this that does not support a definitive finding one way or the other is said to be *underpowered*.

9.9 EXERCISES

9.1 Complete the FDA tutorial on protection of human subjects at http://cme.nci.nih.gov/. E-mail or submit a printed copy of your electronic certificate of completion to your instructor.

9.2 Use a spreadsheet layout like Fig. 9.4 on page 184 to hand calculate the correlation between the survival time of the poorly matched skin graft (x) and the close matched skin graft (y) in the data of exercise 8.6 (you may set it up in a computer spreadsheet). Compute the regression line $\hat{y}_x = a + b \cdot x$. Plot the data and the regression line on a graph. If you produce the graph by hand, use real graph paper and a ruler. You can print your own graph paper at http://www.thinkspot.net/materdei/resources/graphsheet.htm.

9.3 Suppose that a candidate for implantation has been deaf for one year. Describe the posterior predictive distribution of the speech understanding she will achieve with the implant and use tTailArea.xls to compute the probability that she will achieve at least 50% speech understanding.

9.4 Refer to Table 9.4 on page 196, and compute the predicted salary of,
(a) A white associate with a BA and 3 years of experience
(b) A nonwhite agent with an MA and 3 years of experience.

9.5 You think that raises might not be the same for nonwhite and white employees. In Table 9.6 on page 198 add two design variables to allow different slopes for nonwhite and for white employees. Suppose that the average raise turns out to be $55 for nonwhite employees and $63 for white employees, and that all the other regression coefficients are the same as in Table 9.4. Write down the estimated regression equation and compute the predicted salaries and difference between the salaries of nonwhite and white employees with an MS and 5 years of experience. Write out how the model would be specified in PROC MIXED.

9.6   In addition to different slopes, you also think that the reward for an MS might not be the same for nonwhite and white employees. In Table 9.6 on page 198, add two more design variables to allow different MS effects for nonwhite and for white employees. Suppose that the MS effect turns out to be $527 for nonwhites and $1023 for whites. Write down the estimated regression equation, and compute the predicted salaries and difference of predicted salaries for nonwhite and white employees with an MS and 5 years of experience. (Your equation and estimates should incorporate the different slopes.) Write out how the model would be specified in PROC MIXED.

9.10 REFERENCES

Finkelstein, M.O., and Levin, B., *Statistics for Lawyers, Second Edition*, New York, Springer-Verlag, Inc. (2001).

Fryauf-Bertschy H., Tyler R.S., Kelsay, D.M., Gantz,B.J., Woodworth, G.G., "Cochlear Implant Use By Prelingually Deafened Children: The Influences of Age at Implant and Length of Device Use." *Journal of Speech, Language, & Hearing Research*, Vol. 40, No. 1 (1997), pp 183–199.

Kadane, J.B.,ed. (1996), *Bayesian Methods and Ethics in a Clinical Trial Design*, New York: John Wiley and Sons.

Tyler, R.S., Teagle, H.F.B., Kelsay, D.M.R, Gantz, B.J., Woodworth, G.G., and Parkinson, A.J., "Speech Perception by Prelingually dEaf Children after Six Years of Cochlear Implant Use: Effects of Age at Implantation," *Annals of Otology, Rhinology, & Laryngology – Supplement*, Vol. 185 (2000), pp. 82-84.

10

Logistic Regression

10.1 INTRODUCTION

It is often the case that the response variable in a study is binary (a disease is present or absent, the patient survived for 10 years or did not, etc.). Although the analyst wants to determine how much the explanatory variables influence the response, it is not possible to use ordinary linear regression, because the response variable is categorical.

Logistic regression is the standard way to analyze binary response variables. For example, in a fictional study of smoking and health, the subjects are a large sample of adults between the ages of 40 and 85, the response variable is lung cancer, and the explanatory variables are smoking intensity (cigarettes per day), age, and gender. The purpose of logistic regression is to estimate the conditional probability that a person will develop lung cancer based on the information we have about that person: age, smoking intensity, and gender. We'll call the conditional probability

$$p_x = P\left(\text{cancer}|\text{age,smoking,gender}\right) \qquad (10.1)$$

The subscript x stands for the explanatory variables age, smoking intensity, and gender. A *logistic regression model* is a linear regression equation in which the response variable is the log odds,

$$\log\text{odds} = \ln\left(\frac{p_x}{1-p_x}\right) = \beta_1 \cdot x_1 + \cdots \beta_k \cdot x_k \qquad (10.2)$$

The mathematical function $\ln[x/(1-x)]$ is called the *logistic function*, or *logit* for short. Thus, the log odds, $\ln[p/(1\text{-}p)]$, is "the logit of p" or more often, just "the logit." The β's are called *logistic regression coefficients*, and the x's are *design variables* as defined in Chapter 9. The logistic regression model, Equation (10.2), looks like a linear regression model, but there is a difficulty — the response variable (log odds) is not directly observed. All we can actually observe is whether or not a subject has the adverse outcome, lung cancer. The details of how it is possible to estimate regression coefficients under these circumstances can be

209

found in more advanced statistics texts. Here we will be concerned with the practical matters of how to use a statistical computer package to estimate the regression coefficients, how to interpret the output of such an analysis, and how to interpret published reports of logistic regression analyses.

Table 10.1 shows the analysis of the fictional smoking study. There are three design variables and three regression coefficients. The column headed "Estimate" contains the estimated logistic regression coefficients. The estimated logistic regression coefficient of the constant is $\hat{\beta}_1$, its standard error is seb_1, the estimated logistic regression coefficient for the explanatory variable cigarettes per day is $\hat{\beta}_2$, its standard error is seb_2, and so on. If the data set is sufficiently large, then the posterior distribution of each logistic regression coefficient is approximately normal with parameter μ equal to the estimated logistic regression coefficient $\hat{\beta}$, and parameter σ equal to seb. This interpretation is valid if the prior distributions of the regression coefficients are noninformative and if there are no separating hyperplanes. The latter condition is quite technical and difficult to verify. The SAS LOGISTIC procedure (see Section 10.5) halts execution when it appears that this condition is violated; however, with sparse data the best strategy is to estimate the logistic regression model with a Bayesian package such as WinBUGS.

The posterior distribution of the logistic regression coefficient associated with smoking intensity is approximately normal with $\mu = 0.15$ and $\sigma = 0.05$. The 95% CI is $\mu \pm 1.96 \cdot \sigma$, or 0.052 to 0.248. This interval excludes zero, which means that there is at least 97.5% posterior probability that the logistic regression coefficient is positive, which is fairly strong evidence that higher smoking intensity is associated with an increase in the rate of lung cancer. The question is what is the impact of that increase — is it negligible, or could it represent a public health problem? In the next section we shall see that a logistic regression coefficient can be interpreted as the logarithm of an odds ratio, which will enable us to assess the impact of each risk factor.

Table 10.1   Logistic regression analysis of a smoking study.[a]

Variable	Estimate[b]	SE[c]	95% Confidence Interval	
			Lower	Upper
Intercept	−9.200	0.9	−11.0	−7.4
Cigs/day	0.150	0.05	0.05	0.25
Male	0.100	0.07	−0.04	0.24
Age (years)	0.002	.0009	0.0002	0.0038

a. Fictitious data.
b. Estimated logistic regression coefficients.
c. Standard error of logistic regression coefficient.

10.2 INTERPRETING LOGISTIC REGRESSION COEFFICIENTS

In linear regression, a regression coefficient represents the average change in the response variable y produced by a unit change in the explanatory variable associated with that coefficient. In a similar way, a logistic regression coefficient represents the change in the logarithm of the odds (logit) of the adverse outcome produced by a unit change in the explanatory variable. Thus, each additional cigarette smoked per day adds 0.15 to the logit, being male adds 0.1, and each year of age adds 0.002. The logit for subjects with a particular combination of characteristics is built incrementally from a base (the constant β_1), to which is added a contribution from each risk factor. The contribution from a given explanatory variable, such as cigarettes smoked per day, is the product of the regression coefficient and the numerical value of the design variable. Thus 20 cigarettes/day adds an estimated $0.15 \times 20 = 3$ points to the logit, being male adds another $0.1 \times 1 = 0.1$ point and being 45 years old adds $0.002 \times 45 = 0.09$ point, all of which, added to the intercept ($\beta_1 = -9.2$), come to a grand total of 6.02, the estimated logit for a 45-year-old male pack a day smoker (45M20)

$$\widehat{\text{logit}} \quad = \quad \underset{\text{constant}}{-9.2} \quad + \quad \underset{\text{pack a day}}{0.15 \times 20} \quad + \quad \underset{\text{male}}{0.1 \times 1} \quad + \quad \underset{\text{age 45}}{0.002 \times 45}$$

What, exactly, does it mean to have a logit of -6.02? It is, of course the logarithm of the odds on lung cancer; however, it is easier to interpret when converted first to odds and then to probability:

$$\widehat{\text{odds}} \quad = \quad \exp\left(\widehat{\text{logit}}\right) \quad = \quad \exp(-6.01) \quad = \quad 0.002454$$

$$\hat{p}_x \quad = \quad \frac{\widehat{\text{odds}}}{1+\widehat{\text{odds}}} \quad = \quad \frac{0.00245}{1.00245} \quad = \quad 0.002448 \tag{10.3}$$

Thus the estimated lung cancer rate for 45-year-old male pack-a-day smokers is $\hat{p}_{45M20} = 0.002448$, that is about 245 cases per 100,000. For 45-year-old male nonsmokers (45M0) the logit is -9.01, the odds are $\exp(-9.01) = 0.000122$ and the estimated rate of lung cancer is $\hat{p}_{45M0} = \text{odds}/(1+\text{odds}) = 0.000122$, that is about 12 in 100,000 (see row 2 of Table 10.2).

Treating pack-a-day smokers as the exposed group, the relative risk (the rate of lung cancer in the exposed group divided by the rate in the unexposed group) is $0.002448/0.000122 = 20.07$, and the odds ratio is $0.002454/0.000122 = 20.12$. The process of computing log odds, odds, probabilities, relative risks, and odds ratios can be made routine by using design variables as illustrated in Table 10.2. Simply multiply the numerical value of each design variable by the

Table 10.2   Logits, Odds, and Probabilities via Design Variables

Explanatory Vars			Design Variables				logits, odds, rates			
Cigs/ Day	Sex	Age	X_1	X_2	X_3	X_4	Logit	Odds	Rate P_x	Odds Ratio
0	M	45	1	0	1	45	-9.01	0.000122	0.000122	}20.1
20	M	45	1	20	1	45	-6.01	0.002454	0.002448	
20	M	25	1	20	1	25	-6.05	0.002358	0.002352	}1.1
20	F	25	1	20	0	25	-6.15	0.002133	0.002129	
$\hat{\beta}$ values:			-9.2	0.15	0.10	0.002				

logistic regression coefficient at the foot of its column and add the products in each row. Logits are converted into odds by using Equation (10.3).

10.3   ADJUSTED ODDS RATIOS

Each of the odds ratios in Table 10.2 compares two groups of people who differ with respect to one explanatory variable but are otherwise identical. For example, the first odds ratio, 20.1, compares pack-a-day smokers with nonsmokers of the same sex and age (male, age 45). Similarly, the second odds ratio, 1.1, compares males with females, holding smoking and age fixed (pack a day, age 25). They are, in other words, *adjusted odds ratios*, and they quantify the impact of one explanatory variable, *all other things being equal*.

There is a much simpler way to compute an adjusted odds ratio using the mathematical relationship $\ln(A/B) = \ln(A) - \ln(B)$. For example, in Table 10.2 we determined that the log odds on lung cancer for 45 year old, male, pack-a-day smokers is -6.01 and the corresponding log odds for identical nonsmokers is -9.01. The odds are exp(-6.01) and exp(-9.01), respectively, and the adjusted odds ratio for pack-a-day smoking is their ratio,

$$\text{odds ratio} = \frac{\exp(-6.01)}{\exp(-9.01)} = \exp(-6.01 - (-9.01))$$

$$= \exp(3) = \exp\left(\begin{array}{c}\text{increment for}\\ \text{20 cigs. a day}\end{array}\right) \quad (10.4)$$

$$= \exp(0.15 \times 20) = \exp(\beta_2 \cdot \Delta X_2)$$

The expression ΔX_2, represents the difference of the numerical values of the design variable X_2 (cigarettes per day) in the numerator and denominator of the odds ratio; all other design variables (age and gender) are the same in the numerator and denominator and therefore cancel out of the ratio. Since the numerator value is $X_2 = 20$ cigs./day and the denominator value is $X_2 = 0$ the difference is $\Delta X_2 = 20{-}0 = 20$. Boolean design variables like X_3 (gender=male) are a special case. Such design variable are either 1 (male) or 0 (female); consequently the difference ΔX_3 for males vs. females is 1. Thus the adjusted odds ratio for a Boolean design variable is $\exp(\beta)$. Using this method, it is easy to determine that the odds ratio for males vs. females (all other things being equal) is $\exp(\beta_3) = \exp(0.1) = 1.1$, which is identical to the odds ratio for gender computed the hard way in the last two rows of Table 10.2. Unfortunately, there is no shortcut to relative risks, which must be computed as in Table 10.2.

10.4   A CASE STUDY: SURVIVING MELANOMA

Dr. Lynn Schuchter and colleagues (Schuchter et al. 1996) developed a logistic regression equation for predicting 10-year survival in patients with primary melanoma. Excerpts from this study are reproduced with permission of the American College of Physicians. The authors described their study this way,

Patients: 488 patients with primary cutaneous melanoma who had no apparent metastatic disease. Patients were followed prospectively for at least 10 years. An independent validation sample of 142 patients was used to assess the stability of the model.

Measurements: Six clinical and pathologic variables that predict survival and are readily available to the clinician were used to develop a prediction model. The variables were tested for their association with death by using a univariate logistic regression model. Point estimates were generated for the probability of surviving melanoma at 10 years. Variables that were statistically significantly associated with survival were retained for testing in a logistic regression model.

The estimated logistic regression equation was described in as follows:

The logistic regression equation for probability of surviving at least 10 years is $(1+e^{-x})^{-1}$, where $x = -2.245 + 3.929$ (if tumor thickness < 0.76 mm) $+ 2.251$ (if thickness is between 0.76 mm and 1.69 mm, inclusive) $+ 1.066$ (if thickness is between 1.70 mm and 3.60 mm, inclusive) $+ 0.7126$ (if the patient is female) $+ 1.470$ (if tumor site is extremity) $+ 1.086$ (if patient age < 60 years).

The quantity referred to as x in this passage is the sum of products of regression coefficients each multiplied by a design variable, that is, it is the estimated logit:

$$
\begin{aligned}
x &= \widehat{\text{logit}} \\
&= -2.245 \cdot X_1 + 3.929 \cdot X_2 + 2.251 \cdot X_3 + 1.066 \cdot X_4 \\
&\quad +0.7126 \cdot X_5 + 1.470 \cdot X_6 + 1.086 \cdot X_7
\end{aligned}
\tag{10.5}
$$

We'll identify the design variables shortly, but first we need to understand the expression $(1+e^{-x})^{-1}$. Since x is the estimated logit, the authors are reporting that the estimated probability, or rate, of 10 year survival is

$$
\hat{P}(\text{survival}\,|\,\text{risk factors}) = \hat{p}_x = \frac{1}{1+e^{-\widehat{\text{logit}}}}
\tag{10.6}
$$

which is not the same as our expression for the estimated rate

$$
\hat{p}_x = \frac{\widehat{\text{odds}}}{1+\widehat{\text{odds}}} = \frac{e^{\widehat{\text{logit}}}}{1+e^{\widehat{\text{logit}}}}
\tag{10.7}
$$

To see that Equations (10.6) and (10.7) are equivalent to each other, multiply the numerator and denominator of Equation (10.6) by $e^{\widehat{\text{logit}}}$. To make the algebra easier to follow, let L symbolize the estimated logit; then starting with Equation (10.6) we have

$$
\begin{aligned}
\frac{1}{1+e^{-L}} \cdot \frac{e^L}{e^L} &= \frac{e^L}{e^L + e^{-L} \cdot e^L} = \frac{e^L}{e^L + e^{L-L}} \\
&= \frac{e^L}{e^L + e^0} = \frac{e^L}{e^L + 1}
\end{aligned}
$$

The design variables in Equation (10.5) are defined in the quoted passage. For example, the range of tumor thickness is covered by three Boolean design variables: x_2 = (tumor thickness < 0.76), x_3 = (thickness between 0.76 and 1.69 mm), and x_4 = (thickness between 1.70 and 3.60 mm). The reference category is thickness > 3.30 mm) The other design variables (all of them Boolean) are listed in Table 10.3.

The authors' table of estimated logistic regression coefficients, adjusted odds ratios, and confidence intervals (approximate credible intervals) is reproduced in Table 10.4. The adjusted odds ratios are computed as $\exp(\hat{\beta})$, because $\Delta X=1$ for Boolean design variables. For example the adjusted odds ratio for female sex is $\exp(\beta_5) = \exp(0.71) = 2.0$. An approximate 95% CI for an adjusted odds ratio is

$$
\exp\left[\left(\hat{\beta} \pm 1.96 \cdot \text{seb}\right) \cdot \Delta X\right]
\tag{10.8}
$$

Here the quantity ΔX is the difference between the value of the design variable in

Table 10.3 Definitions of design variables in the melanoma analysis.

Design variable	Authors' description
x1	(Constant always 1)
x2	Tumor thickness < 0.76
x3	Thickness between 0.76 and 1.69 mm
x4	Thickness between 1.70 and 3.60 mm
x5	Female sex
x6	Tumor site is extremity[a]
x7	Patient age < 60 years

a. Axis location includes trunk, head and neck, and volar (palm) and subungual (nails) sites.

the group in the numerator of the adjusted odds ratio and the design variable value in the denominator; in this case all design variables are binary (0/1) so $\Delta X=1$. For example, the estimated adjusted odds ratio corresponding to β_2, the regression coefficient for X_2 (lesion thickness < 0.76 mm) compares patients in the thinnest category of lesions with patients in the reference category (thickness > 3.60 mm), all other things being equal. The adjusted odds ratio is 50.8 with 95% CI (18.5, 140), indicating that patients with lesions under 0.76 mm are estimated to have 51 times greater odds of survival than patients with lesions over 3.6 mm, all other things being equal.

If the denominator of the odds ratio is not the reference category, as for example in a comparison between lesion thickness < 0.76 mm and thickness in the range 0.76 – 1.69 mm, the adjusted odds ratio is the ratio of two "reference category" odds ratios: 50.8/9.5 = 5.3.

Table 10.4 Adjusted odds ratios for independent predictors of survival[a]

Variable	Coefficient	Adjusted odds ratio (95% CI)
Lesion thickness		
< 0.76 mm	3.93	50.8 (18.5 to 140)[b]
0.76 – 1.69 mm	2.25	9.5 (4.01 to 22.5)[b]
1.70 – 3.60 mm	1.07	2.9 (1.29 to 6.54)[b]
Female sex	0.71	2.0 (1.2 to 3.6)
Primary lesion on an extremity	1.47	4.4 (2.3 to 8.4)
Age at diagnosis $\leq$ 60 y	1.10	3.0 (1.7 to 5.3)

a. Source Table 1 of Schuchter et al. (1996).

That is, an odds ratio not involving the reference category is a ratio of odds ratios,

$$\frac{\text{odds}_1}{\text{odds}_2} = \frac{\dfrac{\text{odds}_1}{\text{odds}_{\text{ref}}}}{\dfrac{\text{odds}_2}{\text{odds}_{\text{ref}}}} = \frac{\text{OR}_1}{\text{OR}_2} \qquad (10.9)$$

Unfortunately, it is not possible to compute a CI for this derived odds ratio without more information.

Standard errors of the regression coefficients were not reported in Schluchter et al. (1996), but can be deduced from the reported confidence intervals via Equation (7.15). Thus, for example, the standard error of $\hat{\beta}_2$, the regression coefficient for (thickness < 0.76 mm) is,

$$seb_2 = \frac{\ln\left(\dfrac{ucl}{lcl}\right)}{2 \cdot 1.96} = \frac{\ln\left(\dfrac{140}{18.4}\right)}{3.92} = 0.52 \qquad (10.10)$$

Consequently the posterior distribution of β_2 is approximately normal with $\mu = \hat{\beta}_2 = 3.93$ (from Table 10.4) and $\sigma = seb_2 = 0.52$ [from Equation (10.10)]. With this information it is possible to compute the posterior probability of a sentence about the true adjusted odds ratio in the population, for example, the probability that the odds ratio exceeds 20.

The first step in the computation is to notice that since $\text{OR}_2 = \exp(\beta_2)$ it follows that $\beta_2 = \ln(\text{OR}_2)$; therefore,

$$P(\text{OR}_2 > 20|\text{data}) = P(\ln(\text{OR}_2) > \ln(20)) = P(\beta_2 > 2.996)$$

But β_2 has an approximately normal posterior distribution with $\mu = 3.93$ and $\sigma = 0.52$; consequently,

$$P(\beta_2 > 2.996) = P\left(Z > \frac{2.996 - 3.93}{0.52}\right) = P(Z > -1.80) = 0.964$$

Besides reporting adjusted odds ratios, Schuchter et al. (1996) provided a table of estimated probabilities of 10-year survival for various categories of melanoma patients (see Table 10.5). For example, a woman aged 51 with a 0.72-mm lesion on her arm has an estimated 99% probability of surviving 10 years, whereas a 70 year old man with a 4-mm lesion on his back has only a 10% estimated probability of surviving 10 years.

Table 10.5 Estimated 10-year survival probabilities.[a]

Thickness (mm)	Age (yr)	Tumor with Extremity Location — Female Patients	Tumor with Extremity Location — Male Patients	Tumor with Axis Location[b] — Female Patients	Tumor with Axis Location[b] — Male Patients
< 0.76	≤ 60	0.99 (0.98, 1.00)	0.98 (0.96, 0.99)	0.97 (0.93, 0.99)	0.94 (0.88, 0.97)
	> 60	0.98 (0.95, 0.99)	0.96 (0.89, 0.98)	0.92 (0.82, 0.96)	0.84 (0.70, 0.93)
0.76–1.69	≤ 60	0.96 (0.92, 0.98)	0.93 (0.85, 0.97)	0.86 (0.76, 0.92)	0.75 (0.72, 0.84)
	> 60	0.90 (0.80, 0.95)	0.81 (0.64, 0.91)	0.67 (0.50, 0.81)	0.50 (0.33, 0.67)
1.70–3.60	≤ 60	0.89 (0.80, 0.94)	0.80 (0.65, 0.89)	0.65 (0.50, 0.77)	0.48 (0.35, 0.61)
	> 60	0.73 (0.57, 0.85)	0.57 (0.38, 0.75)	0.38 (0.24, 0.55)	0.24 (0.14, 0.37)
> 3.60	≤ 60	0.74 (0.53, 0.87)	0.58 (0.36, 0.77)	0.39 (0.21, 0.60)	0.24 (0.13, 0.40)
	> 60	0.48 (0.28, 0.69)	0.32 (0.16, 0.53)	0.18 (0.08, 0.35)	0.10 (0.04, 0.20)

a. Source: Table 3 of Schuchter, et al. 1996.
b. Axis location includes trunk, head and neck, and volar (palm) and subungual (nails) sites.

Table 10.6 shows how the estimated survival probabilities were calculated. Note that our computed values for 51-year-old females and 70-year-old males agree exactly with Schuchter's values in Table 10.5. Although estimated probabilities can be computed from design variables and logistic regression coefficients, the confidence intervals must be computed with a statistical software package such as SAS PROC LOGISTIC (see Section 10.5).

Table 10.6 Computation of estimated survival probabilities.

	Logistic regression coefficients −2.25 3.93 2.25 1.07 0.71 1.47 1.10							Estimates of:		
	x_1	x_2	x_3	x_4	x_5	x_6	x_7	logit[a]	odds[b]	p[c]
Female, age 51, 72-mm lesion on arm	1	1	0	0	1	1	1	4.96	143	0.99
Male, age 70, 4-mm lesion on back	1	0	0	0	0	0	0	−2.25	0.105	0.10

a. Equation (10.5)
b. Equation (10.6)
c. Equation (10.7)

The authors' results and conclusions suggest that the predictive model will be useful to physicians treating melanoma patients,

Conclusions: A prognostic model that uses four readily accessible variables more accurately predicts outcome in patients with primary melanoma than does tumor thickness alone. This four-variable model can identify patients at high risk for the recurrence of disease, an identification that becomes increasingly important as adjuvant therapies are developed for treatment of melanoma."

According to the Online Medical Dictionary (http://cancerweb.ncl.ac.uk/omd/contents/A.html), an *adjuvant therapy* is "a treatment that is added to increase the effectiveness of a primary treatment. In cancer, adjuvant treatment usually refers to chemotherapy, hormonal therapy or radiation therapy after surgery to increase the likelihood of killing all cancer cells." So it appears that the prognostic logistic regression model could be used to identify patients most in need of more aggressive treatment.

10.5   COMPUTING LOGISTIC REGRESSION WITH SAS SOFTWARE

The data file CSSsubset.sas7bdat, available for downloading from the textbook website, is a subset of the Georgia Charging and Sentencing data base. Details can be found in Baldus et al. (1990). These data were compiled from records of the department of corrections and the parole board to assess the levels of racial discrimination and arbitrariness occurring at different levels within Georgia's capital charging and sentencing system. The data cover approximately 700 variables and a sample of 1000 murder and voluntary-manslaughter cases. Information was obtained for all known penalty trial cases, that is, cases at risk of receiving a death penalty, and for a sample of other types of cases stratified by case type (voluntary-manslaughter conviction, and non penalty trial life sentence) and by state judicial circuit. Numerous measures of defendant blameworthiness were developed as a basis for assessing levels of arbitrariness and discrimination in the capital charging and sentencing system. Variables include race, sex, and socioeconomic class as well as crime codes, jury and bench decisions, final plea, term, and number of counts convicted.

The data file SentStudy.sas7bdat is a subset of this data base that contains these six variables:

Case   Case identifier
Sent   Death sentence (D=death, L = life)
AggCirx Total number of statutory aggravating circumstances present
MitCirx Total number of mitigating circumstances present
Def   Defendant race (B=black, NB = nonblack)
Vic   Victim race (W=any white, NW= all nonwhite)

The SAS procedure for logistic regression is called LOGISTIC. It has many features in common with the linear regression procedure MIXED; however, it does not construct design variables the way PROC MIXED does, which makes interpreting the output somewhat difficult when there are product variables in the model. A so-called *additive effects* model is analyzed in Fig. 10.1. Notice that it is necessary to specify which value of the response variable Sent (sentence) corresponds to the outcome to be predicted, in this case it is D (death). It is also necessary to specify which values of the categorical variables Vic and Def are the reference categories. Unlike the MIXED procedure, LOGISTIC has more than one way of constructing design variables; the option PARAM=REF selects the version of design variables used in this book. The program assumes that the data file SentStudy.sas7bdat was downloaded to the directory C:\temp.

Output of the SAS program includes logistic regression coefficients with standard errors and confidence intervals (approximate credible intervals), and adjusted odds ratios with confidence intervals. The adjusted odds ratio estimates are antilogs of the estimated regression coefficients. 95% Wald Confidence Limits are computed with Equation (10.8) and can be interpreted as approximate CI's

Program

```
LIBNAME BioStat "C:\temp\";
PROC LOGISTIC DATA=BioStat.SentStudy;
    CLASS Def(REF="NB" Vic(REF="NW") / PARAM=REF;
    MODEL Sent(EVENT="D")=Vic Def AggCirx MitCirx;
RUN;
```

Logistic regression coefficient estimates and standard errors.

Parameter		DF	Estimate	StdErr	Wald ChiSq	Pr>ChiSq
Intercept		1	-3.8138	0.4232	81.2187	<.0001
Vic	W	1	0.6874	0.3201	4.6118	0.0318
Def	B	1	-0.3739	0.2646	1.9973	0.1576
AGGCIRX		1	0.9357	0.1008	86.2387	<.0001
MITCIRX		1	-0.4088	0.1135	12.9716	0.0003

Adjusted odds ratio estimates and CI's.

Effect			Point Estimate	95% Wald Confidence Limits	
Vic	W	vs NW	1.989	1.062	3.724
Def	B	vs NB	0.688	0.410	1.156
AGGCIRX			2.549	2.092	3.106
MITCIRX			0.664	0.532	0.830

Figure 10.1 Logistic Regression via SAS PROC LOGISTIC. Adjusted odds ratio estimates are antilogs of regression coefficients.

for the adjusted odds ratios. The PARAM=REF option is mandatory; otherwise the regression coefficients for categorical variables do not agree with the adjusted odds ratios.

PROC LOGISTIC can also create a file of estimated probabilities of the adverse outcome (death) as in Equations (10.5) and (10.6). In addition, it computes confidence intervals for the estimated probabilities, a calculation that cannot be done by hand with the information in the default printout. The procedure for requesting probability estimates and confidence intervals is shown in Fig. 10.3. The requests are for a ladder of cases, from mild crimes with no aggravating circumstances and three mitigating circumstances, to severe crimes with six aggravating circumstances and no mitigating circumstances. The estimates with 95% CI's are shown in Fig. 10.3; not surprisingly, the estimated probability of death increases as the crimes become more serious; however, when the victim is white, the probability is even higher.

Create a request file

```
DATA Request;
    INPUT Def $ Vic $ AggCirx MitCirx;
DATALINES;
B   NW   0   3
B   W    0   3
B   NW   1   2
B   W    1   2
B   NW   3   1
B   W    3   1
B   NW   6   0
B   W    6   0
RUN;
```

Make the predictions

```
LIBNAME BioStat "C:\temp\";
PROC LOGISTIC DATA=BioStat.SentStudy;
    CLASS Def(REF="NB" Vic(REF="NW") / PARAM=REF;
    MODEL Sent(EVENT="D")=Vic Def AggCirx MitCirx;
    SCORE DATA=Request OUT=Predict;
PROC PRINT DATA=Predict;
RUN;
```

Figure 10.2   Computing estimated probabilities of death with 95% CI's.

Printout of estimated probabilities (edited to fit)

Def	Vic	Agg Cirx	Mit Cirx	P_D	LCL_D	UCL_D
B	NW	0	3	0.00443	0.00205	0.00957
B	W	0	3	0.00878	0.00363	0.02110
B	NW	1	2	0.01680	0.00936	0.02997
B	W	1	2	0.03286	0.01724	0.06171
B	NW	3	1	0.14315	0.09208	0.21580
B	W	3	1	0.24937	0.18230	0.33111
B	NW	6	0	0.80637	0.64070	0.90676
B	W	6	0	0.89225	0.81674	0.93897

Figure 10.3   Estimated probabilities, P_D, of the adverse outcome death with 95% confidence intervals, LCL_D to UCL_D, interpretable as approximate 95% CI's.

10.6   EXERCISES

10.1 Using the estimated logistic regression coefficients in Table 10.1 on page 210, calculate the odds ratio for a nonsmoking 60-year-old male vs. an identical 25-year-old male.

10.2 Using the same coefficients, what is the estimated probability that a 70 year old, two-pack-a-day, male will have lung cancer? What is the odds ratio for this person compared to an identical nonsmoker?

10.3 Using the estimated logit model in Equation (10.5) on page 214, calculate the estimated 10-year survival probability for a 55-year-old woman with a 0.85 mm melanoma lesion on her arm. What is the estimated probability if the lesion is on her back? Compute the estimated adjusted odds ratio and adjusted relative risk. Compute the odds ratio two ways — as the ratio of estimated odds and as $\exp(\beta \cdot \Delta X)$. The two methods should give the same result.

10.4 Repeat exercise 10.3 for a man. Is the adjusted odds ratio the same as for a woman? Is the adjusted relative risk the same for men and women?

10.5 Compute adjusted odds ratios and CI's for each of the explanatory variables in Table 10.1 on page 210. Compute the adjusted odds ratio and confidence interval for a 10-year age difference. Compute the adjusted odds ratio and CI for a 10-cigarette/day difference. Why does this equation predict the same odds ratio for two packs a day vs. one pack a day as for one pack a day vs. nonsmoking.

10.6 Compute seb for each β (other than the constant) in Table 10.4 on page 215. Compute the posterior probability that OR_6 exceeds 4. (OR_6 is the adjusted odds ratio for extremity vs. axis location.)

10.7 In Table 10.1 on page 210, compute the estimated adjusted OR and approximate 95% CI for pack-a-day smoking vs. nonsmoking. Compute the approximate posterior probability that the true OR exceeds 10. Hint: the true OR is exp(20·β), not β ; consequently, the required probability is

$$P(e^{20 \cdot \beta} > 10 | \text{ data}) = P(20 \cdot \beta > \ln(10) | \text{ data}) = P\left(\beta > \frac{\ln(10)}{20} \middle| \text{ data} \right)$$

10.8 Referring to Exercise 7.9 on page 138, do you think the odds ratio needs to be adjusted? What confounding variables do you think should be included.

10.7 REFERENCES

Baldus, D, Woodworth, GG, and Pulaski, C., *Equal Justice and the Death Penalty: A Legal and Empirical Analysis. Boston: Northeastern University Press*, 1990.

Schuchter, Lynn, Schultz, Delray J., Synnestvedt, Marie, Trock, Bruce J., Guerry, DuPont, Elder, David E., Elenitsas, Rosalie, Clark, Wallace H., and Halpern, A.C. (the Pigmented Lesion Group), "A Prognostic Model for Predicting 10-Year Survival in Patients with Primary Melanoma," *Annals of Internal Medicine*, Vol. 125 (1 September 1996), pp. 369–375.

11

Hierarchical Models

11.1 META-ANALYSIS

Meta-analysis, or *research synthesis*, deals with the problem of reaching consensus on a particular question, using evidence from multiple independent studies. For example, Morris et al. (1992) reported that there were seven published studies on the question of whether chlorinated water is a risk factor for bladder cancer. Excerpts from this paper are reproduced with permission of the American Public Health Association. Each of the *primary studies*, summarized in Table 11.1 and Table 11.2, reported an estimated adjusted odds ratio that contrasted people who were exposed to chlorinated drinking water with people who for one reason or another mostly drank unchlorinated water. Each study reported an estimated odds ratio and confidence interval adjusted for one or more confounding variables. Six of the seven studies found increased risks (odds ratio > 1), but in only three of these studies did the 95% CI rule out the possibility of equal risk (OR = 1) or even a reduced risk (OR < 1).

Table 11.1 **Studies of chlorinated water and bladder cancer.**

Principal Author	Year	Adjusted OR	LCL[a]	UCL	Method	Quality
Cantor	1987	1.19	1.07	1.32	Logistic	78
Zierler	1988	1.40	1.20	2.10	M-H[b]	71
Wilkins	1986	2.20	0.71	6.82	Logistic	61
Gottlieb	1982	1.18	0.95	1.45	adj[c]	49
Brenniman	1980	0.98	0.77	1.25	adj[c]	46
Young	1981	1.15	0.70	1.89	Logistic	45
Alvanja	1978	1.69	1.07	2.67	adj[c]	43
Consensus		1.21	1.06	1.42		

a. Lower and upper endpoints of the 95% credible interval.
b. M-H is the Mantel-Haenszel method, which produces an approximate logistic regression estimate.
c. The odds ratio was adjusted by some method other than logistic regression.

Table 11.2 Study characteristics.

Investigator	Population	Cases	Controls	Exposure ascertainment	Adjusted for[a]
Cantor	White residents of 10 US regions	2962 newly diagnosed cases of bladder cancer	Community controls matched 2::1 for age, sex, and region by random dialing and medicare files	History of residence and beverage consumption combined with sampling of water utilities.	Age, sex, smoking, occupation, size of place of longest residence
Zierler	MA residents	51,645 kidney, bladder, lung, breast, pancreas, GI cancer deaths	cardiovasular, cerebro-vascular, or pulmonary deaths or lymphatic cancer deaths.	Chlorination / chloramination at address on death certificate.	Age, sex surface / groundwater, size, poverty of county of residence.
Wilkins	31,000 residents of Washington Co, MD	Diagnosis of cancer in 12 years after initial survey.	Unexposed: Deep well users.	Exposed: Users of chlorinated surface water in Hagerstown, MD	Age, sex, smoking, marital status, education.
Gottlieb	Louisiana residents	10,205 kidney, bladder, stomach, liver, colorectal cancer deaths.	Other cancer deaths and noncancer deaths matched 1:1 for age, sex, race, and county.	Chlorinated / unchlorinated water ad address on death cert. and at birthplace.	surface / groundwater, cardiovascular disease, age, sex, race, county
Brenniman	White residents of IL	3 208 GI and urinary tract cancer deaths	Noncancer deaths matched 14:1 for age, sex, and county.	Chlorinated / unchlorinated water at address on death cert.	Urbanicity, population densit, age, sex, county.
Young	White, Female WI residents	8029 cancer deaths	Noncancer deaths matched 1:1 for age, sex, race, and county	Chlorinated / unchlorinated water at address on death cert.	Pop density, occupation, marital status, rural runoff in water, age, sex, race, county
Alvanja	NY State residents	3446 GI and urinary tract cancer deaths	Noncancer and lung cancer deaths matched 1:1 for age, sex, county.	Chlorinated / unchlorinated water at address on death cert.	Urbanicity, occupation, age, sex, county

a. In addition to variables used to match cases and controls.

This chapter shows how to use meta-analysis to combine the information from these seven studies and reach consensus on the interpretation of the evidence that chlorinated water is a risk factor for bladder cancer. The results of a meta-analysis of the seven chlorinated water studies are reported in last row of Table 11.1. The consensus of the seven studies is that the most probable value of the odds ratio is 1.21 (a 21% increase in the odds of bladder cancer) with 95% credible interval 1.06 to 1.42.

11.2 HIERARCHICAL MODELS

A characteristic of meta-analysis data that we have not encoutered up to this point is that it is *hierarchical*, consisting of one layer of sampling above another. To give a simple example, an investigator could make a list of all public hospitals in the United States and randomly select 30 of them for further study. Within each of the 30 selected hosptials, the investiagor could take a sample of 100 patients. So the data would consist of a sample of 30 samples of 100 patients each.

Meta-analysis also has a hierarchical structure. We imagine the population consisting of all the research studies that could have been done on chlorinated water and bladder cancer. Meta-analysis is based on the belief that the seven studies that were actually done can be treated as a random sample from the population of studies. The purpose of each of these seven studies was to estimate the odds ratio in its study population (residents of Massachusetts, for example). In order to estimate the odds ratio, each of the seven studies had to draw a sample of cases and controls from its study population. Consequently, a meta-analysis, like our hypothetical hospitals study, is fundamentally a sample of samples; the idea is diagrammed in Fig. 11.1.

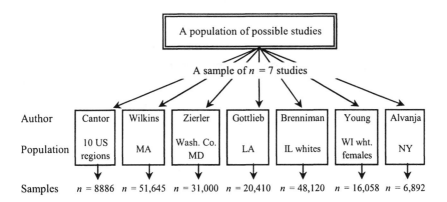

Figure 11.1   Hierarchical structure of a meta-analysis.

Each study population has a true (but unobserved) odds ratio, which it estimated from a sample of subjects. For example, the true odds ratio for Cantor's study population is an unknown quantity; Cantor obtained a sample of 8886 subjects from his study population and computed the estimated odds ratio, 1.19, and its 95% confidence interval, (1.07 to 1.32).

The goal of the meta-analysis is to say something about the box of possible studies at the top of the hierarchy. The reason for that interest is that the box at the top of the hierarchy represents the entire country (or world), not just the seven study areas. There are many things that could be said about the box of studies: the average funding per study, the proportion of female principal investigators, and so on; however, the only aspects of the box that are relevant to public health are the median odds ratio and the variability of the odds ratio from one region of the country to another. (Median rather than mean, in case the distribution of odds ratios is skewed.) Fig. 11.2 is a *model* that captures the essential features of Fig. 11.1 without the extraneous details.

11.3   NORMAL APPROXIMATION

The unknown quantities in this analysis are odds ratios; however, as we learned in Section 7.6.5, the log odds ratio is more convenient to work with, since it typically has a nearly normal distribution. Since the expression "log odds ratio" takes up a lot of space, we'll use the symbol λ (lambda) to to stand for a true, population log odds ratio and $\hat{\lambda}$ to stand for an estimate.

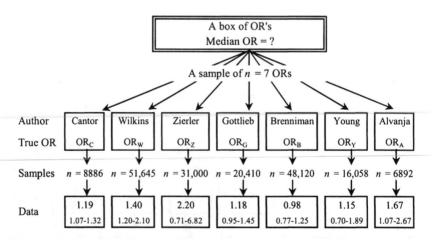

Figure 11.2   A model of the essential features of the meta-analysis. The data are represented by estimated odds ratios and confidence intervals. The goal is to use Bayes rule to compute the posterior distribution of the median of the box of odds ratios given the data.

Estimated log odds ratios and standard errors for Cantor's study are

$$\hat{\lambda} = \ln\left(\widehat{OR}\right) = \ln(1.19) = 0.174, \text{ and}$$

$$se\lambda = \frac{\ln\left(LCL/UCL\right)}{2 \cdot 1.96} = \frac{\ln(1.32/1.07)}{3.92} = 0.054 \tag{11.1}$$

where $\widehat{OR}$, LCL, and UCL are the adjusted odds ratio and lower and upper confidence limits from the first row of Table 11.1. This calculation was repeated for the six other studies and the results are listed in Table 11.3.

An important by-product of normalizing the data is that the mean of a normal distribution is also its median and therefore the antilog, exp( ), of the mean of the log odds ratio is the median of the odds ratio, which is the unknown quantity of interest (see Fig. 11.2 and the top panel of Fig. 8.11).

11.4 EXCHANGEABILITY

How can it be that a mere seven studies can tell us something about the population of all possible studies? What drives the meta-analysis is the assumption that the studies are *exchangeable*. The word *exchangeable* means more or less the same as the legal term *fungible*.[5] Gallons of pasturized 2% milk in the dairy case at a supermarket are fungible: one is as good as another. Not that they are absolutely identical, just that I have no reason to prefer this gallon of milk over the another one (assuming same brand, same expiration date).

Table 11.3   Information in the meta-sample

Studies in the meta sample	What we'd like to observe	What we do observe (estimates and standard erors)	
Author	true log odds ratio λ	estimate	standard error
Cantor	λ_C	0.174	0.054
Zierler	λ_Z	0.336	0.143
Wilkins	λ_W	0.788	0.577
Gottlieb	λ_G	0.166	0.108
Brenniman	λ_B	−0.020	0.124
Young	λ_Y	0.140	0.253
Alvanja	λ_A	0.525	0.233

5. *Fungible*: Of such kind or nature that one specimen or part may be used in place of another in satisfaction of an obligation, *Webster's New Collegiate Dictionary*, 2nd ed.

Exchangeability of studies of chlorinated water means that, other things being equal, one study population and design is as good as another for the purpose of studying the population of odds ratios of possible studies (Fig. 11.2). Exchangeability does not mean that the studies are identical or produce identical results; it means that I have no reason to believe that one population has a higher odds ratio than another and no reason to believe that one study would produce a biased estimate of the odds ratio of its study population. Exchangeability refers to the state of knowledge of the person evaluating the evidence — it means that she can see no meaningful distinction among the studies (at least with respect to population odds ratios). In other words, she would view the seven studies described in the previous section as exchangeable with each other and with other studies that could have been performed if she possessed no information that would cause her to predict that one study would have a systematically higher or lower odds ratio than another.

Exchangeability is not to be taken for granted. Obviously if the studies all used the same subject selection criteria, if they all adjusted for the same confounding variables, and if they differed only in geographic location, then it might be reasonable to presume that they were exchangeable; however, the seven studies in the chlorinated water meta-analysis differ in technique (for example, one is prospective, six are retrospective), adjustment (for example, not all adjust for smoking), and selection criteria (one excluded males; three excluded nonwhites). The identification of a set of exchangeable studies is not trivial and must be made by a person or team with intimate knowledge of the research area.

11.5   THE STATISTICAL MODEL: PRIOR AND LIKELIHOOD

Exchangeability of studies means that for the purpose of estimating the box average, a meta-sample can be thought of as a sample from a box containing the true log odds ratios (λ's) of all possible studies. For example, one possible study (call it study 1) might be a case–control study of death certificates in Los Angeles County, and it's true λ value, λ_1, would be the logarithm of the adjusted odds ratio in that population; λ_2 might be the true log odds ratio for a case–control study of newly diagnosed bladder cancer cases vs. sex- and age-matched controls at the University of Iowa Hospitals and Clinics; and so on for the thousands of other other λ's in the box. The unknown quantity of interest, which we will call θ, is the mean of all the λ's in the box; it is the *consensus log odds ratio.*

The inferential problem is to figure out what the meta-sample data in Table 11.3 tell us about the consensus log odds ratio. The method, as always, is to use Bayes' rule to compute the posterior distributions of the unknown quantities. The two inputs to a Bayes rule calculation are (1) the prior distribution of the unknown quantities (the true log odds ratios of the seven studies) and (2) the likelihood of the observed data given the values of the unknown quantities.

11.5.1   The Prior

The hierarchical model has two levels: a so-called *hyperpopulation* of potential studies and, for each potential study, a real population of people. From the hyper-population, the meta-investigator obtains a sample of k studies ($k = 7$ in the chlorinated water meta-analysis). The meta-investigator models the hyperpopulation as a box of λ values. The values in the box are assumed to be approximately normally distributed; consequently the only unknown quantities are μ_{box} and σ_{box}, which we will call μ_λ and σ_λ. They are, respectively, the most probable log odds ratio value in the population of studies and the variability of the odds ratio from study to study. These so-called *hyperparameters* describe the contents of the box of studies at the top of the hierarchy depicted in Fig. 11.1.

The best information about the hyperparameters μ_λ and σ_λ would have been the actual numerical values of a sample of λ's from the hyper-population. The numbers in that sample are denoted $\lambda_1, \lambda_2, ..., \lambda_k$. In the chlorinated-water-bladder-cancer meta-analysis, there are $k=7$ studies; λ_1 is the true odds ratio for Cantor's study population, λ_2 is the true odds ratio for Zierler's study population, and so on. Thus the statistical model for the top stage of the hierarchical model is

$$\lambda_1, \lambda_2, ..., \lambda_k \text{ is a sample of size } k \text{ from}$$
$$\text{a normal box with parameters } \mu_\lambda \text{ and } \sigma_\lambda \tag{11.2}$$

The meta-investigator typically has no prior information on either μ_λ or σ_λ and represents that ignorance by putting nearly flat prior distributions on μ_λ and $\ln(\sigma_\lambda)$. We will shortly learn how to do this using the WinBUGS package.

11.5.2   Likelihood of the Data

The other input to a Bayes' rule calculation is the likelihood of the data given the values of the unobservable quantities $\lambda_1, \lambda_2, ..., \lambda_k$. The amount of data available to the meta-investigator depends on how much information was provided by the seven primary study investigator (Cantor, Zierler, etc.). Sometimes it is possible to obtain the original raw study data; sometimes all that is available is an adjusted odds ratio and credible interval as is the case in Table 11.1.

In this minimal-data case, the first step is to convert the estimated odds ratios and confidence intervals into estimated log odds ratios, $\hat{\lambda}_1, \hat{\lambda}_2, ..., \hat{\lambda}_k$, and their standard errors, se λ_1, se $\lambda_2, ...,$ se λ_k, as in the third and fourth columns of Table 11.3. The likelihood of these data, in theory, is the probability of observing the particular values in Table 11.3 expressed as a function of the unknown true values. So, for example, if the true value of λ_1 is 0.400, then the observed value, $\hat{\lambda}_1 = 0.175$, would be extremely improbable (because the values are separated by more than 4 se). Consequently, the posterior probability that $\lambda_1 = 0.400$ would

Each $\hat{\lambda}_i$ is approximately normally distributed with $\mu = \lambda_i$ and $\sigma = \mathrm{se}\lambda_i$ (11.3)

be small because posterior probability is proportional to the product of the prior and the likelihood.

Computing the exact likelihood would require having access to the raw data underlying each of the seven studies. Here we are using an approximation to the likelihood that exploits the fact that for large samples and efficient, approximately normally distributed estimates[6], the standard errors can be treated as constants, in which case the likelihood of the data $\hat{\lambda}_1, \hat{\lambda}_2, ..., \hat{\lambda}_k$ is specified as follows:

Smith et al. (1995) make the case that the approximation in Equation (11.3) breaks down if the meta-analysis includes too many primary studies with extremely small sample sizes. As a rule of thumb, the approximation appears to be acceptable if each of the component studies has at least 10 observations in each of the four cells in Table 11.4 (the rule of 10), and produces a posterior distribution virtually indistinguishable from that of the full-data analysis if there are are at least 30 observations in each cell (the rule of 30). It is probably the case that the the approximation is acceptable even if a minority of the studies have one cell with a count as small as 3; however, a reliable rule of thumb for that situation awaits further investigation.

When the cell counts in Table 11.4 are not reported in the primary studies it is still possible to deduce the smallest possible cell count using the inequality

$$\text{minimum cell count} > 1/\mathrm{se}^2 \qquad (11.4)$$

Thus a conservative form of the rule of 10 is $\mathrm{se}^2 < 1/10$, or $\mathrm{se} < 0.32$, and a conservative form of the rule of 30 is $\mathrm{se} < 0.18$. Six of the seven component studies in the chlorinated water meta-analysis satisfy the rule of 10, and four of

Table 11.4   Rule of thumb for the normal likelihood approximation.

		Group	
		Exposed	Unexposed
Outcome	Favorable	$\geq 10^a$	≥ 10
	Unfavorable	≥ 10	≥ 10

a. 10 or more observations in all categories in all studies gives an adequate approximation, 30 or more gives an approximation indistinguishable from the full-data likelihood.

6. *Efficient* estimates are estimates that for large samples are as precise as is possible, given the data. Means of normal samples, logits, log odds ratios, regression coefficients, logistic regression coefficients, adjusted means and differences, and adjusted log odds ratios produced by logistic regression are all efficient estimates.

them satisfy the rule of 30; consequently, the normal approximation should be reasonably accurate.

To recapitulate, we have now have established that Equation (11.3) provides a good approximation to the likelihood of the data. What remains is to use Bayes' rule to compute the posterior distribution of the mean of the box of log odds ratios, and, by reverse transformation, the distribution of the median odds ratio.

11.6 SIMULATION OF THE POSTERIOR DISTRIBUTION

The primary goal of the meta-analysis is to compute the posterior distribution of μ_λ, the population mean log odds ratio; however, it is not possible to write down an algebraic expression even for an approximation of the posterior distribution in a meta-analysis. Since μ_λ is the average of a box of numbers it follows that *if* the meta-investigator could observe the numerical values of the true odds ratios in the primary studies (λ_1, λ_2, ..., λ_7) *then* the posterior distribution of their box average, μ_λ would be approximately t(6) with

$$\mu = \bar{\lambda} = \frac{\lambda_1 + \lambda_2 + \cdots + \lambda_k}{k},$$

$$\sigma = \text{sem} = \frac{s}{\sqrt{k}} = \frac{\sqrt{\frac{\sum_{i=1}^{k}\left(\lambda_i - \bar{\lambda}\right)^2}{k-1}}}{\sqrt{k}} \tag{11.5}$$

[If these formulas seem complicated, note that they can be derived from Fig. 8.2 and Equation (8.1) by pretending that the unobserved true odds ratios, $\lambda_1, \lambda_2, ...,$ λ_k, are observed data, i.e., $x_1 = \lambda_1$, $x_2 = \lambda_1$, ..., $x_n = \lambda_k$, and n=k.)

However, since the true λ's are not directly observable, the meta-investigator cannot compute their average or standard deviation. Instead it is necessary to use a simulation technique called Markov chain Monte Carlo (MCMC) to compute the posterior distribution. MCMC methods work by generating simulated observations of the unknown quantities λ_1, λ_2, ..., λ_k, μ_λ, and σ_λ. To see why that might be useful, suppose that you want to compute $P(\mu_\lambda < 0.30 \mid \text{data})$. Since MCMC is able to supply you with any number of simulated values of the unknown quantity μ_λ, you could ask it to generate perhaps 50,000 values; lets say the values are -0.13, 0.27, 0.49, 0.33, etc. For some of these values the sentence "$\mu_\lambda < 0.30$" is true, for some it is false. Let's say the sentence is true 46,623 times out of 50,000; that is, the proportion of simulations in which it is true is $46,623/50,000 = 0.9325$, or 93.25%. That proportion is the Monte Carlo approximation of $P(\mu_\lambda < 0.30 \mid \text{data})$. Since the approximation is based on simulations, there will be a small margin of error, called the *MC error*. It is analogous to roundoff error and (unlike the standard error) does not represent lack of information; indeed, the MC error can be be made as small as desired by increasing the number of simulations.

Fig. 11.4 on page 235 shows how to set up the WinBUGS computer package to generate simulated values of the unknown quantities λ_1, λ_2, ..., λ_k, μ_λ, and σ_λ and to compute approximate posterior probabilities. This program was used to generate 500,000 simulated values of μ_λ, which are depicted as a histogram in Fig. 11.3 and as a left-tail area table in Table 11.5 on page 233. The histogram is plotted at intervals of 0.01 on the μ_λ scale. The horizontal coordinate of each circle is the center of an interval of width 0.01 on the μ_λ axis, and the vertical coordinate of each circle is the relative frequency (percent) of simulated μ_λ values in that interval. For example, the vertical coordinate at 0.0215 on the horizontal axis is 5.58%, meaning that about 5.58% of the posterior probability is in the interval 0.21 to 0.22. Probability estimates near 0.1 (i.e. 10%) are accurate to a Monte Carlo error of about ±0.0008; probability estimates near 0.01 (i.e. 1%) are accurate to about ±0.0003. To halve the error of a Monte Carlo approximation it is necessary to quadruple the number of simulated μ_λ values.

Table 11.5 shows the left tail areas for the posterior density curve in Fig. 11.3. It was computed by calculating the proportion of simulated μ_λ values to the left of a given point on the horizontal axis. For example, exactly 2976 of the 500,000 simulated μ_λ values were negative; consequently, the Monte Carlo estimate of the tail area to the left of 0.00 on the μ_λ axis is 2976/500,000 = 0.005952 or about 0.6%; so there is about 99.4% posterior probability that chlorinated drinking water is associated with an increased risk of bladder cancer. This table enables us to compute posterior probabilities of other sentences about the consensus odds ratio. For example, suppose that someone thinks that 30% increased odds would reperesent a serious public health problem; a relevant

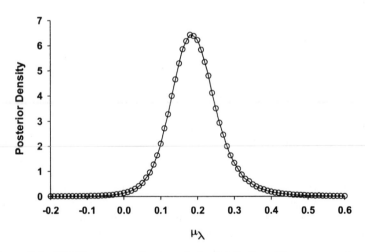

Figure 11.3   MCMC approximation to posterior distribution of the consensus log odds

Table 11.5   Left Tail Area of the Posterior Distribution of the Consensus log Odds Ratio (μ_λ) for the Morris et al. Bladder Cancer Meta-Analysis[a].

μ_λ	-0.09	-0.08	-0.07	-0.06	-0.05	-0.04	-0.03	-0.02	-0.01	-0.00
-0.30	0	0	0	0	0	0	.0001	.0001	.0001	.0001
-0.20	.0001	.0001	.0001	.0001	.0001	.0001	.0002	.0002	.0002	.0002
-0.10	.0003	.0003	.0003	.0004	.0004	.0005	.0006	.0006	.0008	.0009
-0.00	.0011	.0013	.0015	.0018	.0021	.0026	.0032	.0038	.0048	.0059

μ_λ	0.00	0.01	0.02	0.03	0.04	0.05	0.06	0.07	0.08	0.09
0.00	.0059	.0075	.0095	.0122	.0157	.0204	.0266	.0349	.0460	.0606
0.10	.0791	.1028	.1330	.1697	.2130	.2630	.3194	.3801	.4435	.5080
0.20	.5716	.6319	.6877	.7372	.7813	.8192	.8513	.8781	.9002	.9180
0.30	.9326	.9443	.9539	.9539	.9619	.9737	.9780	.9814	.9843	.9868
0.40	.9887	.9905	.9919	.9931	.9941	.9950	.9956	.9962	.9967	.9972
0.50	.9976	.9979	.9981	.9984	.9986	.9988	.9989	.9990	.9991	.9992
0.60	.9993	.9994	.9995	.9996	.9996	.9996	.9996	.9997	.9997	.9998
0.70	.9998	.9998	.9998	.9998	.9999	.9999	.9999	.9999	.9999	.9999
0.80	.9999	.9999	.9999	.9999	.9999	.9999	.9999	.9999	.9999	.9999
0.90	.9999	.9999	1	1	1	1	1	1	1	1

a. This table was computed from the seven bladder cancer studies listed in Table 11.1 it is not valid for any other meta-analysis.

calculation for that person is the probability that the odds ratio is less than 1.3: $P(OR < 1.3 \mid \text{data}) = P(\ln(OR) < \ln(1.3)) = P(\mu_\lambda < 0.262) = 0.856$, or 85.6%, by interpolation. This sort of table is not routinely produced by MCMC software and in practice other techniques are used to compute left tail areas (see the WinBUGS tip in Exercise 11.4).

The table can also be used to compute percentiles and credible intervals. For example, the the interval between the 2.5th and 97.5th percentiles is a 95% credible interval. The 2.5th percentile is the point on the μ_λ axis where the left tail area is 0.025, which is somewhere between $\mu_\lambda = 0.05$ (tail area = 0.204) and $\mu_\lambda = 0.06$ (tail area = 0.266). It is necessary to interpolate to find the percentile $P_{2.5} = .057$. The 97.5th percentile is $P_{97.5} = 0.353$, so the 95% credible interval for the consensus log odds ratio μ_λ is 0.057 to 0.353, and the 95% credible interval for the consensus odds ratio is $\exp(0.057)$ to $\exp(0.353)$; that is, 1.06 to 1.42. In practice, MCMC software such as WinBUGS computes percentiles and credible intervals.

This meta-analysis has been presented in great detail in order to introduce the principles of hierarchical models and meta-analysis. However, in practice the analysis can be done simply and quickly with the WinBUGS statistical package.

Fig. 11.4 shows how to set up WinBUGS to compute the 95% credible interval, and three posterior tail areas; output from the program is in Table 11.6 on page 235. The node mu.lambda is μ_λ, the mean, or consensus, log odds ratio, and the other nodes are explained in the footnote. The 95% confidence interval for μ_λ is 1.059 to 1.42, which agrees with the hand calculation using the tail area table (Table 11.5). The antilog (exp) of the mean log odds ratio is the median odds ratio. Three tail areas for the odds ratio are reported in Table 11.6; for example the mean of node ORlt1.3 is the Monte Carlo estimate of the posterior probability $P(\text{OR} < 1.3 \mid \text{data})$. According to Table 11.5, the computed value is 0.861 accurate to ± 3 in the third decimal place ("MC error" column), which is close to the less accurate interpolated value, 0.856, that we computed from Table 11.5. The column labeled "sd" is the posterior distribution of a node; it is the analog of the standard error (se) in conventional statistical output. For approximately normal posterior distributions the mean and sd are the parameters μ and σ of the posterior distribution; however, in this analysis neither the odds ratio nor the log odds ratio, μ_λ, has an approximately normal posterior distribution. Output that is either irrelevant or misleading has been printed in strikeout type and should be ignored.

Instructions for entering and running WinBUGS analyses such as this are in Appendix B. An exact meta analysis of odds ratios is possible if the numbers of adverse and favorable outcomes in the numerator and denominator of the odds ratio are known for each study. For an example of how to set up this analysis, see the "Blocker" example in WinBUGS Examples, Vol. I, under the WinBUGS Help menu.

```
MODEL {
  # HyperPrior: No Prior Knowledge of Consensus ln(OR)
    mu.lambda ~ dnorm(0,.001)
    tau.lambda ~ dgamma(.001,.001)
    sigma.lambda <- 1/sqrt(tau.lambda)
  # Prior: Distribution of lambda's for individual studies
    for (i in 1:k) {lambda[i] ~ dnorm(mu.lambda,tau.lambda)}
  # Approximate Normal Likelihood of Observed Data
    for (i in 1:k) {
      tau.hat[i] <- 1/(se[i]*se[i])
      lambda.hat[i] ~ dnorm(lambda[i],tau.hat[i])
    }
  # Posterior Tail Probs for the Consensus Odds Ratio
    OR <- exp(mu.lambda)
  # P(OR > 1.0 | Data):
    ORgt1.0 <- 1-step(1.0-OR)
  # P(OR > 1.1 | Data):
    ORgt1.1 <- 1-step(1.1-OR)
  # P(OR < 1.3 | Data):
    ORlt1.3 <- step(1.3-OR)
  # ... requests for more tail areas if desired ...
}
DATA list(k=7,
  lambda.hat=c(0.174,0.336,0.788,0.166,-0.020,0.140,0.525),
          se=c(0.054,0.143,0.577,0.108,0.125,0.253,0.233))
```

Figure 11.4 WinBUGS program for the bladder cancer meta-analysis.

Table 11.6 **WinBUGS output for the bladder cancer meta-analysis.**

Node[a]	Mean[b]	sd	MC error[c]	2.5%[d]	Median[e]	97.5%[d]	Start	Sample[f]
OR	1.215	0.09234	9.754E-4	1.059	1.207	1.42	1001	50000
mu.lambda	0.192	0.07397	7.809E-4	0.05718	0.1884	0.3508	1001	50000
ORgt1.0	0.994[g]	0.07697	4.325E-4	1.0	1.0	1.0	1001	50000
ORgt1.1	0.9305[g]	0.2544	0.001983	0.0	1.0	1.0	1001	50000
ORlt1.3	0.8611[g]	0.3459	0.003381	0.0	1.0	1.0	1001	50000

a. As used here, *node* designates a simulated unknown quantity.
b. The average of the simulated values of this node, thus the mean of a true–false node is a probability.
c. Accuracy of the simulation (here about three digits); to halve the error quadruple the simulations.
d. Lower and upper endpoints of the 95% credible interval, meaningless for posterior probabilities.
e. 50th percentile; thus $P(OR < 1.207 \mid data) = 0.50$.
f. Number of simulated observations from the posterior distribution.
g. Estimated posterior probabilities, thus $P(OR > 1 \mid Data) \cong 0.994$, or 99.4%, and $P(OR \leq 1.3 \mid data) \cong 0.861$, or 86.1%.

11.7 BORROWING STRENGTH

The data in Table 11.7 are the numbers of pediatric cardiac surgeries and deaths in 12 hospitals. One of these, hospital A experienced no deaths in 47 surgeries; however, assuming that the hospitals are comparable (exchangeable), there is a strong presumption that hospital A simply was lucky with the first 47 surgeries and that eventually its death rate will come to resemble that of its peer institutions — something on the order of 4 to 14 deaths per hundred surgeries. On the other hand, we cannot ignore the fact that there were no deaths in that hospital. What is needed, therefore, is some way to incorporate both sources of information — the observed data in hospital A combined with the experience at similar institutions. The process is informally called *borrowing strength* since the comparatively small sample from hospital A is strengthened by observations from similar hospitals. The data can be thought of as a sample from a box of exchangeable hospitals. Exchangeability is the key to the analysis presented here; consequently it would not be appropriate if the box contained a heterogeneous mix of suburban and small city hospitals, inner city hospitals, and university hospitals.

Table 11.7   Pediatric cardiac surgery deaths.[a]

Hospital	No. of operations n_i	No. of deaths x_i	Crude death rate
A	47	0	0.000
B	148	18	0.122
C	119	8	0.067
D	810	46	0.057
E	211	8	0.038
F	196	13	0.066
G	148	9	0.061
H	215	31	0.144
I	207	14	0.068
J	97	8	0.082
K	256	29	0.113
L	360	24	0.067

a. Source: WinBUGS help menu, *Examples Vol I*, "Surgical: Instutional

Fig. 11.5 displays the hierarchical structure of the mortality rate data. The 12 hospitals in the study are, in effect, a sample from a population, or box, of exchangeable hospitals. Each hospital in the box has a true death rate, p, and the true death rates for the 12 hospitals in the study are sampled from a box of rates. However, as in the meta-analysis example, instead of being able to observe the true rates in the sample of hospitals, the investigator observes the number of deaths (x) in a finite number of surgeries (n) in each hospital.

Fig. 11.6 shows how to set up the calculation in WinBUGS. We have used a beta distribution (see Section 6.6) to represent the prior distribution of pediatric surgical mortality rates in the box of hospitals. That means that the distribution the true rates is completely specified by the two positive-valued shape-determining hyperparameters α and β of the beta distribution (see Section 6.6). If the hospital rates are very similar to each other then α and β will be quite large, and if the rates are quite dissimilar then the parameters α and β will be small. Since we do not know what the true rates are, we certainly do not know whether they are similar to each orther or vary substantially from hospital to hospital; therefore, the hyperparemeters α and β are unknown quantities about which we know essentially nothing. In the WinBUGS program, α and β are designated a and b. We express virtually complete ignorance about about a and b by means of *gamma* distributions (see the next section).

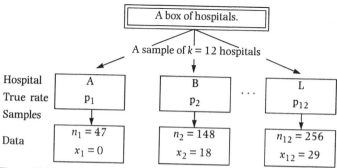

Figure 11.5   Hierarchical structure of the hospital data.

```
MODEL PedMort {
    #Hyperprior for the Box of Rates
        a~dgamma(.001,.001)
        b~dgamma(.001,.001)
    #Prior Distribution of the True Rates
    for (i in 1:k) {
        #Prior distribution of  Hospital i's True Rate
        p[i] ~ dbeta(a,b)
        #Likelihood of Hospital i's Data
        x[i] ~ dbin(p[i],n[i])
    }
}
DATA list(k=12,
    n = c(47,148,119,810,211,196,148,215,207,97,256,360),
    x = c( 0, 18,  8, 46,  8, 13,  9, 31, 14, 8, 29, 24))

INITIAL VALUES list(a=1,b=1)
```

Figure 11.6   WinBUGS program for the hospital study.

Since five of the hospitals experienced fewer than 10 deaths, a normal likelihood approximation might not be accurate; for that reason, it seemed advisable to use the exact *binomial* likelihood [see Equation (5.4)].

The WinBUGS program in Fig. 11.6, like any Bayes' rule calculation, involves specifying the prior distribution and the likelihood. The prior specification consists of two parts. First, we specify that we know next to nothing about the distribution of mortaility rates among hospitals,

$$a \sim dgamma(.001,.001) \quad \text{and} \quad b \sim dgamma(.001,.001)$$

which means that $\ln(a)$ and $\ln(b)$ have nearly flat prior distributions over the entire number line. Second, we specify the prior distribution of true death rates of the hospitals are a sample from a box with a beta(a, b) distribution,

$$p[i] \sim dbeta(a,b)$$

The likelihood specification states that the deaths, x[i], among the n[i] surgeries in hospital *i* are, in effect, a sample from a box of surgical patients with mortality rate p[i],

$$x[i] \sim dbin(p[i],n[i])$$

The distinction between this hierarchical model analysis, which is sometimes called *Bayes empirical Bayes*, and the meta analysis of the previous sections is that we are not interested in the box average μ_χ, which was the objective of the meta analysis. Instead we are interested in more accurate estimates of the death rates for individual hospitals. Each hospital's rate is estimated from the data for that hospital *strengthened* by the data from the other hospitals.

Edited WinBUGS output is presented in Table 11.8. Either the posterior mean or the posterior median could be used in place of a hospital's crude death rate as a strengthened estimate of the true mortality rate in that hospital. These estimates combine the direct information from each hospital with *collateral* information provided by its peer institutions. The collateral information has very little effect on the estimate for hospital D, which performed over 800 surgeries (crude rate = 0.057, strengthened estimate = 0.059). However, the crude estimate for hospital A, which performed only 47 surgeries, is comparatively imprecise and is therefore influenced by the experience in the other hospitals (crude rate = 0.000, strengthened estimate = 0.048).

Why is the strengthened estimate preferable? Doesn't it fudge the data? Why should we not take the crude rate at face value? The answer is that there are two sources of data, first the experience in hospital A, and second the knowledge (in the sense of *justified* belief) that hospital A is exchangeable with the other hospitals. Thus the strengthened estimate is more honest; it is consistent with our belief that hospital A is exchangeable with the others in the sense that it possesses no characteristics that would meaningfully distinguish it from them. If that is what we believe, then the strengthened estimate, 0.048, is indeed more honest in the sense of more accurately reflecting what we believe to be the truth. Both Bayesian and conventional statistical practice agree on that point (see Efron and Morris 1977). As a corollary, if prior to seeing the mortality data we had other information that distinguished hospital A from the others in some way relevant to pediatric mortality, then the results of the hierarchical model analysis would not honestly reflect our knowledge and would for that reason be incorrect.

Table 11.8   WinBUGS output for the pediatric mortality rate data.

Hospital	μ mean	σ sd	median	95% CI 2.5%	97.5
A	0.0497	0.0218	0.0490	0.0109	0.0944
B	0.1037	0.0216	0.1016	0.0675	0.1515
C	0.0714	0.0180	0.0702	0.0397	0.1107
D	0.0592	0.0080	0.0589	0.0444	0.0758
E	0.0505	0.0137	0.0499	0.0260	0.0791
F	0.0698	0.0151	0.0690	0.0428	0.1022
G	0.0671	0.0164	0.0662	0.0379	0.1023
H	0.1226	0.0215	0.1213	0.0847	0.1684
I	0.0705	0.0150	0.0696	0.0435	0.1022
J	0.0797	0.0202	0.0780	0.0448	0.1245
K	0.1029	0.0172	0.1017	0.0726	0.1398
L	0.0689	0.0119	0.0684	0.0471	0.0937

11.8   THE GAMMA FAMILY OF DISTRIBUTIONS

The gamma family of densities includes a variety of shapes from nearly normal to highly skewed. Unlike the beta distribution, introduced in Section 6.6, which covers the interval from 0 to 1, or the normal, which covers the entire number line, the gamma distribution covers the positive half of the number line from 0 to infinity. Therefore the gamma density is a useful way to describe knowledge about a positive-valued unknown quantity.

The gamma family has two parameters, which we will designate a and s. Parameter a determines the *shape* of the density curve and parameter s determines the *scale*, or spread, of the density curve. The gamma density with parameters a and s is denoted dgamma(a, s) in WinBUGS. Fig. 11.7 illustrates how the shape and scale parameters influence the appearance of the curve. The bottom two panels show that changing the scale does not affect the shape. The scale parameter determines how concentrated the distribution is. Increasing the scale parameter makes the distribution more concentrated; for example, the distribution with the smaller scale parameter (s = 0.6) spreads its probability between 10 and 50 on the number line while the distribution with the larger scale parameter (s = 6) concentrates is probability between 1 and 5).

As a rule of thumb, the gamma density is approximately normal when the shape parameter exceeds 25; however, it is always slightly positively skewed. For shape parameter values in the range $0 < a \leq 1$, the distribution is highly skewed with the mode (peak) at zero. For a > 1, the mode of the gamma density is

$$\text{mode} = \tilde{\mu} = \frac{a-1}{s} \tag{11.6}$$

If the shape parameter a is 2 or smaller, the curve is not bell-shaped; but for values of a bigger than 2, the curve has points of inflection at

$$\tilde{\mu} \pm \sqrt{\frac{\tilde{\mu}}{s}} = \tilde{\mu} \pm \tilde{\sigma} \tag{11.7}$$

The gamma(0.001,0.001) distribution in the upper right panel of Fig. 11.7 is often used to represent a state of almost complete ignorance about a positive-valued unknown quantity. This might seem a bit odd, since that distribution curve seems to be concentrated near zero. The explanation is this: if unknown quantity V has a gamma(0.001,0.001) distribution, then its logarithm, $\ln(V)$, is approximately uniformly distributed over the entire number line. Consequently gamma(0.001,0.001) expresses almost complete ignorance about the logarithm of the unknown quantity.

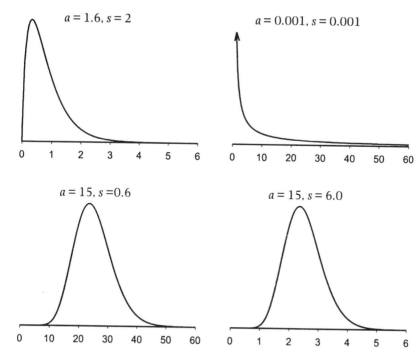

Figure 11.7   Some gamma distributions.

11.9 EXERCISES

11.1 Use Table 11.5 on page 233 to compute $P(OR < 1.2)$ and $P(OR < 1.6)$. Interpolate to to obtain the first and third quartiles and median of μ_λ. Convert these into quartiles and median of the posterior distribution of OR.

11.2 Table 11.9 lists the 11 U.S. studies of environmental tobacco smoke (ETS) as a risk factor for lung cancer that were used in the EPA's report, *Respiratory Health Effects of Passive Smoking*. For each study, convert the odds ratios and confidence intervals to log odds ratios and standard errors using Equation (11.1) on page 227. Use Equation (11.4) on page 230 to determine which studies obey the rule of thumb in Table 11.4 on page 230.

Table 11.9   U.S. Studies of ETS.

Study ID	Odds Ratio	LCL	UCL
BROW	1.50	0.48	4.72
BUFF	0.68	0.32	1.41
BUTL (coh)	2.01	0.61	6.73
CORR	1.89	0.85	4.14
FONT	1.28	1.03	1.60
GARF	1.27	0.91	1.79
GARF (coh)	1.16	0.89	1.52
HUMB	2.00	0.83	4.97
JANE	0.79	0.52	1.17
KABA	0.73	0.27	1.89
WU	1.32	0.59	2.93

11.3 Enter $k = 11$ and the log odds ratios and standard errors that you computed in exercise 11.2 into the DATA list in the WinBUGS program in Fig. 11.4 on page 235 and compute an approximate 95% credible interval for the consensus log odds ratio and the odds ratio. Modify the tail area requests to compute $P(OR > 1 \mid data)$ and $P(OR > 1.05 \mid data)$.

The EPA report, which did not use a Bayesian analysis, stated that the consensus odds ratio was 1.19 with 90% confidence interval 1.04 to 1.35 (to compute the 90% credible interval in WinBUGS, request the 5th and 95th percentiles in Fig. B.9 on page 312). Do you agree with the EPA interval? If not, is the EPA interval too long or too short? Does this strengthen or weaken the evidence in support of regulating ETS?

11.4 Run the WinBUGS anlysis of pediatric death rates using the program in Fig. 11.4. Use the step( ) function to compute the posterior probability that the death rate for hospital A exceeds 0.075 (i.e., $P(p[1] > 0.075)$).

WinBUGS Tip: Use the *step* function to compute left or right tail areas. Suppose that U is the unknown quantity (node) of interest and you want to compute $P(U \leq 1.3)$. In the WinBUGS program define a node that indicates when the sentence "$U \leq 1.3$" is true: Ule1.3 <– step(1.3 – U). The mean (average) of this node in the output window is the required tail probability.

Here's how it works: If V is any node then step(V) equals 1 if $V \geq 0$ and equals 0 if $V < 0$. Consequently step(a – U) equals 1 if $a - U \geq 0$; i.e, if $a \geq U$.

How to:

Compute $P(U \leq a)$: new_node_name <– step(a – U)
Compute $P(U < a)$: new_node_name <– 1–step(U – a)
Compute $p(U \geq b)$: new_node_name <– step(U – b)
Compute $P(U > b)$: new_node_name <– 1–step(b – U)

11.10   REFERENCES

Efron, B., and Morris, C. "Stein's paradox in statistics." *Scientific American*, Vol. 236, No. 5 (1977), pp. 119–127.

Morris, R.D., Audet, A.M., Angelillo , I.F., Chalmers , T.C., and Mosteller, F, "Chlorination, Chlorination By-products, and Cancer: a Meta-analysis," *American Journal of Public Health, Vol 82, No. 7 (Jul. 1992), pp. 955-963*. Erratum, Vol. 83, No. 9 (Sep. 1993), pp. 1257.

D J Spiegelhalter, D.J., Thomas, A., and Best, N.G., "Surgical: Institutional Ranking," in *WinBUGS 1.4/Help Menu/Examples*, Vol. 1, MRC Biostatistics Unit, 2002.

Smith,T.C., Spiegelhalter, D.J., and Thomas, A., "Bayesian approaches to random-effects meta-analysis: A comparative study," *Statistics in Medicine*, Vol. 2 (1995), pp. 2685-2699.

United States Environmental Protection Agency. *Respiratory Health Effects of Passive Smoking: Lung Cancer and Other Disorders*, EPA/600/6-90/006F, Dec. 1992.

12

Time to Event Analysis

12.1 INTRODUCTION

In its simplest form, time-to-event analysis, or *survival analysis*, is the study of length of life: the period between birth and death. More generally it is the study of the length of time between two events sometimes called *intake* and *endpoint*. The length of the time interval between these two events is called the *survival time*. For example, Table 12.1 is a *life table* that summarizes the survival times of 45,000 people diagnosed with colon cancer; this example was suggested by Greene et al. (2002) but the data presented here are fictional. This group of 45,000 patients is a *cohort*, that is, a group of people (or things) with meaningful similarities (in this case the same diagnosis) who are followed for a period of time. The *follow-up time* can be a few days (as in the case of recovery of consciousness after a closed head injury) or can be several decades.

In this example, each patient's intake time was the date of his/her diagnosis and his/her endpoint time was the date of death; consequently, survival time is the length of time between diagnosis and death. For large data sets such as this one, survival times are often *grouped* into convenient intervals (weeks, months, quarters, etc.). This is the case in Table 12.1, where the survival times have been grouped into half-year intervals. Thus, interval 1 is the first six months after diagnosis, interval 2 is the next six months, and so on.

Survival analysis was developed by actuaries and by statisticians, which means, unfortunately, that there are two systems of terms and symbols for the same concepts. Statistical symbols and terminology will be mostly used in this chapter.

12.2 LIFE TABLES

Table 12.1 is arranged in the form of a *life table* or *mortality table*. The basic data are the initial size of the cohort (45,000) and the numbers of deaths in each period; all other numbers in the table were derived from these. Thus the number of survivors at the beginning of a period is the number of patients alive at the beginning of the previous period minus the number of deaths during that period. The death rate in each period is the number of deaths in that period divided by the number of people alive at the beginning of the period. For example, there were

245

Table 12.1   Life table of 45,000 colon cancer patients[a]

Period	Months	No. alive at start of period	No. of deaths during period	Death rate during period	Hazard rate (failures/month)
i	x to $x+6$	l_x	d_x	$q_x = d_x/l_x$	$h_x \cong -\ln(1-q_x)/6$
1	00–06	45000	3600	0.080	0.0139
2	06–12	41400	3519	0.085	0.01489
3	12–18	37881	3410	0.090	0.01579
4	18–24	34471	3102	0.090	0.01579
5	24–30	31369	2855	0.091	0.01599
6	30–36	28514	2453	0.086	0.0150
7	36–42	26061	2111	0.081	0.0141
8	42–48	23950	1725	0.072	0.0125
9	48–54	22225	1401	0.063	0.0108
10	54–60	20824	1125	0.054	0.0093

a. Source: hypothetical data.

37,881 patients still alive at the end of period 2, but 3401 of them died during period 3, a death rate of $3401/37,881 = 0.090$, or 9%.

Statisticians use the symbol t (for time) rather than x to stand for survival times. The number alive at time t, expressed as a fraction of the original cohort, is called the *survival function* and is expressed as a fraction of the size of the original cohort, For example, the proportion of patients still alive at time 0 is $S(0) = 1$ and

$$S(t) = \frac{l_t}{l_0} \tag{12.1}$$

at the end of twelve months is $S(12) = 37,881/45,000 = 0.842$.

Survival operates like compound interest in reverse — there was an "opening balance" of 45,000 patients, which was reduced by a certain percentage in each period. Eight percent died in the first six-month period and 92%, or 41,400, survived; 8.5% of these survivors died in the second period and 91.5%, or 37,811, survived; and so on. That is, 92% of the original cohort survived to the end of the first period; 91.5% of that 92%, or 81.2% or the original cohort, survived to end of the second period; 91% of 91.5%, of 92%, or 76.6% of the cohort, was alive at the end of the third period; and so on.

Recursive computing equations for the survival function are:

$$
\begin{aligned}
S_0 &= 1 \\
S_1 &= S_0 \cdot (1-q_1) &= (1-q_1) \\
S_2 &= S_1 \cdot (1-q_2) &= (1-q_1)\cdot(1-q_2) \\
&\;\;\vdots &&\;\;\vdots \\
S_k &= S_{k-1} \cdot (1-q_k) &= (1-q_1)\cdot(1-q_2)\cdots(1-q_k)
\end{aligned}
\tag{12.2}
$$

Here S_k is the surviving portion of the original cohort at the end period k, and q_i is the death rate in the ith period.

The net interest paid on a loan or paid by a savings account depends on how often the interest is *compounded*. Suppose you put $100 in a savings account that pays 2% per calendar quarter. That's $2 every three months, which you'd think would amount to $8 per year; however, your closing balance at the end of one year would actually be $100×(1.02)×(1.02)×(1.02)·(1.02) = $108.24, so the annual interest would be a bit higher than 8%, actually 8.24%.

Death rates, like compound interest rates, depend on how often the rate is compounded. Actuaries use double subscripting to represent death rates for intervals of different lengths. For example, the symbol $_6q_{12}$ is the death rate for the 6-month period ending at month 12. Even if the death rate were constant over a 12 month period, the 1-month death rate would be more than one-sixth of the 6-month death rate, and the 12-month death rate would be less than twice the 6-month rate. What ties all these rates together is the *hazard function* (also called the *failure rate* or *force of mortality*). The hazard function is a sort of instantaneous death rate — intuitively, the death rate in an infinitely short period of time. Strictly speaking, the hazard function is the *density* or *force* of death, and just as a physical force increases velocity (rate of travel), the hazard function (force of mortality) increases the rate of death. The technical definition of the hazard function is the derivative of the negative logarithm of the survival function, $h(t) = -d\ln(S(t))/dt$; however, for short periods (or longer ones if the hazard rate is constant) the relationship between the hazard rate $h(t)$ and the death rate $_Tq_t$ over a time period of length T units is approximated by,

$$
_Tq_t \cong 1 - \exp\left(-T \cdot h(t)\right). \tag{12.3}
$$

For example, if the hazard rate is a constant $h(t) = 0.174$ deaths per person per year, then the death rate $_6q_t$ over a 6 month period ending at time t is $1 - \exp(-0.5×0.174) = 0.084$, or 8.4%. Conversely, if the death rate for a time period of length T, is $_Tq_t$, then the hazard rate is

$$
h(t) \cong \frac{-\ln(1 - {_Tq_t})}{T} \tag{12.4}
$$

For example, 6 months is half a year (0.5 years), and the 6 month death rate in period 10 is $_6q_{60} = 0.054$ (see Table 12.1); consequently the hazard rate in period 10 is approximately $h(60) = -\ln(0.946)/6 = 0.0093$, or 0.93% per month. Note that all of these calculations presume that the hazard rate stays nearly constant within each period.

12.3   SURVIVAL ANALYSIS

It is unusual to have complete survival-time data on any sample of subjects, because that would require the investigator to wait until there were no survivors left. Most often there are only a few hundred subjects and the survival data are almost invariably incomplete. For example, Table 12.2 lists the disease-free survival times of a hypothetical study of 85 patients after a particular form of cancer surgery. The example was inspired by Beenken et al. (2003) but the data are fictional. The intake point is a patient's date of surgery and the endpoint is the date of recurrence of the cancer. The study design called for following the patients for 21 years (252 months); however, some patients were still disease-free at the end of the planned observation period and some patients were *lost to follow-up*, perhaps because they moved without leaving a forwarding address or died from some other cause prior to any recurrence of the cancer. Patients lost to follow-up

Table 12.2   Disease-free survival times.[a]

Lymph Node Involvement					
4+ nodes		1–3 nodes		0 nodes	
1	152	1	128	17	>185
3	>152	4	>147	18	>186
3	157	6	>149	28	>194
5	>229	6	>154	42	>195
9	>252	9	>216	51	>197
10	>252	12	>217	>100	>205
11		13	>223	101	>212
11		16	>229	106	>219
23		50	>231	>116	221
23		53	>239	117	>221
36		58	>242	121	>252
38		63		>125	>252
49		65		>126	>252
65		80		>140	>252
121		88		153	>252
128		97		>160	>252
>139		>123		>164	>252

a. Source: hypothetical data.

might have experienced a recurrence of the cancer at some time in the future, but the investigator would not know if or when this happened. Such observations are said to be *right-censored*, and in Table 12.2 are marked with a greater-than symbol (>) indicating that the disease-free survival time is known only to exceed the follow-up time for that patient.

As this example suggests, studies of medical interventions often define "survival time" not as time to death but as the time to failure of the therapy or as the time to the emergence of some *surrogate* endpoint indicating that the therapy is likely to fail. For example, Ziegler et al. (2003) used the appearance of islet autoantibodies as a surrogate for diabetes. In studies of anti-HIV medications, reappearance of detectable HIV virus particles is sometimes used as a surrogate for full-blown AIDS. Survival analysis is not confined to living creatures and has been widely applied to things as diverse as light bulbs (time to burnout) and service providers (time between errors), to mention but two examples.

12.3.1   The Kaplan–Meier Survival Function Estimate

Probably the most widely used descriptive display of survival data is the Kaplan–Meier survival function. It is a graph (or table) of the estimated survival rate at any given time after the beginning of the observation period. The Kaplan–Meier algorithm, in effect, applies the failure rate among noncensored patients to the censored patients. Table 12.3 shows the Kaplan–Meier calculation for the 23 patients in Table 12.2 with the worst prognosis prior to surgery (4+ nodes involved). The calculation uses a modification of the "compound interest" method of Equation (12.2) on page 247.

The intervals are bounded by times at which one or more patients failed or were censored. The number at risk is the number of survivors at the beginning of the period. The number of survivors (number at risk) at the beginning of a period is reduced by the number who failed or were censored at the end of that period. for example, 6 patients were alive at 139 months; one of these died and one was censored at 152 months, leaving 4 known survivors. The failure rate in each interval is the number of failures divided by the number at risk (censored patients are not counted as failures because, by definition, they were known to be alive at the time they were censored). The survival rate at the end of a period is the survival rate at the beginning times the complement of the failure rate.

Kaplan–Meier survival curves for the disease-free survival times in Table 12.2 are graphed in Fig. 12.1. The graph conveys a great deal of information. For example, by reading up from 10 years on the horizontal axis as in Fig. 12.2 we determine that the 10-year survival rates are 39% (4+ nodes), 43% (1–3 nodes), and 85% (0 nodes). Reading horizontally at 50% survival (0.50 on the vertical axis), we determine that the median survival times are 3.8 years for patients with 4+ nodes, 8 years for patients with 1–3 nodes, and more than 20 years for patients with no nodes. Clearly, the lower the survival curve, the shorter the median survival time.

Table 12.3   Kaplan-Meier survival curve for patients with 4 + nodes involved.

interval	number at risk	number failed	number censored	failure rate	survival rate
x to x_{next}	l_x	d_x	c_x	$q_x = d_x/l_x$	$S_{next} = S_x \cdot (1 - q_x)$
n/a	n/a	n/a	n/a	n/a	1.0000
0-1	23	1	0	0.0435	0.9565
1-3	22	2	0	0.0909	0.8696
3-5	20	1	0	0.0500	0.8261
5-9	19	1	0	0.0526	0.7826
9-10	18	1	0	0.0556	0.7391
10-11	17	2	0	0.1176	0.6522
22-23	15	2	0	0.1333	0.5652
23-36	13	1	0	0.0769	0.5217
36-38	12	1	0	0.0833	0.4783
38-49	11	1	0	0.0909	0.4348
49-65	10	1	0	0.1000	0.3913
65-121	9	1	0	0.1111	0.3478
121-128	8	1	0	0.1250	0.3043
128-139	7	0	1	0.0000	0.3043
139-152	6	1	1	0.1667	0.2536
152-157	3	1	0	0.2500	0.1902
157-229	2	0	1	0.0000	0.1902
229-252	2	0	2	0.0000	0.1902

The LIFETEST procedure of the SAS statistical package computes and plots Kaplan–Meier survival functions and in addition reports standard errors for the survival function, and estimates and confidence intervals for mean, median and quartile survival times. Table 12.4 is a condensed version of the LIFETEST

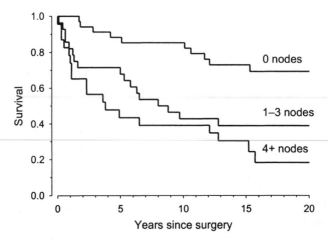

Figure 12.1   Kaplan–Meier curves. Node status influences survival.

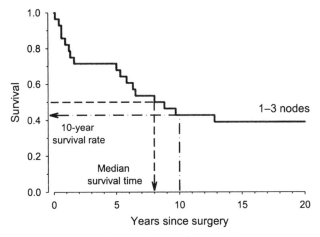

Figure 12.2 Survival rates and quantile survival times.

output. In addition to providing estimates and confidence intervals, the program reports p values for differences between pairs of survival curves. Recall that in conventional statistical practice, a p value purports to quantify the weight of evidence against a null hypothesis. For the survival curves shown in Fig. 12.1 the null hypothesis is that the three subpopulations, or strata, of patients defined by node status have the same survival function. The p value most commonly used in conventional statistical reports comes from the *log-rank test*; in this case the reported p value was 0.0006, which conventional statistical practice would take as strong evidence against the null hypothesis; however, both conventional and

Table 12.4 Kaplan–Meier estimates from the SAS LIFETEST procedure.

	0 nodes		1-3 nodes		4+ nodes	
Survival rate	est[a]	95% CCI[b]	est	95% CCI	est	95% CCI
at 5 years	0.853	(0.733, 0.973)	0.607	(0.427, 0.787)	0.435	(0.233, 0.637)
at 10 years	0.760	(0.615, 0.905)	0.429	(0.245, 0.613)	0.391	(0.191, 0.591)
at 20 years	0.615	(0.413, 0.817)	0.390	(0.208, 0.572)	0.190	(0.018, 0.362)
Survival time[c]:						
Q1	121	$(51, \infty)$	14.5	(6, 65)	10	(3, 36)
Median	–	$(221, \infty)$	84	$(53, \infty)$	38	(11, 128)

a. Maximum likelihood estimate.
b. Conventional confidence interval.
c. Quantiles (first quartile and median) of the failure time distribution, see Fig. 12.2.

Bayesian statisticians question this interpretation. In the next section we examine the question of whether the conventional confidence intervals in Table 12.4 can be interpreted as approximate credible intervals and whether the p value can be interpreted as an approximate posterior probability.

12.3.2   Interpretation of Kaplan–Meier Statistics

The Kaplan–Meier method is an example of a *nonparametric* statistical analysis. In a nonparametric analysis, the investigator has no prior knowledge of the form of the survival function. In contrast, in a *parametric* statistical analysis, the investigator knows (based perhaps on extensive prior experience) that the survival curve has a specific functional form with a few unknown *parameters* to be determined by the data. Probably the most common functional form for the survival function is the Weibull survival function

$$S(t) = \exp\left(-\gamma \cdot t^{\alpha}\right) \tag{12.5}$$

The functional form of the survival function is completely known because the graph of $\ln(-\ln(S(t)))$ vs. $\ln(t)$ is a straight line with unknown slope α and unknown intercept $\ln(\gamma)$.

$$\ln\left(-\ln\left(S\left(t\right)\right)\right) = \ln\left(\gamma\right) + \alpha \cdot \ln\left(t\right) \tag{12.6}$$

The fundamental shape is a straight line, and the parameters α and γ simply determine the slope and intercept.

We have already seen other parametric statistical analyses, involving a single rate parameter p (Chapter 7) or two parameters, μ_{box} and σ_{box} (Chapter 8), and have seen that what is involved is computing the posterior distribution of an unknown quantity, such as a difference or relative risk. We will see in the next section that parametric analyses of survival times are done in a similar way. However, nonparametric analyses are another matter. In a nonparametric survival analysis the unknown "quantity" is not a single number (such as a box mean or a success rate or a difference) but is the *entire graph* of the survival curve. Consequently, prior and posterior "distributions" refer not to possible values of a single unknown quantity, but to possible graphs of the survival function. For example, the left panel of Fig. 12.3 shows four of the uncountably infinite number of possible graphs of the true survival curve for patients with 1–3 nodes. Each possible graph has some posterior probability of being the true graph.

One can conceptualize the distribution of a finite number of graphs as a box of plastic chips similar to Fig. 8.1 on page 142, but each chip is engraved with the graph of a possible survival function rather than a single number. The probability of a particular graph is the proportion of chips in the box bearing that graph;

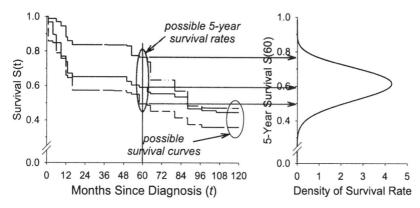

Figure 12.3   A "box" of survival curves induces the distribution of the 5-year survival

however, in contrast to Fig. 8.1, even if we knew what was engraved on each and every chip, it would still not be possible to draw a distribution of survival functions because there is no natural way to arrange the stacks of chips along the the number line. Nevertheless, it is still possible to draw the distribution of a *single* feature of the graphs, for example, the 5-year survival rate — the height of the graph at 5 years (60 months). The idea is simple: imagine that someone went through the box of graphs, worked out the height of each graph at 5 years (as in Fig. 12.3) and engraved that number on the back of the chip. Now the box of curves has become a box of numbers, and it is possible to draw the distribution of those numbers.

The fundamental inferential output of the Kaplan–Meier procedure consists of a table and graph of the estimated survival function, denoted $\hat{S}(t)$. The tabular output gives values at steps in the graph and also reports the standard error of the survival rate at those points. For example in Table 12.5 the estimated survival rate at 1 year (12 months) for patients with 1-3 nodes involved, is 0.7857, and its standard error is 0.0775. The survival estimate at a time not listed in the table is equal to the survival estimate at the next smaller time value in the table. For example, the estimated survival rate at 7 months is the 6 month estimate 0.8571. These estimates and standard errors can be used to construct approximate posterior distributions and for that reason, many aspects of a conventional Kaplan–Meier analysis are interpretable as posterior probabilities

The argument for the validity of this interpretation is based on the work of Susarla and Van Ryzin (1976), who demonstrated that the Kaplan–Meier estimate $\hat{S}(t)$ of the survival function at time t, is the mean (μ) of the posterior distribution of the true survival function $S(t)$ at time t. The Kaplan–Meier estimate has this interpretation provided the investigator uses no prior knowledge of the shape of the survival function S(t). So, for convenience, we will say that the

**Table 12.5   Kaplan–Meier survival function estimates and standard errors
and Susarla–VanRyzin posterior σ's for patients with 1–3 node involvement.**

Month	K-M Survival[a]	K-M standard error	Susarla & Van Ryzin σ	Month	K-M Survival	K-M standard error	Susarla & Van Ryzin σ
t	$\hat{S}(t)$	se	$\sigma(t)$	t	$\hat{S}(t)$	se	$\sigma(t)$
0	1.0000	0	0	53	0.6429	0.0906	0.0890
1	0.9643	0.0351	0.0345	58	0.6071	0.0923	0.0907
4	0.9286	0.0487	0.0478	63	0.5714	0.0935	0.0919
6	0.8571	0.0661	0.0650	65	0.5357	0.0942	0.0926
9	0.8214	0.0724	0.0711	80	0.5000	0.0945	0.0928
12	0.7857	0.0775	0.0762	88	0.4643	0.0942	0.0926
13	0.7500	0.0818	0.0804	97	0.4286	0.0935	0.0919
16	0.7143	0.0854	0.0839	128	0.3896	0.0928	0.0911
50	0.6786	0.0883	0.0867				

a. For a time not in the table use the next smaller survival time in the table; for example, the 5-year (60 month) survival estimate is the 58-month estimate, 0.6071.

Kaplan–Meier method is the result of applying Bayes' rule with a total-ignorance prior.

Susarla and Van Ryzin also obtained an equation for computing the standard deviation (σ) of the posterior distribution of the survival function at a given time t. As Table 12.5 indicates, the Kaplan–Meier standard errors appear to be a good approximation of the exact σ of the posterior distribution. For large samples, the normal approximation to the posterior distribution will be accurate; however, for small samples and for times at which the survival rate is close to 1.0 (for example, at 1 month) or close to 0.0, it is better to use a beta approximation (see Section 6.6, page 100) with parameters α and β given by

$$\alpha = \tilde{n} \cdot \mu \text{ and } \beta = \tilde{n} \cdot (1 - \mu), \text{ where } \tilde{n} = \frac{\mu \cdot (1 - \mu)}{\sigma^2} - 1 \qquad (12.7)$$

For example, at $t = 1$ month, the posterior distribution of the survival rate has parameters $\mu = 0.9643$ and $\sigma = 0.0345$; consequently, the parameters of the beta approximation are $\alpha = 26.9$ and $\beta = 0.997$ [see Equation (12.8)].

$$\tilde{n} = \frac{0.9643\times(1-0.9643)}{0.0345^2} - 1 = 27.9$$
$$\alpha = 27.92\times0.9643 \qquad = 26.9$$
$$\beta = 27.92\times0.0357 \qquad = 0.997$$

(12.8)

Inserting these parameters in the spreadsheet BetaTailAreaTable.xls, we find that posterior 95% credible interval for the 5 year survival rate is 0.872 to 0.999. The approximate normal 95% credible interval ($\mu \pm 1.96\cdot\sigma$, or 0.897 to 1.032), is easier to compute but is not very accurate. Using the Kaplan–Meier standard error reported by the LIFETEST procedure in place of the exact Susarla–Van Ryzin value produces nearly the same results.

12.3.3 Comparing Two Survival Curves

Points on a survival curve are rates. For example, $S(60)$ is the survival rate at 5 years after diagnosis. Therefore, many of the methods of Chapter 7 are applicable. Since it is more common to report relative risks than differences, we show in Fig. 12.4 how to obtain the approximate posterior distribution of the relative risk (RR) of failure in the first 5 years for patients with 0 nodes compared to patients with 1–3 nodes.

Technical note: The calculation in Fig. 12.4 ultimately flows from the beta approximation in Equation (12.7), which represents the posterior distribution as if it were the result of observing $\alpha-1$ successes and $\beta-1$ failures. The idea is to use the method of Fig. 7.3 on page 124 but with n replaced by $\alpha+\beta-2$ and x (the number of adverse events) replaced by $\beta-1$. This idea is the basis for Fig. 12.4but the calculation has been re-expressed in terms of the Kaplan–Meier estimates and standard errors rather than the derived quantities α and β. The right tail area calculation in the last row is based on Equation (7.14) on page 126 with b=1.

12.3.4 Unanticipated Consequences of Complete Ignorance

Kaplan–Meier estimates and standard errors flow from the assumption of complete ignorance about the form of the survival function. Although "complete ignorance" is easily said, it is less easily modeled when the thing about which we are ignorant is the entire survival function, rather than a single number such as the difference between two box averages.

To understand this point, notice that although the Kaplan–Meier survival curve is indeed the Bayes' rule estimate of the survival curve assuming complete prior ignorance, it is a step function with flat stretches interrupted by abrupt drops. The flat stretches represent time intervals where (so the graph claims) there is absolutely no risk of death, and the abrupt drops represent discrete time points where there is a highly concentrated risk of death. For example, the

Kaplan–Meier survival curve in Fig. 12.2 is flat between years 2 and 5. Since the graph represents the investigator's posterior knowledge about the survival curve, it attributes to the investigator the absolute belief that in the population of patients with 1-3 node involvement, nobody will die between years 2 and 5. It seems doubtful that any rational person would actually put much probability on that statement. For that reason, any reasonable definition of complete ignorance should start with the belief that the survival curve decreases continuously without abrupt drops. Although various ways of doing this have been proposed, there is at this time (2004) no user-friendly software suitable for nonspecialists; readers should consult this textbook's website for more recent developments.

Posterior Distributions of Failure Time Quantiles. Posterior uncertainty about the survival function can be converted into the posterior distribution of a quantile failure time. The most commonly reported quantiles are the first- and third-quartile (Q1 and Q3) and median survival times; for example, the first-quartile survival time is the time at which a quarter of the patients have "failed." Fig. 12.5 shows how our posterior uncertainty about the survival curve for the

	Lower risk (L) 0 nodes	Higher risk (H) 1-3 nodes
Parameters of Posterior distributions of survival rates at t=60 months		
$\hat{S}(t) = \mu$	$\mu_L = 0.8529$	$\mu_H = 0.6071$
$se \cong \sigma$	$\sigma_L = 0.0607$	$\sigma_H = 0.0923$
Approximate Posterior Distributions of log Relative Risk		
$\widehat{RR} = \left(1 - \hat{S}_H\right)/\left(1 - \hat{S}_L\right)$	2.671	
$\mu \cong \ln\left(\widehat{RR}\right)$	0.982	
$\sigma = \sqrt{\left(\dfrac{se_L}{1-\hat{S}_L}\right)^2 + \left(\dfrac{se_H}{1-\hat{S}_H}\right)^2}$	0.475	
Approximate credible intervals and tail areas for ln(RR) and RR		
95% Credible Interval for ln(RR): $\mu \pm 1.96 \cdot \sigma$	0.052	1.913
95% CI for RR: $\exp(\mu - 1.96 \cdot \sigma)$ to $\exp(\mu + 1.96 \cdot \sigma)$	1.05	6.77
$P(RR > 1) = P(\ln(RR) > 0) = P(Z > (0-\mu)/\sigma) = P(Z > -2.31) = 0.989$		

Figure 12.4 Log-normal approximation of the posterior distribution of the relative risk of failure within the first 5 years.

population of patients with 1–3 nodes (lower graph) induces a posterior distribution of the first quartile survival time for these patients (the time at which 75% of patients are still alive). One of the unanticipated consequences of using a total ignorance prior is that the resulting posterior distribution of the Q1 failure time is concentrated at those discrete points where the Kaplan–Meier estimate steps down. The distribution has a conspicuous hole between months 24 and 50, meaning that the an investigator professing total prior ignorance is forced into the absurd position of claiming to be absolutely certain that the quartile failure time cannot possibly be anywhere in the interval from 18 to 50 months. It seems strange to be so firmly convinced on the evidence of so few observations.

Thus, in a technical sense it is possible to compute a 95% posterior credible interval for the Q1 failure time; however, it would be better described as a credible donut because it has a hole in the middle. This example makes us reluctant to recommend treating Kaplan–Meier confidence intervals of quartile and median failure times as if they were credible intervals. Since steps in the Kaplan–Meier survival function occur at observed failure (or censoring) times, the donut effect will be less severe for large data sets.

It is important to understand that the source of the donut problem is the presumption that the investigator has no prior belief about the continuity or smoothness of the survival function, a presumption that produces quite incredible posterior distributions for quantile survival times. There are many ways to cure the problem by incorporating continuity and smoothness into the nonparametric analysis; however, they are well beyond the scope of this text.

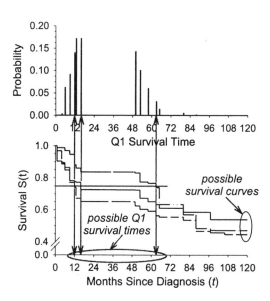

Figure 12.5 The "box" of survival curves (lower graph) induces the distribution of first quartile (Q1) survival times (upper graphs). The distribution is not continuous.

12.3.5   Interpreting *p* values

Statistical programs that implement the Kaplan–Meier method generally also compute *p* values, and by far the most commonly of these is produced by the *log-rank test*. The two-group log-rank statistic can be interpreted as a comparison of the hazard rate in one population (let's say patients with 0 nodes) with the hazard rate in a second population (let's say patients with 1–3 nodes). What complicates matters is that there is not a single hazard rate; instead, the hazard rate varies with time. For example, in Table 12.1 the hazard is not constant over the 5-year follow-up, but peaks at 16 per 1000 per month and at the end of 5 years has dropped to 9 per 1000 per month. In humans, the hazard rate at age 21 is about 14 per 10,000 per year and rises to about 14 per 1000 by age 61 (increasing by a factor of 10 about every 30 years). So in order to compare "the" hazard rates in two populations it is necessary to use some sort of average.

The log-rank statistic is proportional to an estimate of the difference Δ of the *weighted average hazard rates* between the two groups, where the weights are proportional to the *harmonic mean* number of person-years (or person-months) of exposure to each hazard rate in the two groups combined. (The harmonic mean of a list of numbers is the reciprocal of the average of the reciprocals.) This idea is not unreasonable — weight the hazard by the number of people exposed to the hazard. However, no one to date has obtained the posterior distribution of Δ and for that reason it is unknown whether the *p* value for the log-rank statistic can be converted into a good approximation of the posterior probability that Δ is positive. How that question is eventually resolved is to some extent irrelevant, since it is more common to report *hazard ratios* rather than hazard differences.

12.3.6   Computing Kaplan–Meier Estimates

Fig. 12.6 is the SAS program that computed Fig. 12.1 and Table 12.4 using the data in Table 12.2. The first step is to arrange the data in three columns:

Nodes: A categorical variable indicating the node category:
                    4 = 4+ lymph nodes involved
                    2 = 1 to 3 lymph nodes involved
                    0 = no lymph nodes involved

Time: The survival time of each patient

Censored: A categorical variable indicating a censored observation:
                    1 = yes (patient was alive at the end of the observation period)
                    0 = no (observation period ended at the patient's death)

For example, in the 4+ node group, survival time 128 is represented by data line "4 128 0," and survival time ">139" is represented by data line"4 139 1."

Other data sets will have a different categorical variable (or variables) that defines subgroups of patients (gender, for example) but will always have variables Time and Censored. The STRATA keyword lists one or more categorical variables that define subgroups of subjects. The syntax of the TIME keyword is

TIME survival_time_variable*censoring_indicator_variable(value);

where the value in parentheses is the code for a censored observation, in this case value 1 indicates a censored observation.

```
Data Table12_2;
    INPUT Nodes Time Censored;
DATALINES;
4   1   0
4   3   0
...
4   1280
4   1391
...
2   1   0
2   4   0
...
2   2421
0   17  0
...
0   2521
0   2521
;;;;
PROC LIFETEST DATA=Table12_2 PLOTS=(S);
    STRATA Nodes;
    TIME Time*Censored(1);
RUN;
```

Figure 12.6   SAS program to compute Fig. 12.1 and Table 12.4.

Output from the program consists of a graph similar to Fig. 12.1, a series of tables containing the information summarized in Table 12.4, and a table of the survival function of each subgroup, with confidence intervals. The Kaplan-Meier survival function estimate for stratum 3 (patients with 4+ nodes involved) is shown in Fig. 12.7; for example, the estimated survival rate at 12 months, $S(12)$, is 0.6522 with standard error 0.0993, which can be used to compute approximate posterior credible intervals using the beta method described in the paragraphs containing Equations (12.7) and (12.8). The estimated median survival time is 50 months with confidence interval 11 to 128 months. The

```
                        The LIFETEST Procedure
                        Stratum 3: Nodes = 4
                      Product-Limit Survival Estimates
```

			Standard	Number	Number
Time	Survival	Failure	Error	Failed	Left
0.000	1.0000	0	0	0	23
1.000	0.9565	0.0435	0.0425	1	22
3.000	.	.	.	2	21
3.000	0.8696	0.1304	0.0702	3	20
5.000	0.8261	0.1739	0.0790	4	19
9.000	0.7826	0.2174	0.0860	5	18
10.000	0.7391	0.2609	0.0916	6	17
11.000	0.6522	0.3478	0.0993	8	15
23.000	0.5652	0.4348	0.1034	10	13
36.000	0.5217	0.4783	0.1042	11	12
38.000	0.4783	0.5217	0.1042	12	11
49.000	0.4348	0.5652	0.1034	13	10
65.000	0.3913	0.6087	0.1018	14	9
121.000	0.3478	0.6522	0.0993	15	8
128.000	0.3043	0.6957	0.0959	16	7
139.000*	.	.	.	16	6
152.000	0.2536	0.7464	0.0924	17	5
152.000*	.	.	.	17	4
157.000	0.1902	0.8098	0.0884	18	3
228.000*	.	.	.	18	2
252.000*	.	.	.	18	1

NOTE: The marked survival times are censored observations.

```
                  Quartile Estimates
                        95% Confidence Interval
        Percent    Estimate    [Lower      Upper)
           75       157.000    49.000         .
           50        38.000    11.000     128.000
           25        10.000     3.000      36.000
```

Figure 12.7   Kaplan–Meier estimate and quartile survival times.

confidence interval is, technically, an approximate 95% credible interval for a total ignorance prior, but recall the warnings in Section 12.3.4 about the "incredible" consequences of total ignorance.

A near-total-ignorance analysis can be set up in WinBUGS as well and can answer research questions that conventional analysis cannot; however, the setup is rather tricky and requires the assistance of an experienced WinBUGS user (see "Leuk" in WinBUGS Examples, Vol I).

12.4   REGRESSION METHODS

The previous section was concerned with comparisons among three groups of cancer patients, those with 0, 1–3, or 4+ nodes involved. It did appear that there were substantial differences in the survival of the three groups. However, there may have been other differences between the groups in addition to node status. For example, suppose that survival time after diagnosis is influenced by race (W, N), node status (0, 1–3, 4+), and number of cigarettes smoked per day

(0, 1–10, 11–20, 21–30, 31+). There are 30 possible configurations of these explanatory variables ($2 \times 3 \times 5$), and a Kaplan–Meier analysis would require estimating 30 survival curves — one curve for white nonsmokers with 0 nodes, one curve for nonwhite nonsmokers with 0 nodes, and so on. Such an undertaking is clearly not feasible without an extremely large sample. For example even with 300 subjects, at least one of the groups would have 10 or fewer members, too few to produce a reliable estimate of the survival curve; and even if there were sufficient numbers of subjects to estimate 30 survival curves, the task of reporting and interpreting the results would be daunting.

For these reasons it is useful to have regression-like methods for adjusting survival rate estimates to remove the effect of confounding variables and to quantify the effect of explanatory variables. The most commonly used method is Cox *proportional hazards regression*, which is a *semiparametric* analysis. There is, in addition, a large and versatile arsenal of fully *parametric* methods such as *Weibull regression*.

12.4.1 Proportional Hazards Regression

Cox (1972) proposed that the effects of explanatory and confounding variables be represented as a multiplier of a *baseline hazard function* called $h_0(t)$. The investigator is assumed to have no prior knowledge about the shape of the baseline hazard function; however, the influence of explanatory and confounding variables is a log-linear regression.

To begin with a simple example, suppose that, for a particular cancer diagnosis, the patient's survival time is influenced by whether there is any lymph node involvement at the time of diagnosis (no or yes) and by the number of cigarettes smoked per day. The design variables are x_1, a Boolean variable indicating node involvement ($1 = $ yes, $0 = $ no), and x_2, the number of cigarettes smoked per day. The baseline category is nonsmokers with no node involvement. The hazard function for the baseline category is called $h_0(t)$, where t is time since diagnosis, and the hazard function for other categories is denoted $h_x(t)$, where x stands for the bundle of explanatory variables (node status, cigarettes per day).

The proportional hazards model relates the hazard function for patients in category x to the hazard function for patients in the baseline category,

$$h_x(t) = h_0(t) \cdot \exp\left(\beta_1 \cdot x_1 + \beta_1 \cdot x_2\right), \tag{12.9}$$

where x_1 and x_2 are design variables. The *hazard ratio* is the ratio of two hazard rates.

The hazard ratio is the ratio of death rates in vanishingly short time intervals and is therefore interpretable as the odds ratio or relative risk of dying in the next instant of time for people in category x compared to people in the baseline category (nonsmoking patients with no node involvement).

$$\frac{h_x(t)}{h_0(t)} = \exp\left(\beta_1 \cdot x_1 + \beta_1 \cdot x_2\right) = HR_x \qquad (12.10)$$

The Cox model separates the estimation of the baseline hazard function from the estimation of the influence of explanatory and confounding variables. The baseline hazard portion of the model is nonparametric in the sense that the investigator assumes no prior knowledge of its form; however, the influences of the explanatory variables are described in a *parametric* linear regression model with regression coefficients β_1 and β_2. The Cox model is therefore said to be *semiparametric* — a hybrid of nonparametric and parametric components.

The interpretation of proportional hazards regression coefficients is similar to the interpretation of logistic regression coefficients, with the hazard ratio taking the place of log odds ratio (see Section 10.2, page 211). To help illustrate how to interpret proportional hazards regression coefficients, let us assume that the estimated coefficients are as shown in Table 12.6. The column labeled "adjusted hazard ratio" contains the antilogs of the regression coefficients; for example, the regression coefficient for node involvement is $\beta_1 = 0.6$, and therefore its adjusted hazard ratio is $\exp(0.6) = 1.82$. The interpretation is that at the same point in time (t), with all other things being equal, patients with some node involvement have an 82% increased death rate compared to patients with no node involvement.

In publications it is the practice to compute the adjusted hazard ratio as the antilog of the regression coefficient whether the design variable is Boolean or not; however, for a continuous explanatory variable like the number of cigarettes per day, the hazard ratio depends on the difference between the numbers of cigarettes per day in the two groups being compared. For example, $\exp(\hat{\beta}_2 \times 10) = \exp(0.2) = 1.22$ is the adjusted hazard ratio for 10-cigarette-a-day smokers compared to nonsmokers with the same node status and is also interpretable as the hazard ratio for 20 cigarette a day smokers compared to 10 cigarette a day smokers with the same node status.

The hazard function $h(t)$ is the instantaneous death rate at time t, and therefore higher values of the hazard function produce more deaths and therefore fewer survivors at any given time. The specific mathematical relationship is Where $S_0(t)$ is the survival function corresponding to the baseline hazard function $h_0(t)$ and HR_x is the hazard ratio of category x relative to the baseline category defined in Equation (12.10).

Table 12.6   Proportional hazards regression coefficient estimates.

Explanatory variables x	Estimates $\hat{\beta}$	Adjusted hazard ratio HR
Node involvement	0.60	1.82
No. of cigarettes/day	0.02	1.02

$$S_x(t) = \left[S_0(t)\right]^{HR_x} \qquad (12.11)$$

To illustrate how the hazard ratio modifies the survival rate, suppose that the baseline survival rate at 60 months is $S_0(60) = 0.70$, that is, the survival rate for nonsmoking patients with no node involvement. Now consider a patient with no node involvement who smoked 20 cigarettes a day. His estimated hazard ratio relative to the baseline category is

$$\widehat{HR} = \exp\left(\hat{\beta}_1 \cdot x_1 + \hat{\beta}_2 \cdot x_2\right) = \exp\left(0.60 \times 0 + 0.02 \times 20\right) = 1.49 \qquad (12.12)$$

Suppose we want to compute his estimated probability of surviving 10 or more years after diagnosis, in other words, the estimated survival rate at 120 months for pack-a-day smokers with no node involvement, $\hat{S}_x(120)$. According to Equation (12.2) it is the baseline survival function at 120 months raised to the 1.49 power. So, if we had an estimate of the baseline survival rate at 120 months, we could compute the survival function for any other category of patients. For example, if the estimated baseline survival rate at 120 months is $\hat{S}_0(120) = 0.97$, then the survival rate for pack-a-day smokers with no node involvement is

$$\hat{S}_x(120) = \left(0.97\right)^{1.49} = 0.956 \qquad (12.13)$$

Notice that if the hazard ratio is greater than 1.0, indicating an increased death rate, the survival rate for category x is lower than that in the baseline category, and if the hazard ratio is less that 1.0, indicating a decreased death rate, the survival rate increases relative to the baseline category.

In some cases, the baseline category has no members; for example, if one explanatory variable is the patient's body mass index (BMI), the baseline category will consist of people with BMI equal to zero, who would have to be either massless or infinitely tall. It such cases it is more convenient to express survival in terms of a more convenient reference category. Suppose, for example, that a convenient reference category is people who smoked 5 cigarettes a day and had some node involvement. The estimated 10-year survival rate for this category is symbolized $\hat{S}_{ref}(120)$, and the estimated survival rate for any other category is

$$\hat{S}_x(120) = \left[\hat{S}_{ref}(120)\right]^{HR_x/HR_{ref}} \qquad (12.14)$$

This equation could be applied to any time point and therefore provides a way to compute the graph of the entire survival function for any category of patients; however, since the baseline survival function is estimated nonparametrically, it will have the undesirable features of the Kaplan–Meier survival function. In other words, it will be a discontinuous function with steps at observed survival times. Survival functions for nonbaseline categories inherit these discontinuities, and therefore quantile survival times for baseline or nonbaseline categories will have spiky posterior distributions similar to the top panel of Fig. 12.5. Consequently, in studies that observe only a small number of failures, reported confidence intervals for quantile failure times should be interpreted with considerable caution.

A Case Study: Metastatic Prostate Cancer. Halabi et al. (2003) followed 1101 patients with metastatic hormone-refractory adenocarcinoma of the prostate for a ten-year observation period. Excerpts from that paper are reproduced with permission of the author and the Journal of Clinical Oncology.

A proportional hazards model was used to develop a prognostic model for predicting survival. The response variable was the total survival time, defined as the time between entry into the study and death. The explanatory variables are listed in Table 12.7, along with their approximate ranges and medians. Thus in this study the average patient was in performance status 1, and had Gleason sum ≥ 8, LDH = 220, alkaline phosphatase = 150, PSA = 100, no Visceral disease, and hemoglobin = 12; we will use this as the reference category in Equation (12.14).

The authors provided a graphical *nomogram* (their Figure 3) that allows the reader to compute the estimated survival rate for any category of patients at 12 months and at 24 months. For the patients in the reference category as defined in the previous paragraph, those values are

$$\hat{S}_{\text{ref}}(12) = 0.60$$
$$\hat{S}_{\text{ref}}(24) = 0.24$$

$$(12.15)$$

Table 12.7 **Explanatory variables in the prostate cancer survival model.**

Explanatory variable[a]	Values	Median[b]
Performance status	0,1,2	1
Gleason sum	<8 or ≥ 8	<8
LDH	6 to 4000	220
Alkaline phosphatase	10 to 2500	150
PSA	1 to 5000	100
Visceral disease	No or Yes	No
Hemoglobin	7 to 17	12

a. Defined in Halabi et al. (2003).
b. Average median or model category in Halabi's learning and validation samples.

Proportional hazards regression coefficients, hazard ratios (antilogs of the regression coefficients), and 95% confidence intervals for the hazard ratios are reported in Table 12.8. From this table we can use Equation (12.14) to compute the estimated survival rate at 12 and 24 months for any category of patients, and to compute the hazard ratio between any two categories of patients. Moreover, the confidence intervals in Table 12.8 can be reverse-engineered using Equation (7.15) on page 128 to obtain approximate posterior distributions of the regression coefficients and hazard ratios for *single* explanatory variables. What we cannot compute with the given information is the posterior distribution of the survival rate at a particular time, or the posterior predictive distribution of the survival time of an individual patient, or the posterior distribution of a hazard ratio involving more than one explanatory variable, although these could be computed from the raw data using WinBUGS.

To illustrate the sort of thing that can be done with the information in Equation (12.15) and Table 12.8, we'll compute the estimated 12- and 24-month survival rates for patients at the third quartile on all explanatory variables; namely, performance status=1, Gleason sum $\geq$ 8, LDH=409, alkaline phosphatase=355, PSA=336, no visceral disease, and hemoglobin=11. Let's call this category of patients Q3. In order to do the calculation we first need to calculate the hazard ratio for category Q3 as well as for the reference category; each relative to the baseline category. Hazard ratios relative to baseline are

Table 12.8 Proportional hazards regression coefficients.[a]

Explanatory variable[b]	Regression coefficient	Adjusted hazard ratio	95% CCI[c]
Performance status:			
0		1	Baseline[d]
1	0.392	1.48	1.31 to 1.67
2	0.784	2.19	1.94 to 2.47
Gleason sum:			
< 8		1	Baseline
8–10	0.335	1.40	1.20 to 1.62
ln(LDH)	0.312	1.37	1.21 to 1.55
ln(alkaline phosphatase)	0.211	1.23	1.12 to 1.36
ln(PSA)	0.093	1.10	1.05 to 1.15
Visceral disease:			
No		1	Baseline
Yes	0.161	1.17	0.95 to 1.46
Hemoglobin	−0.082	0.92	0.87 to 0.97

a. Source: Halabi, et al. (2003) Table 3.
b. See above for definitions of the explanatory variables.
c. Conventional confidence interval (approximate flat prior credible interval).
d. The baseline category is: performance status 0, Gleason sum <8, no visceral disease, and all continuous explanatory variables equal to 0.

Table 12.9   Computing hazard ratios for two patient categories.

			Patient categories	
Explanatory variables	Design variables	Regression coefficients	Reference (ref)	3rd quartile (Q3)
			Design variable values	
Perf. status:				
1	x_1	0.392	1	1
2	x_2	0.784	0	0
Gleason sum:				
8–10	x_3	0.335	0	1
ln(LDH)	x_4	0.312	ln(220)=5.39	ln(409)=6.01
ln(alk. phos.)	x_5	0.211	ln(150)=5.01	ln(355)=5.87
ln(PSA)	x_6	0.093	ln(100)=4.61	ln(336)=5.82
Visceral disease:				
Yes	x_7	0.161	0	0
Hemoglobin	x_8	−0.082	12	11
		Sum of $\beta \times X$:	2.57	3.48
		Hazard ratio = exp(sum):	HR_{ref}=13.15	HR_{Q3}=32.50

calculated in Table 12.9 using Equation (12.10) with eight design variables and regression coefficients.

We have now assembled all the ingredients of Equation (12.14) and can calculate survival rates at 12 and 24 months for patients at the third quartile of each of the explanatory variables,

$$\hat{S}_{Q3}(12) = \left[\hat{S}_{ref}(12)\right]^{HR_{Q3}/HR_{ref}} = (0.60)^{32.50/13.15} = 0.28$$

$$\hat{S}_{Q3}(24) = \left[\hat{S}_{ref}(24)\right]^{HR_{Q3}/HR_{ref}} = (0.24)^{32.50/13.15} = 0.03$$

                (12.16)

Many papers that report proportional hazards regression analyses provide some means of obtaining estimates of the survival function. Here we have done the calculation "the hard way" to try to provide some insight into how the proportional hazards model uses the explanatory variables to modulate the survival rate. Frequently, the investigator will provide tables or graphs of the survival function at the first and third quartiles and median of the estimated hazard ratio or for clinically important combinations of the explanatory variables; examples are Table IV in Giles et al. (2003) and Figure 5 in Halabi et al. (2003).

Another way that proportional hazards regression is used is to identify clinically important risk factors. For example, any risk factor that produces a 10% or more increased hazard might be judged clinically important. In that case the investigator would be interested in calculating the posterior probability $P(\text{HR} > 1.10 \mid \text{Data})$. Assuming that the investigator places flat priors on the regression coefficients and that the subjects were followed long enough to observe a substantial number of deaths (failures), the 95% confidence intervals can be interpreted as approximate 95% credible intervals and reverse engineered as in Equation (7.15) to obtain the parameters of a normal approximation to the posterior distribution.

For example, the median PSA was 100, and the third quartile was about 3 times larger, 336. Does a change that large produce a 10% or more increased hazard rate? This is a question about the true hazard ratio for patients with PSA=336 vs. patients with PSA=100, all other factors being equal. The expression for that hazard ratio is $\exp(\beta_6 \cdot \Delta X_6)$, where β_6 is the true but unknown regression coefficient for ln(PSA), and, according to Table 12.9, $\Delta X_6 = \ln(336) - \ln(100) = 1.21$. For convenience, we'll call this the *PSAQ3 hazard ratio*, or HR_{PSAQ3}, with the understanding that it reflects the effect of changing the PSR from its median to its third quartile. To complete the posterior probability calculation, we need the posterior distribution of β_6, which, by Equation (7.6) on page 117, is approximately normal with

$$\mu = \hat{\beta}_6 = 0.093$$

$$\sigma = \text{seb}_6 = \frac{\ln(UCL/LCL)}{2 \times 1.96} = \frac{\ln(1.15/1.05)}{3.92} = 0.023$$

where UCL and LCL are the upper and lower endpoints of the 95% confidence interval for the adjusted hazard ratio, not the regression coefficient. Thus,

$$
\begin{aligned}
P\left(\text{HR}_{\text{PSAQ3}} \geq 1.1\right) &= P\left(\exp\left(\beta_6 \cdot 1.21\right) \geq 1.1\right) = P\left(\beta_6 \cdot 1.21 \geq \ln(1.1)\right) \\
&= P\left(\beta_6 \geq 0.0788\right) = P\left(Z \geq \frac{0.0788 - \mu}{\sigma}\right) \\
&= P\left(Z \geq -0.62\right) = 0.732
\end{aligned}
$$

The conclusion is that there is about a 73% posterior probability that roughly tripling the PSR from 100 to 336 will increase the hazard rate (i.e., the instantaneous death rate) by 10%.

12.4.2 Computing Proportional Hazards Regressions

Fig. 12.8 is the output of a re-analysis of the much-analyzed multiple myeloma survival data in Table 2 of Krall et al. (1975). In this analysis there are two explanatory variables: gender and blood urea nitrogen or BUN. Other analysts have found that the logarithm of BUN produces a better model specification and we will follow their lead; however, we have opted to use different slopes (regression coefficients for log BUN) for males and females. The explanatory variables and design variables are as follows:

Variable	Description
Time	Survival time
D	Death (the censoring variable; D=0 indicates that the patient was still alive at the end of the observation period)
BUN	Blood urea nitrogen at diagnosis
HG	Hemoglobin at diagnosis (not used in the analysis)
SEX	Gender of the patient
lnBUN	log BUN
FEMALE	Boolean variable indicating (SEX="F")
MALE	Boolean variable indicating (SEX="M")
lnBUN_FEM	lnBUN if SEX="F", else 0
lnBUN_MALE	lnBUN if SEX="M", else 0

The last four design variables are computed in the DATA step:

```
DATA Krall;
    Input T D BUN HG SEX $;
    /* Compute the Design Variables */
    lnBUN=LOG(BUN);
    FEMALE=(SEX="F"); MALE=(SEX="M");
    lnBUN_FEM=lnBUN*FEMALE;
    lnBUN_MALE=lnBUN*MALE;
DATALINES;
1.25    1    165    9.4    M
1.25    1     87    12     M
2       1     33    9.8    M
    ... 59 data lines not shown ...
53      0     13    12     M
57      0     18    12.5   M
77      0     12    14     M
;;;;
RUN;
```

The bare-bones syntax of the proportional hazards regression procedure (PROC PHREG) is

```
PROC PHREG DATA=data_set_name;
    MODEL survival_time*censoring(value)=design_vars;
```

In this case the data file (see the DATA step) is called KRALL, variable T is the survival time, and variable D = 0 indicates a censored observation; consequently, the commands are,

```
PROC PHREG DATA=KRALL;
    MODEL T*D(0)=FEMALE MALE lnBUN_FEM lnBUN_MALE;
RUN;
```

These commands produce a table of estimated regression coefficients, standard errors, and adjusted hazard ratios with 95% confidence intervals:

Variable	DF	Parameter Estimate	Standard Error	ChiSq	Pr>ChiSq	Hazard Ratio	95% Hazard Ratio Confidence Limits	
FEMALE	1	2.81925	1.77392	2.5258	0.1120	16.764	0.518	542.437
MALE	0	0	.	.	.	.	.	.
lnBUN_FEM	1	0.24335	0.43525	0.3126	0.5761	1.276	0.543	2.993
lnBUN_MALE	1	1.19928	0.35821	11.2091	0.0008	3.318	1.644	6.695

Figure 12.8   Proportional hazards regression coefficient estimates from PROC PHREG.

The table lists p values under the heading Pr > Chi-Square, which in this case can be converted into approximate posterior tail areas using Equation (7.11) on page 122. The parameter estimate, $\hat{\beta}$, and standard error, can be used as the μ and σ of the normal approximation to the flat-prior posterior distribution of the true regression coefficient. The column headed ChiSq is the square of μ/σ, that is, $(\hat{\beta}/se)^2$, and can be ignored.

The program becomes a bit more complicated when the analyst wants to produce estimated survival curves for selected categories of patients. To do this it is necessary to create a second data file containing the design variable values for the categories or patients for which estimated survival curves are desired. The following DATA step shows how to set up a request for survival curves for three categories: males with BUN=6, males with BUN=21, and females with BUN=21.

```
DATA CategoryRequests;
    INPUT BUN SEX $;
    lnBUN=LOG(BUN);
    FEMALE=(SEX="F"); MALE=(SEX="M");
    lnBUN_FEM=lnBUN*FEMALE; lnBUN_MALE=lnBUN*MALE;
DATALINES;
    6 M
    21 M
    37 F
RUN;
```

The program that produces estimated survival curves for these three categories is as follows:

```
PROC PHREG DATA=KRALL;
   BASELINE OUT=SurvivalCurves
      COVARIATES=CategoryRequests SURVIVAL=S
      LOWER=LCL UPPER=UCL/ NOMEAN;
   MODEL T*D(O)=FEMALE MALE lnBUN_FEM lnBUN_MALE /
      RISKLIMITS;
PROC PRINT DATA=SurvivalCurves;
RUN;
```

This program creates a SAS data file called SurvivalCurves. The portion of that file corresponding to 12- and 24-month survival estimates is reproduced in Fig. 12.9. The estimated 12 month survival rate is the same as the 11 month rate because there were no failures or censored observations between 11 and 13 months. The estimated 12 month survival rate for males with BUN=6 is 92.2%, which drops to 69.4% for males with BUN=21; females with BUN=21 have a slightly lower 12 month survival rate than males with the same BUN.

FEMALE	MALE	lnBUN_FEM	lnBUN_MALE	T	S	LCL	UCL
			(Males with BUN=6)				
0	1	0.00000	1.79176	11	0.92182	0.83832	1.00000
0	1	0.00000	1.79176	24	0.84151	0.69646	1.00000
			(Males with BUN=21)				
0	1	0.00000	3.04452	11	0.69371	0.57409	0.83825
0	1	0.00000	3.04452	24	0.46060	0.32410	0.65459
			(Females with BUN=21)				
1	0	3.61092	0.00000	11	0.71615	0.59228	0.86593
1	0	3.61092	0.00000	24	0.49276	0.34201	0.70997

Figure 12.9 Estimated 12 and 24-month survival rates (S) and 95% confidence intervals (LCL, UCL) for three categories of multiple myeloma patients.

The program and data for this example are available for downloading as file CoxPHReg.SAS from the textbook website.

12.4.3   Weibull Regression

The difference between Cox proportional hazards regression and Weibull regression is that in the latter the baseline hazard function is modeled parametrically and smoothly. This has the advantage of smoothing the posterior distribution of quantiles, but introduces the risk that the incorrect model was used for the baseline hazard function, because the Weibull hazard function has the property of increasing (or decreasing) as a *power* of time,

$$h_0\left(t\right) = b \cdot t^{b-1} \tag{12.17}$$

and not all populations have that characteristic. For example, as we will learn in Section 12.5, adult humans experience an *exponentially* increasing risk of death (hazard function). Consequently, using Weibull models in extremely long-term human survival studies is probably not appropriate, although they have been successfully applied in comparatively short-term studies (follow-up periods of at most a decade, say).

Combining Equations (12.8) and (12.17) produces the Weibull proportional hazards regression model

$$
\begin{aligned}
h_x(t) &= b \cdot t^{b-1} \cdot \exp\left(\beta_1 \cdot x_1 + \beta_2 \cdot x_2 + \cdots + \beta_k \cdot x_k\right) \\
&= b \cdot t^{b-1} \cdot \mathrm{HR}_x
\end{aligned}
\tag{12.18}
$$

where HR_x is the hazard ratio relative to the baseline category. Weibull regression coefficients have the same interpretation as Cox regression coefficients or logistic regression coefficients, and we will not go over that material again.

Unlike the Kaplan–Meier and Cox survival functions, the Weibull survival function can be expressed as an algebraic function

$$
S_x(t) = \exp\left(-t^b \cdot \mathrm{HR}_x\right)
\tag{12.19}
$$

12.4.4 Weibull Regression with WinBUGS

This section is a reanalysis of the proportional hazards regression model of Sections 12.4.1 and 12.4.2; however, in this analysis the baseline hazard function is a Weibull model, rather than being estimated nonparametrically. The advantage of this approach is that the survival function estimates are continuous and smooth and have no unrealistic abrupt drops. The data, as before, are the multiple myeloma survival from Krall et al. (1975). The continuous explanatory variable is the logarithm of blood urea nitrogen at diagnosis (BUN) and the four regression coefficients are intercepts and slopes for male patients and female patients.

The LIFREG procedure has a Weibull regression option; however, this analysis was done with the WinBUGS program rather than SAS to show how to use WinBUGS and because it is capable of computing posterior probabilities that answer a much wider range of research questions. In particular, we wanted to compute posterior credible intervals for contrasts between male and female survival rates. Detailed instructions for setting up the data and design variables for WinBUGS analysis are presented in Appendix Section B.9.

The part of the WinBUGS output corresponding the SAS proportional hazards regression output in Table 12.8 is reproduced in Table 12.10. The male slopes in the SAS and WinBUGS analyses are close to 1.2 in both analyses and the

Table 12.10   B. Output[a]: Posterior Moments and Quantiles of Regression Coefficients.

Description	Node[b]	Mean	sd	MC error	2.5%	Median	97.5%
Female intercept	beta[1]	-4.84	1.44	0.05	-7.64	-4.84	-1.97
Female slope	beta[2]	0.21	0.43	0.015	-0.69	0.24	0.98
Male intercept	beta[3]	-8.00	1.42	0.06	-10.92	-7.96	-5.56
Male slope	beta[4]	1.26	0.36	0.015	0.57	1.26	1.99
Weibull parameter b	shape	1.16	0.13	0.004	0.93	1.15	1.42

a. Based on 60,000 simulations.
b. "Node" is WinBUGS' term for an unknown quantity in the model.

female slopes are close to 0.2. The intercepts are quite different between the two analyses, but that is in part an artifact of the way the baseline hazard function is defined in the two analyses; in fact the difference between the male and female intercepts is about 3 in either analysis.

Recall that the hazard function, $h_x(t)$ is the force of mortality, or instantaneous death rate, at time t, and that a hazard ratio (the ratio of the hazard functions for two different categories of patients) has the same meaning as a relative risk or odds ratio. Inserting the posterior means of parameter b and the regression coefficients from Table 12.10 into Equation (12.18) gives estimated male and female hazard functions,

$$\hat{h}_{male}(t) = 1.16 \cdot t^{1.16-1} \cdot \exp\left(-8.00 + 1.26 \cdot \ln\left(BUN\right)\right)$$
$$\hat{h}_{female}(t) = 1.16 \cdot t^{1.16-1} \cdot \exp\left(-4.84 + 0.21 \cdot \ln\left(BUN\right)\right) \qquad (12.20)$$

Consequently the female/male hazard ratio is

$$\widehat{HR}_{f-m} = \frac{\hat{h}_{female}}{\hat{h}_{male}} = \frac{\exp\left(-4.84 + 0.21 \cdot \ln\left(BUN\right)\right)}{\exp\left(-8.00 + 1.26 \cdot \ln\left(BUN\right)\right)} = \exp\left(3.16 - 1.05 \cdot \ln\left(BUN\right)\right) \quad (12.21)$$

The clinical relevance of this equation is that men and women with the same BUN value at diagnosis do not have the same prognosis. At the lowest value, BUN=6, the estimated hazard ratio is $\exp(3.16 - 1.05 \cdot \ln(6)) = 3.59$, indicating that the death rate for women is 3.6 times higher than for men. However, at the highest value, BUN=172, the estimated hazard ratio is 0.11, indicating that the death rate for women is about 1/9 of that for men.

In order to make probability statements about the female-vs.-male hazard ratio it is necessary to include the ratio as a computed *node*[7] in the WinBUGS program (see Appendix B for an explanation of WinBUGS terminology, and Appendix Section B.9 in particular for details on setting up the calculation of

HR_{f-m} at selected BUN values). WinBUGS output concerning the male vs. female contrast is listed in Table 12.11. The output consists of estimates of the survival function at 24 months, $S(24)$, separately for men and women at the minimum, maximum, median, and quartile BUN values observed in the data set.

In appears that BUN is powerful prognostic factor, as evidenced by the fact that Male survival rates drop from 86% to 1% as the BUN value increases from its minimum value to its maximum value. However, it appears that the BUN level has different prognostic implications for men and women, as can be seen by comparing female and male hazard rates or cumulative death rates at different BUN values. For example, the relative risk (RR) of death within 24 months of diagnosis for women vs. men is 3.11 at BUN=6, a 211% increased risk of death for women, but is 1 or less for BUN values above the median (21), indicating a lower risk of death for women than for men with the same BUN value.

An important message to take away from this analysis is that Bayesian methods are capable of much more flexible and focused analyses of the particular

Table 12.11 Posterior distributions of survival rates and contrasts.

Explanatory Variables		Posterior Distribution				
		Moments		Quantiles		
Sex	BUN	μ	σ	2.5%	median	97.5%
		Female/Male hazard ratios				
	6	5.099	5.187	0.754	3.545	18.56
	14	1.619	0.710	0.651	1.481	3.359
	21	1.025	0.331	0.521	0.978	1.807
	37	0.583	0.238	0.228	0.546	1.146
	172	0.201	0.274	0.009	0.114	0.918
		Female/Male cumulative relative risks at 24 months				
	6	3.852	3.119	0.779	2.972	12.09
	14	1.438	0.495	0.713	1.358	2.617
	21	1.007	0.233	0.615	0.984	1.531
	37	0.707	0.182	0.370	0.701	1.078
	172	0.622	0.266	0.116	0.647	0.999
		Male and female survival rates at 24 months				
M	6	0.863	0.073	0.682	0.878	0.963
F	6	0.607	0.167	0.233	0.628	0.867
M	14	0.681	0.080	0.514	0.685	0.825
F	14	0.568	0.094	0.378	0.570	0.746
M	21	0.536	0.071	0.396	0.536	0.675
F	21	0.543	0.081	0.382	0.543	0.698
M	37	0.283	0.074	0.152	0.278	0.441
F	37	0.499	0.115	0.285	0.496	0.728
M	172	0.010	0.034	0.000	0.000	0.101
F	172	0.385	0.261	0.008	0.359	0.885

7. "Node" is WinBUGS's term for an unknown quantity. It is hoped, the reader will not be confused by the fact that *lymph* nodes were mentioned earlier in this chapter.

scientific issues in a given study than are conventional statistical methods. This advantage does not come from more flexible software, but is characteristic of Bayesian inference. In Bayesian inference, it is possible to compute the posterior distribution of any function of any combination of unknown quantities in the model. On the other hand, conventional statistical methodology has no general theory for constructing a confidence interval for a function of unknown quantities.

12.5 THE HUMAN HAZARD FUNCTION

How long humans live, of course, depends on many factors. Table 12.12 is a fragment of the life table based on US death rates for all races and genders combined in the year 1999. Hazard rates are not listed, but are practically the same as death rates; for example, the death rate at age 62 is 0.01427, and the hazard rate is $-\log(1-0.01427) = 0.01437$. Graphs of the hazard and survival functions are shown in Fig. 12.10. The hazard function has been graphed on the log scale and on that scale appears nearly linear from about age 30 to 100. In other words,

$$\ln\big(h(t)\big) \cong a + b \cdot t \tag{12.22}$$

Exponentiating each side produces the hazard function,

$$h(t) \cong A \cdot \exp(b \cdot t), \quad \text{where } A = \exp(a) \tag{12.23}$$

Thus adult humans, at least in industrialized nations, have an exponentially increasing hazard function. This hazard model is called the *Gompertz* model and is often used when a mathematical representation of adult human life-spans is needed as a component of a larger mathematical model.

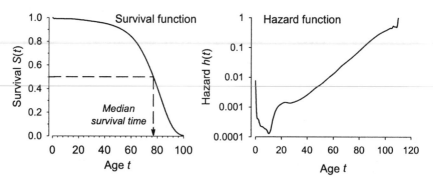

Figure 12.10   Survival and hazard functions derived from 1999 U.S. death rates.

Table 12.12   1999 U.S. life table.

Age	Death rate	Survivors	Deaths	Survival function
x to $x+1$	q_x	l_x	d_x	$S_{t+1}=S_t\cdot(1-q_x)$
0-1	0.00772	100000[a]	772	1.0000
1-1	0.00056	99228	56	0.9923
2-3	0.00040	99172	40	0.9917
19-20	0.00120	98502	118	0.9850
20-21	0.00128	98384	126	0.9838
21-22	0.00136	98258	134	0.9826
61-62	0.01427	83113	1186	0.8311
62-63	0.01565	81927	1282	0.8193
63-64	0.0173	80645	1395	0.8065
108-109	0.50000	4	2	0.00004
109-110	0.50000	2	1	0.00002
110-	1.00000	1	1	0.00001

a. Actuarial life tables are generally computed for a hypothetical population of 100,000 by applying contemporary mortality rates to a single cohort.

12.6 EXERCISES

12.1 Using the data in Table 12.2, calculate the Kaplan–Meier survival function estimate for patients with 1–3 lymph nodes involved.

12.2 Using the statistics in Table 12.4, calculate the approximate posterior probability that the 10-year survival rate for the 1–3-node group is less than 0.50,
   (a) using the normal approximation to the log survival rate,
   (b) using the beta approximation.

12.3 Using the statistics in Table 12.4, calculate the approximate posterior probability that
   (a) the 10-year survival rate is higher in the 1–3-node group than in the 4-node group, $S_{1-3}(10) > S_4(10)$.
   (b) the cumulative death rate within the first 10 years for the 4-node group is more than double the rate in the 0-node group:
        $RR = [1 - S_4(10)]/[1 - S_{1-3}(10)] > 2$.

12.4 According to Table 12.4, the most probable value of the first-quartile survival time is 10 years and the 95% confidence interval is 3 to 36 years. Give two reasons why the normal approximation to the posterior distribution with $\mu = 10$ and $\sigma = (36 - 3)/(2 \cdot 1.96)$ is not valid.

12.5 Download the file "Table12.2.txt" from the textbook website, and analyze the data in Table 12.2 via Cox proportional hazards regression. There is no continuous explanatory variable, but there are three Boolean design variables identifying node status,

   X1=(nodes=4); X2=(nodes=2); X3=(nodes=0);

   Run PHREG with these design variables and including a request for survival rates at 12 and 24 months for all three groups. Answer the following questions based on your output.
   (a) What is the baseline category?
   (b) What are the estimated survival rates at 12 and 24 months?
   (c) Compute hazard ratios and approximate 95% credible intervals for 1–3 vs. 0 nodes and 4 vs. 0 nodes. Hint (use your answer in part a).

12.6 (Difficult — confer with your instructor before undertaking this problem.) Download the data file Krall.txt, which contains an assortment of design variables. Make a subfile containing t.obs[ ] and t.cen[ ] and the design variables that were used in Appendix Section B.9, but add the design variable for hemoglobin (HG). In the request file ask for four 10-year survival rates for hemoglobin values 11 and 12 at BUN=37 for men and women (four requests). Report the posterior moments and quantiles for these four survival rates. Comment on the effect of hemoglobin. You may find it helpful to compute relative risks of death for HG=11 vs. HG=12 separately for men and women. You must not use the ten requests in Table B.4 but instead ask for the four listed above in the following order: male HG=11, male HG=12, female HG=11, female HG=12. In Fig. B.17 remove everything after "#Contrasts of interest" and instead compute relative risks of HG=11 vs. HG=12 for men and women

RR.hg.male=(1-Sr[1])/(1-Sr[2])

and

RR.hg.fem=(1-Sr[3])/(1-Sr[4]).

12.7 REFERENCES

Beenken, S.W., Urist, M.M., Zhang, Y., Desmond, R., Krontiras, H., Medina, H., and Bland, K.I., "Axillary lymph node status, but not tumor size, predicts locoregional recurrence and overall survival after mastectomy for breast cancer," *Annals of Surgery*, Vol. 237, No. 5 (May 2003), pp. 732-738, discussion pp. 738-739.

Cox, D.R., "Regression models and life tables," Journal of the Royal Statistical Society, series B, Vol. 34 (1972), pp. 187-220.

Giles, FJ., Bekele, BN., O'Brien, S., Cortes, JE., Verstovsek, S., Balerdi, M. Yared, M., Zhou, X., Kantarjian, HM., Keating, MJ., Thall, P. and Albitar, M., "A prognostic model for survival in chronic lymphocytic leukaemia based on p53 expression," *British Journal of Haematology*, Vol. 121 (2003), pp. 578-585.

Greene, F. L., Stewart, A.K., Norton, H.J., "A New TNM Staging Strategy for Node-Positive (Stage III) Colon Cancer: An Analysis of 50,042 Patients," *Annals of Surgery*, Vol. 236, No. 4 (Oct. 2002), pp. 416-421.

Halabi, S. Small, EJ. Kantoff, PW., Kattan, MW., Kaplan, EB., Dawson, NA., Levine, EG., Blumenstein, BA., Vogelzang, NJ., "Prognostic model for predicting survival in men with hormone-refractory metastatic prostate cancer," *Journal of Clinical Oncology*, Vol. 21, No. 7 (2003), pp. 1232-1237.

Krall, JM., Uthoff, VA., and Harley, JB., "A Step-up Procedure for Selecting Variables Associated with Surivival," *Biometrics*, Vol. 31, No. 1 (1975), 49-57.

Max-Planck-Gesellschaft, "The Human Lifetable Database," http://www.lifetable.de, 2002.

Susarla, V. and Van Ryzin, J., "Nonparametric Bayesian estimation of survival curves from incomplete observations," *Journal of the American Statistical Association*, Vol. 71 (1976), pp. 897-902.

Ziegler, A.-G., Schmid, S., Huber, D., Hummel, M., Bonifacio, E., "Early Infant Feeding and Risk of Developing Type 1 Diabetes-Associated Autoantibodies," *Journal of the American Medical Association*, Vol. 290, No. 13 (2003), pp. 1721-1728.

13

Decision Analysis

13.1 EVIDENCE AND DECISIONS

The aim of much of the research of biomedical scientists is to learn about nature. To that end, this book has shown how to use posterior distributions to quantify what has been learned from observations. However, some biomedical research is conducted for the purpose of making a decision. For example, clinical trials of a drug or medical device are designed to provide information to a regulatory agency for the purpose of deciding whether to approve or disapprove the drug or device for human use. DNA fingerprints provide evidence to help jurors decide on guilt or innocence. Genetic counseling provides data-based risk probabilities to help couples make informed decisions.

The role of statistical science in supporting decision making is to quantify the evidence upon which the decision is based; but although evidence is one of the raw ingredients of decision making, it is not the only ingredient. The other component of decision making — evaluation of consequences — is not, strictly speaking, in the domain of statistical science (see Fig. 13.1). Nevertheless, statisticians and biomedical professionals need to have some understanding of decision analysis in order to design studies that are adequate to the task of supporting decisions. This chapter starts with a brief introduction to the main ideas of decision analysis and then gives examples of how these ideas are used to plan experiments.

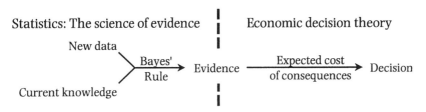

Figure 13.1 Decisions are based on evidence and expected consequences.

13.2 ACTS AND CONSEQUENCES

A simple example will help clarify the framework of decision analysis. Suppose
that there are three drugs available to treat a non-life-threatening, slow-growing
fungal infection. There are two strains of the fungus: Strain R (rare) occurs in
only 10% of patients, and strain F (frequent) occurs in 90% of patients. There are
three available drugs (A, B, and C) to treat the infection. Drug A costs $20 but
works well only for patients with strain R (90% cure rate) and not with strain F
(10% cure rate), drug B costs $5 and works reasonably well on either strain (40%
cure rate for R, 60% cure rate for F), and drug C cures 100% of infections with
either strain but is extremely expensive at $200. A diagnostic test is available to
reliably identify the strain of the infection at a cost of $5. Decisions to be made by
the physician are (1) whether or not to order a test to identify the strain, and (2)
whether to initially treat with drug A or B, followed by drug C if the initial
treatment fails.

Fig. 13.2 shows the physician's *decision tree* in *extensive form*. Each *node*[8] in
the tree is either a decision point (rectangle) or a point at which some previously
unknown fact is revealed (oval). The decision problem is solved by *backward
induction*, which consists of working from right to left, (1) replacing each
probabilistic node by its expected cost and (2) replacing each decision node by
the minimum-expected-cost option available to the decision maker at that node.

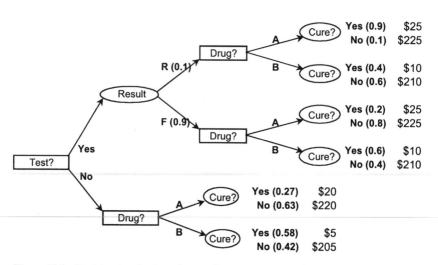

Figure 13.2 Decision tree for drug therapy. Decision points are boxed; probabilistic nodes
are circled; probabilities are in parentheses. Final costs include cost of testing ($5) and
costs of drugs (A: $20, B: $5, C: $200). Drug C is used only if A or B fails.

---

8. Here is yet a third use of the word "node"; in this case it is a branch point in a decision tree.

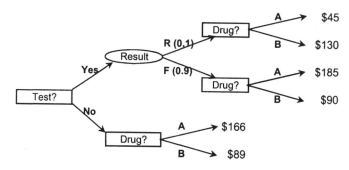

Figure 13.3   First stage of backward induction. Each of the rightmost random nodes has been replaced by its expected value.

In Fig. 13.3 the rightmost probabilistic nodes have been replaced by their expected costs computed as the sums of products of probabilities times dollar values [see Equation (3.14) on page 42]. For example, the expected value of the upper right "cure" node is EV = $0.9 \times \$25 + 0.1 \times \$225 = \$45$. This calculation must be applied separately to each probabilistic (oval) node as shown in Fig. 13.3.

The next step is to make the best (lowest cost) decision at each of the choice points (boxes) at the right side of the tree. For example, the best decision at the upper right drug choice point is to administer drug A, because the expected cost of administering drug A is $45 whereas the expected cost of administering drug B is $130. This procedure is applied to each of the exposed choice points as shown in panel II of Fig. 13.4. At this point the probabilistic node labeled "result" has been exposed and must be replaced by its expected value EV = $0.1 \times \$45 + 0.9 \times \$90 = \$85.50$ as shown in panel III of Fig. 13.4. This decision analysis indicates that it is better to test than not to test, and after testing, it is better to administer drug A to patients who test positive for strain R and to administer drug B to patients who do not. If testing cost $10 rather than $5, then the best decision would have been not to test and administer drug B to all patients.

Although this example is somewhat contrived, the methods illustrated here are widely regarded as the rational approach to decision making. In the biomedical literature indexed by Ovid Medline®, the search phrase "decision analysis or decision theory" produced over 2600 hits through late 2003. These included such decision problems as whether a woman with a prior cesarean section should attempt labor in a subsequent pregnancy or opt for a second cesarean section (Mankuta et al. 2003), the cost-effectiveness of CT screening for lung cancer (Wisnivesky et al. 2003), and whether to use dialysis, kidney transplantation, or pancreas transplantation to treat diabetes patients with

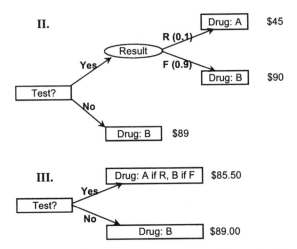

Figure 13.4   Second and third stages of backward induction. II: Each of the rightmost choice points has been replaced by the lowest-cost option. III: The probabilistic result node has been replaced by its expected value. Testing for the rare strain is the better choice.

kidney failure (Knoll and Nichols 2003). See Stangl and Berry (2003) for more examples of decision theory in medicine.

13.3   UTILITY OF NON-MONETARY CONSEQUENCES

Decision analyses often must take into account both monetary consequences and non-monetary ones such as quality of life. How does one assign a dollar value to loss of mobility, for example? In such cases it is necessary to quantify all consequences, whether monetary or not, on a common numerical scale called *utility*. Units of measurement on the utility scale are called *utiles*, just as units of money are called dollars (or pounds or euros). The task of expressing all sorts of consequences in terms of utiles begins with making a list of all of the possible consequences in the decision tree (for example, in an analysis of alternative treatments for diabetes some consequences might be $100,000 in out-of-pocket medical expenses, loss of a limb, loss of 10 years of life, blindness, etc.). The first step is to identify the decision maker's worst imaginable outcome (perhaps blindness) and his or her best imaginable outcome (perhaps 25 more years of good health). The best outcome is assigned a utility value of 100 utiles and the worst is assigned a utility value of 0. Every other outcome, such as $50,000 in unreimbursed medical expenses is assigned a utile value by comparing it to a gamble between the best and worst outcomes.

For example, would you incur $50,000 in unreimbursed medical expenses to avoid a 50 : 50 gamble on good health vs. blindness? If the answer is yes, would you pay $50,000 to avoid a smaller risk of blindness, say a 40 : 60 gamble? As the odds of blindness are reduced, there comes a point at which the chance of blindness is so small that the decision maker is unwilling to spend $50,000 to avoid the risk. Perhaps he/she would be willing to pay $50,000 to avoid a 5% chance of blindness but not to avoid any smaller risk of blindness. That means that incurring $50,000 in unreimbursed medical expenses is worth 95 utiles. Similarly, the decision maker might be willing to lose a kidney to avoid a 10% chance of blindness, meaning that losing a kidney is worth 90 utiles. We'll use the symbol ℧ to identify a cost (or gain) expressed in utiles, so for example on a scale where blindness is worth ℧0 and 25 years of good health is worth ℧100, $50,000 in unreimbursed medical expenses is worth ℧95 and losing one kidney is worth ℧90.

The utile values of all possible consequences in a particular decision problem are used in place of dollar values to determine the best decisions, using the backward induction method illustrated in Section 13.2. Thus the fundamental principle of rational decision theory is to maximize *subjective expected utility* (SEU). For a rigorous derivation of subjective probability using utiles rather than dollars to define fair prices see Savage (1972).

13.4 DECIDING ON THE RIGHT SAMPLE SIZE

Statisticians and biomedical scientists are rarely involved in large-scale decision analysis problems; however, there is one decision problem that is faced by all investigators: "How big should my sample be?" The fully Bayesian (i.e. coherent) approach to that question is to frame it as a decision problem. However, despite decades of writing on the topic, no such methods are in general use, because of the perception that it is difficult if not impossible to assign costs to consequences. For that reason, we will present an example of the full decision-theoretic approach to sample-size determination and then review ad hoc approaches. For a recent review of Bayesian sample size selection, see Lindley (1997) and Adcock (1997).

13.4.1 Noninferiority Studies

In many cases it is necessary to demonstrate experimentally that a new treatment (surgery, drug, medical device, etc.) is not clinically inferior to currently available treatments. Inferiority is often defined as having a substantially lower success rate than existing treatments for the same condition. Success might be defined as a cure or, for chronic conditions, as a lack of complications following treatment. Since the treatment is new, its true success rate is an unknown quantity, symbolized p. On the other hand, the success rate of currently available treatments, the clinically expected success rate, is a known quantity symbolized p_1; for example, the clinically expected success rate might be 96%. Clinicians who might use the new device are assumed to be willing to tolerate a small deviation, called the *clinical tolerance* (symbol: Δ). For example, the clinical tolerance might be 4%, meaning that the new treatment is *inferior* if its success rate is less than 92%, *noninferior* if its success rate exceeds 92%, and *superior* if its success rate exceeds 96%. The goal of a noninferiority study is to decide whether the new procedure is noninferior ($p > p_1 - \Delta$) or inferior ($p < p_1 - \Delta$)

Fig. 13.5 shows the relevant sentences (hypotheses). In conventional statistical terminology the sentence "the new treatment is inferior" ($p < 0.92$) is called the *null hypothesis* (symbol: H_0) and the sentence "the new treatment is non inferior" ($p > 0.92$) is called the *alternative hypothesis* (symbol: H_1). A better terminology might be to call the first sentence the *default hypothesis* or *presumptive hypothesis*, because in the absence of additional evidence it is presumed to be true. The second sentence might appropriately be called the *research hypothesis* or *burdened hypothesis*, because the investigator has the burden of developing research evidence for that hypothesis. In plain terms, the new treatment is presumed inferior until proven otherwise. We will shortly see that the burden of proof arises from the fact that an incorrect decision in favor of the research hypothesis can be much more serious than an incorrect decision if favor of the default hypothesis. In conventional statistical terminology, the former is called a Type I error and the latter is called a Type II error. Committing a Type I error means that an inferior treatment is approved for human use, whereas committing a Type II error means that what would have been an acceptable or even a superior treatment was not approved for human use. Intuitively, the way to avoid a costly Type I error is to presume that the treatment is inferior and to insist on a high standard of proof that it is not inferior; this section fleshes out the mathematical details of that intuition.

Data for a noninferiority study are obtained by treating a sample of n patients with the new treatment. The information derived from the study is completely summarized by the posterior distribution of the success rate p given the data.

In principle the sample size could be any number between 1 and infinity; however, it is unlikely that data from a single subject would be persuasive, and it would not be feasible to apply a new experimental procedure to more than a few

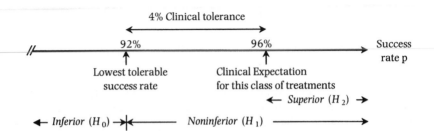

Figure 13.5 Hypotheses in a noninferiority study of a new treatment. The issue is whether the new treatment is inferior (intolerably lower than clinical expectation) or not.

hundred subjects for a complex procedure or to a few thousand for a simple procedure such as mass inoculation. The first decision, therefore, is to choose a sample size n typically somewhere between 1 and 1000. After the n subjects have beet treated with the new procedure and the numbers of treatment failures (f) and successes (s) have been counted, the second step is to decide whether the new procedure is inferior or noninferior, based on the information provided by the sample. Thus there are two decision points: the choice of a sample size n and the so-called *terminal decision* whether the new treatment is inferior or noninferior.

In order to analyze this decision problem, it is necessary to approximate the consequences (costs or losses) that would result from an inappropriate terminal decision. Two simple approximations to the cost structure have been proposed are *linear* and *threshold* cost structures. In the threshold cost structure shown in Table 13.1 there are two kinds of costs: If the terminal decision is that the treatment is noninferior, then it will likely be approved for human use. However, if clinical experience eventually shows that the newly approved treatment is inferior, then there will be a cost of k_2 utiles; this cost has monetary and non-monetary components, such as compensation of injured patients, loss of reputation, and in some cases abandoning that line of business. On the other hand, if the final decision is that the treatment is inferior, then it will not be approved for human use. However, if the treatment would have proved to be superior to existing treatments, then there is a loss (Uk_1) of income from marketing the treatment and also a loss of benefit to patients who might have received the superior treatment. Although neither of these costs can be exactly known in advance, it is possible to estimate them in the same way that it is possible for a homeowner to estimate the cost of replacing a home destroyed by a fire even though that eventuality might never happen.

There is a third cost to be considered: the *cost of information*. Clearly, it would be nice to be able to continue studying the treatment until the success rate is known to a high degree of precision — perhaps by applying the treatment to 10,000 patients. This approach is not possible, for several reasons. First, the new treatment is experimental and it is ethically imperative to place as few patients at risk as possible; second, insurance carriers typically do not reimburse patients for experimental treatments, a cost that must be paid by the investigator; and third, there is the time value of information. The sooner the inferior–noninferior

Table 13.1   Threshold loss structure.

		True status of the treatment	
		Noninferior (H_1)	
	Inferior (H_0)	Neither	Superior (H_2)
Decision	($p < 0.92$)	($0.92 \leq p < 0.96$)	($0.96 \leq p$)
Inferior	0	0	Uk_1
Noninferior	Uk_2	0	0

decision is made, the sooner the treatment can be approved for general use (or abandoned). The cost, from all these sources, of treating a single experimental patient is designated Ʉc.

Fig. 13.6 shows a fragment of the decision tree. The first decision point (selection of the sample size) is labeled "n?". The branches leading out of that node represent all of the sample sizes that could be selected from $n=1$ to $n=1000$; however, only one of these ($n=100$) is shown in the diagram. The probabilistic node labeled "Failures?" represents the results of treating $n=100$ experimental patients; paths leading out of that node represent possible data that might be observed (from no failures to 100 failures). Only two of the 101 branches are shown in detail. The second decision point, labeled "Inferior?", is the terminal decision — whether the treatment is inferior or noninferior. The second probabilistic node, labeled p, represents the true but unknown success rate that the treatment would achieve if used in general practice; the paths leading out of this node lead to costs. For example, if the decision is "no," meaning not inferior, but the treatment is noninferior, then there is only the cost of treating $n=100$ experimental patients ($c \times n = c \times 100$); however, if the treatment is inferior ($p < 0.92$) then the cost (k_2) of erroneously approving an inferior treatment is added to the cost of information ($c \times n$).

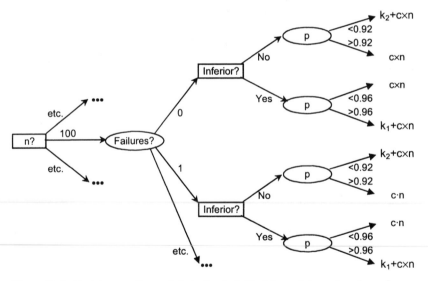

Figure 13.6   Decision tree for noninferiority study. Decision points are boxed, random points are circled. Final costs include per-subject cost (c) and costs of wrong decisions: (k_1) if an inferior treatment is pronounced noninferior, (k_2) if a superior treatment is pronounced inferior.

13.4.2   The Optimal Terminal Decision,

The first step in the decision problem is to compute the expected cost of each decision (yes or no) and select the decision with the lower expected cost. To make the problem concrete, suppose that $k_2/k_1 = 12$ and $c/k_1 = 0.0003$. Furthermore suppose that the decision maker's prior distribution of the true success rate p, based perhaps on a meta-analysis of similar treatments, is approximately beta(14, 1), the equivalent of observing 14 successes and 1 failure in 15 patients. Now suppose that the analyst decided to treat $n=100$ patients and that there were 99 successes and 1 failure, which means that the posterior distribution of p, given the experimental data and the prior information, is beta(113,2). There are too few failures for the normal approximation to be valid and it will be necessary to use the beta calculator (Fig. 6.13 on page 103). Using the beta calculator, we find that the posterior probability that the treatment is inferior is $P(\text{Inferior} \mid \text{data}) = P(p<0.92 \mid \text{data}) = 0.000812$. The probability that the treatment is noninferior is the complement, $1-0.000812 = 0.999188$.

Fig. 13.7 focuses on the decision node following the observation of 1 failure in a sample of $n=100$ experimental patients The probabilities in panel I were computed from the posterior beta(113,2) distribution and the expected costs (ECost) in panel II are computed using the expected value formula (Equation (3.14) on page 42). In this case the lower-cost decision is to conclude

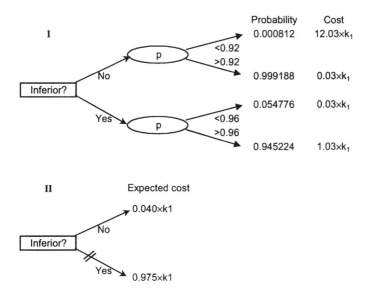

Figure 13.7   Terminal decision analysis following one failure in 100 patients. The decision with the lowest posterior expected cost is to conclude that the treatment is noninferior (Expected cost $= 0.04 \times k_1$). The path not taken is crossed off.

that the treatment is noninferior. Clearly, the decision depends only on the relative costs $k_2/k_1 = 12$ and $c/k_1 = 0.0003$.

This sort of calculation is repeated for each possible number of failures from 0 to 100, as partly shown in Fig. 13.8. Details of the calculation for 1 failure are explained above. If there 3 or fewer failures, the lower cost decision is "no" (i.e., the treatment is not inferior) but for 4 or more failures, the best decision is "yes" (i.e., the treatment is inferior). Three failures is said to be the *critical value*, meaning that if there are more than 3 failures, then the treatment is judged to be inferior. It might in truth not be inferior, but a decision has to be made and this decision rule (inferior if 3 or more failures) is the best procedure given this prior distribution and cost structure. A different cost structure and/or different prior information might raise or lower the critical value.

13.4.3   The Preposterior Distribution of Future Observations

The next step is to compute the cost of the decision to treat a sample of $n=100$ patients rather than some other number. Since that decision leads to the "Failures?" node in Fig. 13.8, what is required is to compute the expected value of that node. Since it is a probabilistic node, its expected value is the sum of probabilities times values (see Equation (3.14) on page 42). The values in question are displayed in Fig. 13.8 (the value associated with 0 failures is 0.031,

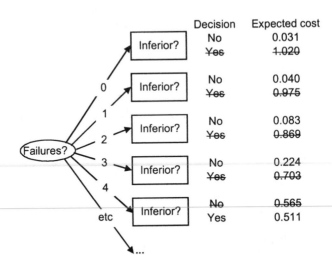

Figure 13.8   Posterior expected costs and best terminal decisions for $n=100$ patients. If there are 3 or fewer failures, the best decision is that the treatment is not inferior; with 4 or more failures, the best decision is that the treatment is inferior.

the value associated with 1 failure is 0.040, etc.). However, the probabilities are obtained as a by-product of Bayes' rule and require a bit of work. The reader should take a moment to return to Section 5.2, page 66, and in particular the top panel of Fig. 5.7 on page 69. Up to this point we have mainly been interested in the posterior probabilities; however, what is required in this case is the probability of data, $P(D)$, which is found at the foot of the "Joint" column in Fig. 5.7. This is the prior probability of the data before the experiment is run and for that reason is called a *predictive probability*. The computing formula for the predictive probability of getting s successes and f failures, when the decision maker's prior knowledge of the success rate is equivalent to s_0 successes and f_0 failures is

$$P(s,f) = \cfrac{\cfrac{(s+s_0-1)\cdot(s+s_0-2)\cdots s_0}{s\cdot(s-1)\cdots 1} \cdot \cfrac{(f+f_0-1)\cdot(f+f_0-2)\cdots f_0}{f\cdot(f-1)\cdots 1}}{\cfrac{(n+n_0-1)\cdot(n+n_0-2)\cdots n_0}{n\cdot(n-1)\cdots 1}} \qquad (13.1)$$

where n is the sample size ($n = s + f$) and n_0 is the effective prior sample size ($n_0 = s_0 + f_0$). In this example, we assume that the decision maker's prior knowledge of the true success rate was equivalent to observing 14 successes and 1 failure, so, for example, the predictive probability of 99 successes and 1 failure is

$$P(s = 99, f = 1) = \cfrac{\cfrac{112\cdot 111\cdots 14}{99\cdot 98\cdots 1}\cdot\cfrac{1}{1}}{\cfrac{114\cdot 113\cdots 15}{100\cdot 99\cdots 1}} = \frac{100\cdot 14}{114\cdot 113} = 0.1087 \qquad (13.2)$$

This equation was used to compute the probabilities in Table 13.2, which shows the first six and last three terms in the calculation of the prior expected cost of a decision based on data from $n=100$ patients. Computation of the posterior expected costs was illustrated in Fig. 13.7, and their position in the decision tree was shown in Fig. 13.8. The prior expected cost is 0.611; however, this number is meaningful only in the context of the prior expected costs for other sample sizes; therefore, the last step in the decision analysis is to graph the prior expected cost vs. sample size and identify the sample with the smallest prior expected cost.

13.4.4 Optimal Sample Size

We now know that the prior expected cost of basing the terminal decision on a sample of $n=100$ patients is 0.126. In order to select the most economical sample size (the one with lowest prior expected cost) it is necessary to repeat the above calculations for every feasible sample size (perhaps 1 to 1000) and select

Table 13.2   Prior expected cost of decision based on data from n=100 Patients

Number of failures	Predictive probability (P)	Posterior expected costs (V)	$P \times V$
0	0.123	0.031	0.004
1	0.109	0.040	0.004
2	0.096	0.083	0.008
3	0.085	0.224	0.019
4	0.075	0.511	0.038
5	0.066	0.336	0.022
⋮	⋮	⋮	⋮
98	0.000	0.030	0.000
99	0.000	0.030	0.000
100	0.000	0.030	0.000
		Prior expected cost = sum of products:	0.126

the sample size with the lowest cost. A SAS program called RatePrePost.sas that does this calculation is available for downloading from the textbook website. Inputs to the program are the prior distribution, the range of sample sizes to be considered, the cost of approving an inferior treatment relative to the cost of failing to approve a superior treatment ($K=k_2/k_1$), and the cost of treating one patient relative to the cost of failing to approve a superior treatment ($C=c/k_1$). General instructions for running the program are in the program file. RatePrePost.sas was used to compute prior expected costs in the noninferiority example discussed above for sample sizes between 100 and 300 (see Fig. 13.9). The computer-generated value at $n=100$ agrees exactly with the hand computations culminating in Table 13.2 above, and the minimum cost is seen to occur at a sample size of $n=202$ patients. The critical value is 9 or fewer failures in 202 patients (a 95.5% or better success rate); if there are more than 8 failures, the treatment will be declared noninferior (see Table 13.3).

It ought to be clear from this example that selecting the "best" sample size is not simple. The analyst is required to (1) specify a prior distribution, (2) specify costs of making incorrect decisions (or at least the ratio of the costs), and (3) specify the cost of obtaining an observation. In the author's experience biomedical researchers find it easy to specify a prior distribution, fairly easy to compute the cost of obtaining an observation, and very difficult to specify potential costs of wrong decisions. Furthermore, when the objective of the research is to learn about nature rather than to make a decision, there is still some interest in selecting an adequate sample size. For these reasons, less formal sample size criteria have been proposed.

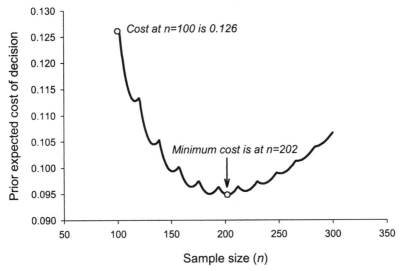

Figure 13.9 Prior expected cost of decisions as a function of sample size. The cost at $n=100$ is computed in Table 13.2. The best (minimum-cost) sample size is $n=202$.

13.5 INFORMAL SAMPLE SIZE SELECTION

The most commonly used informal sample size criteria are (1) controlling the expected width of the posterior credible interval and (2) controlling the probability of making an incorrect decision. The expected width criterion is presented here.

Table 13.3 Critical values[a] for the noninferiority example

n	Critical Value	n	Critical Value
100-102	3	194-211	9
102-120	4	212-229	10
121-139	5	230-247	11
140-157	6	248-265	12
158-175	7	266-283	13
176-193	8	284-300	14

a. The maximum number of failures allowed for a conclusion of noninferiority.
This table is derived from the output of RatePrePost.sas.

13.5.1 Controlling Credible Interval Width

We'll consider situations in which there is a good normal approximation to the posterior distribution. This includes credible intervals for means and differences of means, success rates, differences of success rates, log relative risks, log odds ratios (provided there are at least 10 successes and failures in all samples), and logistic, linear, and proportional hazards regression coefficients. The generic form of a 95% normal credible interval for an unknown quantity θ is $\hat{\theta} \pm 1.96 \times SE$, where the estimate $\hat{\theta}$ is the μ of the posterior distribution of θ, and its standard error (SE) is the σ of the posterior distribution; consequently, the width of the posterior credible interval is $2 \times 1.96 \times SE$.

It is of course desirable to have as short a credible interval as is feasible, since that narrows the range of possible values of the unknown quantity. Suppose that the investigator would like a credible interval of width W_0. In other words, the goal is to select a sample size for which $2 \times 1.96 \times SE < W_0$. That goal has been interpreted in two ways: either (1) select a sample size which guarantees than $2 \times 1.96 \times SE < W_0$ is true with high prior probability, or (2) select a sample size for which the prior expected value of the width is less than W_0. The problem is that the sample size must be selected before the data are obtained (because the data come from the sample), and SE is a function of the data. SE, in short, is an unknown quantity. However, like any unknown quantity, it has a prior distribution, called the predictive distribution or *preposterior distribution*. The term preposterior is a combination of "prior" and "posterior" and refers to the fact that we are making a prior prediction of a quantity that will be known only after (posterior to) observing the data.

The expected length criterion is not coherent and can produce irrational and counterintuitive results, particularly if the unknown quantity is not a so-called *natural parameter*. Natural parameters for Bernoulli rates and contrasts are as follows:

Unknown quantity	Symbol	Natural parameter
Bernoulli rate	p	$\theta = \ln\left(\text{odds}\right) = \ln\left(\dfrac{p}{1-p}\right)$
Two Bernoulli rates	p_T, p_C	$\theta = \ln\left(\text{odds ratio}\right) = \ln\left(\dfrac{p_T/(1-p_T)}{p_C/(1-p_C)}\right)$

(13.3)

Means, differences of means, adjusted means, adjusted mean differences, and regression coefficients are themselves natural parameters and require no reexpression.

13.5.2 Estimating a Bernoulli Success Rate

Even though the investigator might think he or she has substantial prior knowledge about the success rate of a new treatment, it is a good idea to select a sample size that would yield data likely to be persuasive to a decision maker with a different prior. Suppose therefore that the investigator's prior distribution for the unknown success rate p is approximated by a beta(f_0, s_0) distribution (i.e. it is equivalent to n_0 virtual prior observations consisting of f_0 failures and s_0 successes) and that the decision maker's prior is approximated by the beta(f_1, s_1). The restriction on the length of the posterior credible interval should be imposed on the natural parameter; that is, on the log odds (or logit) scale, $\ln(p/(1-p))$ rather than the scale of p. On that scale, the standard error after observing n real observations consisting of s successes and f failures, the decision maker's standard error would be

$$SE \cong \sqrt{\frac{1}{s+s_1} + \frac{1}{f+f_1}} \qquad (13.4)$$

The prior expected value of SE cannot be computed exactly, but a computable upper bound is available in the spreadsheet EWidth.xls, which is available for downloading from the textbook website. Fig. 13.11 shows how to use this program to calculate the investigator's expectation of the width of the decision maker's posterior 95% credible interval. The sample size $n=1323$ was found by trial and error to be the smallest n for which the expected width of the posterior CI was 0.74 or less. The value 0.74 is the clinical tolerance expressed on the log odds scale (see Fig. 13.10).

The sample size is over six times larger than the optimal sample size found in Fig. 13.9. There are several reasons for this. First, the investigator did not have

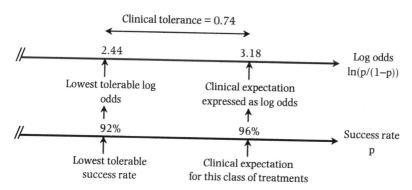

Figure 13.10   Clinical tolerance in the noninferiority study expressed in terms of log-odds. The sample size will be selected to produce a credible interval for the log odds with expected width less than 0.74.

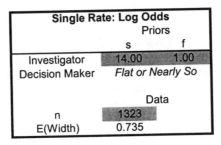

Figure 13.11   EWidth.xls: An investigator with prior knowledge equivalent to observing 14 successes and 1 failure needs a sample of 1323 patients to reduce the prior expected CI width to clinical tolerance 0.74.

precise prior information about the success rate; if, for example, his prior information had been equivalent to 2 failures and 24 successes, then the required sample size would drop to about 600. The second reason the sample size is so large is that the length criterion (0.73) is quite stringent. As a rule of thumb setting the width to $\sqrt{2}$ times the clinical tolerance produces roughly the sample size recommended for conventional statistical inference. For example, in Pagano and Gauvreau (2000, Section 14.5), the formula for n using the commonly recommended design values $\alpha = 0.025$ and $\beta = 0.20$ yields $n=303$. On the other hand, the expected length criterion with $W_0 = \sqrt{2} \cdot 0.74 \approx 1.04$ and a prior effective sample of 2 failures and 24 successes requires a sample size of about 350 (see Fig. 13.12).

Since the expected length criterion is incoherent, setting it to $\sqrt{2}$ times the clinical tolerance seems as good a choice as any and would provide some continuity with conventional practice. However, conventional practice is somewhat ritualized and is certainly not a gold standard. Therefore it is desirable that practitioners become used to thinking in terms of minimizing preposterior

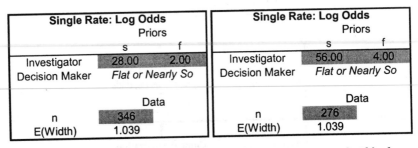

Figure 13.12   From EWidth.xls: calculation of investigator's prior expected width of decision maker's 95% credible interval for a Bernoulli log odds (logit). In each panel the investigator believes the most likely value of the rate is 14/15 or 93.3%. On the left the investigator is less certain about his prior knowledge and as a consequence must use a larger sample. size to achieve the desired expected width, 1.04.

expected utility as the criterion for rational sample size selection as was illustrated in Section 13.4.

When the decision maker does not have strong prior beliefs (f_1 and are s_1 small) and the investigator has a very strong prior opinion about the success rate (f_0 and s_0 are large), the prior expectation of the standard error reduces to

$$E(\text{SE}) \cong \sqrt{\frac{1}{n \cdot p_0 \cdot (1 - p_0)}}, \quad \text{where} \quad p_0 = \frac{s_0}{n_0} \qquad (13.5)$$

Thus, if the expected width, $2 \times 1.96 \times E(\text{SE})$, is to be smaller than some upper bound, W_0, then we must have,

$$2 \cdot 1.96 \cdot \sqrt{\frac{1}{n \cdot p_0 \cdot (1 - p_0)}} \leq W_0 \qquad (13.6)$$

which, after a little algebra, reduces to

$$n \geq \frac{15.4}{p_0 \cdot (1 - p_0) \cdot W_0^2} \qquad (13.7)$$

For example if $W_0 = 1.03$, and $p0 = 0.9333$, then n must be at least 235 to obtain an expected width of 1.03.

This discussion should make it clear that the sample size suggested by an ad hoc criterion such as expected CI width is somewhat sensitive to exactly how the calculation is tweaked, and since the criterion is ad hoc there is no principle by which to select the desired expected width W_0. Equation (13.7) with $W_0 = \sqrt{2} \cdot CT$ (where CT is the clinical tolerance) gives roughly the sample sizes recommended in conventional statistical practice. However, conventional practice is certainly not a gold standard, and apart from a general agreement that decision theory is the only gold standard for sample size selection, it is probably fair to say that Bayesian statisticians have not reached anything like consensus on how to implement it in practice. For a recent review of the debate, see Adcock (1997) and Lindley (1997).

13.6 EXERCISES

13.1 In Section 13.2, assume that drug B costs $50 and that the diagnostic test costs $30. Determine the best treatment strategy, and compute its prior expected cost.

13.2 Repeat the calculation assuming that the tests costs $35. Did the strategy change? Explain why. Since the test involves spending money to purchase data, your answer should speak to the value of your purchase as opposed to its cost.

13.3 Why is it necessary to introduce the idea of utility in complex decision problems such as deciding whether a woman with a prior cesarean section should attempt labor in a subsequent pregnancy or opt for a second cesarean section? Suppose that the decision maker would be willing to spend $30,000 to avoid a 3% risk of the worst outcome. What is the utility of the best and worst outcomes and the utility of $30,000?

13.4 In Fig. 13.9 and Table 13.3 the clinical tolerance is the interval 0.92 to 0.96, the relative cost of approving an inferior treatment is $K = k_2/k_1 = 12$, and the relative cost of obtaining data from one subject is $C = c/k_1 = 0.0003$. Intuitively, it would seem that a more liberal clinical tolerance, say from 0.90 to 0.96, should require a smaller number of patients. (Certainly, if the clinical tolerance were from 0.00 to 1.00, no patients at all would be needed.) Verify this by running RatePrePost.sas with $K = 12$, $C = 0.0003$, but the more liberal clinical tolerance. What is the optimal n and what is the critical value?

13.5 Referring to Exercise 13.4, intuition suggests that if subjects are cheaper to run, then the investigator will run more subjects. Verify this by running RatePrePost.sas with tolerance interval 0.92 to 0.96, $K = 12$, but with a lower cost per subject, $C = 0.0002$. What is the optimal n and what is the critical value?

13.6 Referring to Exercise 13.4, compute the clinical tolerance (CT) in terms of log odds as shown in Fig. 13.10. Suppose that the desired expected width of the posterior 95% credible interval is $W_0 = \sqrt{2} \cdot CT$ and suppose that the decision maker has a flat prior and the analyst has a strong prior belief that the success rate is $p_0 = 0.93$. Compute the smallest sample size needed to achieve the expected width using Equation (13.7).

13.7 Referring to Exercise 13.6, suppose that the decision maker's prior is flat, beta(1, 1), and the analyst does not have a strong opinion; say his prior is beta(1, 5). Intuition would suggest that his vagueness will result in a larger sample size. Use EWidth.xls to discover the smallest sample size needed to achieve the expected width criterion you calculated in exercise 13.6.

13.8 No matter how confident the analyst is, it is the decision maker who must be convinced. Use EWidth.xls to demonstrate this point. Hint: Keep the planned sample size the same ($n = 281$), and continue doubling the investigator's prior s and f in Fig. 13.12. After a certain point the expected width doesn't change much. How does that make the point that the investigator's certainty doesn't relieve him of the burden of providing data to convince the decision maker.

13.7   REFERENCES

Adcock, CJ., "Sample size determination: a review," *The Statistician*, Vol. 46, No. 2 (1997), pp. 261–283.

Knoll GA., and Nichol G., "Dialysis, kidney transplantation, or pancreas transplantation for patients with diabetes mellitus and renal failure a decision analysis of treatment options," *Journal of the American Society of Nephrology*, Vol. 14, No. 2 (Feb. 2002), pp. 500–515.

Lindley, DV., "The choice of sample size," *The Statistician*, Vol. 46, No.2 (1997), pp. 129–138.

Mankuta, D.D., Leshno, M.M., Menasche, M.M., Brezis, M.M., "Vaginal birth after cesarean section trial of labor or repeat cesarean section? A decision analysis," *American Journal of Obstetrics & Gynecology*, Vol. 189, No. 3, (Sep. 2003),   pp. 714–719.

Pagano, M. and Gauvreau, K. (2000), *Principles of Biostatistics, 2nd ed.*, Pacific Grove, CA: Duxbury.

Savage, LJ. (1972) *The Foundations of Statistics*, 2nd rev. ed. New York: Dover.

Stangl, D. and Berry, D., "Bayesian Statistics in Medicine: Where Are We and Where Should We Be Going?" Duke University, Institute of Statistics and Decision Sciences, Discussion Paper Series, paper 97-20, http://isds.duke.edu/cgi-bin/d.pl, 2003.

Wisnivesky, J.P., Mushlin, A.I., Sicherman, N., Henschke, C., "The cost-effectiveness of low-dose CT screening for lung cancer preliminary results of baseline screening," *Chest*, Vol. 124, No. 2 (Aug. 2003), pp. 614–21.

Appendix   A

Tables

A.1 Left tail areas of the standard normal distribution

Z	0.00	0.01	0.02	0.03	0.04	0.05	0.06	0.07	0.08	0.09
−3.9	0.0000	0.0000	0.0000	0.0000	0.0000	0.0000	0.0000	0.0000	0.0000	0.0000
−3.8	0.0001	0.0001	0.0001	0.0001	0.0001	0.0001	0.0001	0.0001	0.0001	0.0001
−3.7	0.0001	0.0001	0.0001	0.0001	0.0001	0.0001	0.0001	0.0001	0.0001	0.0001
−3.6	0.0002	0.0002	0.0001	0.0001	0.0001	0.0001	0.0001	0.0001	0.0001	0.0001
−3.5	0.0002	0.0002	0.0002	0.0002	0.0002	0.0002	0.0002	0.0002	0.0002	0.0002
−3.4	0.0003	0.0003	0.0003	0.0003	0.0003	0.0003	0.0003	0.0003	0.0003	0.0002
−3.3	0.0005	0.0005	0.0005	0.0004	0.0004	0.0004	0.0004	0.0004	0.0004	0.0003
−3.2	0.0007	0.0007	0.0006	0.0006	0.0006	0.0006	0.0006	0.0005	0.0005	0.0005
−3.1	0.0010	0.0009	0.0009	0.0009	0.0008	0.0008	0.0008	0.0008	0.0007	0.0007
−3.0	0.0013	0.0013	0.0013	0.0012	0.0012	0.0011	0.0011	0.0011	0.0010	0.0010
−2.9	0.0019	0.0018	0.0018	0.0017	0.0016	0.0016	0.0015	0.0015	0.0014	0.0014
−2.8	0.0026	0.0025	0.0024	0.0023	0.0023	0.0022	0.0021	0.0021	0.0020	0.0019
−2.7	0.0035	0.0034	0.0033	0.0032	0.0031	0.0030	0.0029	0.0028	0.0027	0.0026
−2.6	0.0047	0.0045	0.0044	0.0043	0.0041	0.0040	0.0039	0.0038	0.0037	0.0036
−2.5	0.0062	0.0060	0.0059	0.0057	0.0055	0.0054	0.0052	0.0051	0.0049	0.0048
−2.4	0.0082	0.0080	0.0078	0.0075	0.0073	0.0071	0.0069	0.0068	0.0066	0.0064
−2.3	0.0107	0.0104	0.0102	0.0099	0.0096	0.0094	0.0091	0.0089	0.0087	0.0084
−2.2	0.0139	0.0136	0.0132	0.0129	0.0125	0.0122	0.0119	0.0116	0.0113	0.0110
−2.1	0.0179	0.0174	0.0170	0.0166	0.0162	0.0158	0.0154	0.0150	0.0146	0.0143
−2.0	0.0228	0.0222	0.0217	0.0212	0.0207	0.0202	0.0197	0.0192	0.0188	0.0183
−1.9	0.0287	0.0281	0.0274	0.0268	0.0262	0.0256	0.0250	0.0244	0.0239	0.0233
−1.8	0.0359	0.0351	0.0344	0.0336	0.0329	0.0322	0.0314	0.0307	0.0301	0.0294
−1.7	0.0446	0.0436	0.0427	0.0418	0.0409	0.0401	0.0392	0.0384	0.0375	0.0367
−1.6	0.0548	0.0537	0.0526	0.0516	0.0505	0.0495	0.0485	0.0475	0.0465	0.0455
−1.5	0.0668	0.0655	0.0643	0.0630	0.0618	0.0606	0.0594	0.0582	0.0571	0.0559
−1.4	0.0808	0.0793	0.0778	0.0764	0.0749	0.0735	0.0721	0.0708	0.0694	0.0681
−1.3	0.0968	0.0951	0.0934	0.0918	0.0901	0.0885	0.0869	0.0853	0.0838	0.0823
−1.2	0.1151	0.1131	0.1112	0.1093	0.1075	0.1056	0.1038	0.1020	0.1003	0.0985
−1.1	0.1357	0.1335	0.1314	0.1292	0.1271	0.1251	0.1230	0.1210	0.1190	0.1170
−1.0	0.1587	0.1562	0.1539	0.1515	0.1492	0.1469	0.1446	0.1423	0.1401	0.1379
−0.9	0.1841	0.1814	0.1788	0.1762	0.1736	0.1711	0.1685	0.1660	0.1635	0.1611
−0.8	0.2119	0.2090	0.2061	0.2033	0.2005	0.1977	0.1949	0.1922	0.1894	0.1867
−0.7	0.2420	0.2389	0.2358	0.2327	0.2296	0.2266	0.2236	0.2206	0.2177	0.2148
−0.6	0.2743	0.2709	0.2676	0.2643	0.2611	0.2578	0.2546	0.2514	0.2483	0.2451
−0.5	0.3085	0.3050	0.3015	0.2981	0.2946	0.2912	0.2877	0.2843	0.2810	0.2776
−0.4	0.3446	0.3409	0.3372	0.3336	0.3300	0.3264	0.3228	0.3192	0.3156	0.3121
−0.3	0.3821	0.3783	0.3745	0.3707	0.3669	0.3632	0.3594	0.3557	0.3520	0.3483
−0.2	0.4207	0.4168	0.4129	0.4090	0.4052	0.4013	0.3974	0.3936	0.3897	0.3859
−0.1	0.4602	0.4562	0.4522	0.4483	0.4443	0.4404	0.4364	0.4325	0.4286	0.4247
−0.0	0.5000	0.4960	0.4920	0.4880	0.4840	0.4801	0.4761	0.4721	0.4681	0.4641

A.1 Left tail areas of the standard normal distribution

Z	0.00	0.01	0.02	0.03	0.04	0.05	0.06	0.07	0.08	0.09
0.0	0.5000	0.5040	0.5080	0.5120	0.5160	0.5199	0.5239	0.5279	0.5319	0.5359
0.1	0.5398	0.5438	0.5478	0.5517	0.5557	0.5596	0.5636	0.5675	0.5714	0.5753
0.2	0.5793	0.5832	0.5871	0.5910	0.5948	0.5987	0.6026	0.6064	0.6103	0.6141
0.3	0.6179	0.6217	0.6255	0.6293	0.6331	0.6368	0.6406	0.6443	0.6480	0.6517
0.4	0.6554	0.6591	0.6628	0.6664	0.6700	0.6736	0.6772	0.6808	0.6844	0.6879
0.5	0.6915	0.6950	0.6985	0.7019	0.7054	0.7088	0.7123	0.7157	0.7190	0.7224
0.6	0.7257	0.7291	0.7324	0.7357	0.7389	0.7422	0.7454	0.7486	0.7517	0.7549
0.7	0.7580	0.7611	0.7642	0.7673	0.7704	0.7734	0.7764	0.7794	0.7823	0.7852
0.8	0.7881	0.7910	0.7939	0.7967	0.7995	0.8023	0.8051	0.8078	0.8106	0.8133
0.9	0.8159	0.8186	0.8212	0.8238	0.8264	0.8289	0.8315	0.8340	0.8365	0.8389
1.0	0.8413	0.8438	0.8461	0.8485	0.8508	0.8531	0.8554	0.8577	0.8599	0.8621
1.1	0.8643	0.8665	0.8686	0.8708	0.8729	0.8749	0.8770	0.8790	0.8810	0.8830
1.2	0.8849	0.8869	0.8888	0.8907	0.8925	0.8944	0.8962	0.8980	0.8997	0.9015
1.3	0.9032	0.9049	0.9066	0.9082	0.9099	0.9115	0.9131	0.9147	0.9162	0.9177
1.4	0.9192	0.9207	0.9222	0.9236	0.9251	0.9265	0.9279	0.9292	0.9306	0.9319
1.5	0.9332	0.9345	0.9357	0.9370	0.9382	0.9394	0.9406	0.9418	0.9429	0.9441
1.6	0.9452	0.9463	0.9474	0.9484	0.9495	0.9505	0.9515	0.9525	0.9535	0.9545
1.7	0.9554	0.9564	0.9573	0.9582	0.9591	0.9599	0.9608	0.9616	0.9625	0.9633
1.8	0.9641	0.9649	0.9656	0.9664	0.9671	0.9678	0.9686	0.9693	0.9699	0.9706
1.9	0.9713	0.9719	0.9726	0.9732	0.9738	0.9744	0.9750	0.9756	0.9761	0.9767
2.0	0.9772	0.9778	0.9783	0.9788	0.9793	0.9798	0.9803	0.9808	0.9812	0.9817
2.1	0.9821	0.9826	0.9830	0.9834	0.9838	0.9842	0.9846	0.9850	0.9854	0.9857
2.2	0.9861	0.9864	0.9868	0.9871	0.9875	0.9878	0.9881	0.9884	0.9887	0.9890
2.3	0.9893	0.9896	0.9898	0.9901	0.9904	0.9906	0.9909	0.9911	0.9913	0.9916
2.4	0.9918	0.9920	0.9922	0.9925	0.9927	0.9929	0.9931	0.9932	0.9934	0.9936
2.5	0.9938	0.9940	0.9941	0.9943	0.9945	0.9946	0.9948	0.9949	0.9951	0.9952
2.6	0.9953	0.9955	0.9956	0.9957	0.9959	0.9960	0.9961	0.9962	0.9963	0.9964
2.7	0.9965	0.9966	0.9967	0.9968	0.9969	0.9970	0.9971	0.9972	0.9973	0.9974
2.8	0.9974	0.9975	0.9976	0.9977	0.9977	0.9978	0.9979	0.9979	0.9980	0.9981
2.9	0.9981	0.9982	0.9982	0.9983	0.9984	0.9984	0.9985	0.9985	0.9986	0.9986
3.0	0.9987	0.9987	0.9987	0.9988	0.9988	0.9989	0.9989	0.9989	0.9990	0.9990
3.1	0.9990	0.9991	0.9991	0.9991	0.9992	0.9992	0.9992	0.9992	0.9993	0.9993
3.2	0.9993	0.9993	0.9994	0.9994	0.9994	0.9994	0.9994	0.9995	0.9995	0.9995
3.3	0.9995	0.9995	0.9995	0.9996	0.9996	0.9996	0.9996	0.9996	0.9996	0.9997
3.4	0.9997	0.9997	0.9997	0.9997	0.9997	0.9997	0.9997	0.9997	0.9997	0.9998
3.5	0.9998	0.9998	0.9998	0.9998	0.9998	0.9998	0.9998	0.9998	0.9998	0.9998
3.6	0.9998	0.9998	0.9999	0.9999	0.9999	0.9999	0.9999	0.9999	0.9999	0.9999
3.7	0.9999	0.9999	0.9999	0.9999	0.9999	0.9999	0.9999	0.9999	0.9999	0.9999
3.8	0.9999	0.9999	0.9999	0.9999	0.9999	0.9999	0.9999	0.9999	0.9999	0.9999
3.9	1.0000	1.0000	1.0000	1.0000	1.0000	1.0000	1.0000	1.0000	1.0000	1.0000

A.2 Quantiles of the standard t distribution family

ν	Left tail area[a]											
	0.005	0.01	0.025	0.05	0.1	0.25	0.75	0.9	0.95	0.975	0.99	0.995
2	−9.92	-6.96	-4.30	-2.92	-1.89	-0.82	0.82	1.89	2.92	4.30	6.96	9.92
3	-5.84	-4.54	-3.18	-2.35	-1.64	-0.76	0.76	1.64	2.35	3.18	4.54	5.84
4	-4.60	-3.75	-2.78	-2.13	-1.53	-0.74	0.74	1.53	2.13	2.78	3.75	4.60
5	-4.03	-3.36	-2.57	-2.02	-1.48	-0.73	0.73	1.48	2.02	2.57	3.36	4.03
6	-3.71	-3.14	-2.45	-1.94	-1.44	-0.72	0.72	1.44	1.94	2.45	3.14	3.71
7	-3.50	-3.00	-2.36	-1.89	-1.41	-0.71	0.71	1.41	1.89	2.36	3.00	3.50
8	-3.36	-2.90	-2.31	-1.86	-1.40	-0.71	0.71	1.40	1.86	2.31	2.90	3.36
9	-3.25	-2.82	-2.26	-1.83	-1.38	-0.70	0.70	1.38	1.83	2.26	2.82	3.25
10	-3.17	-2.76	-2.23	-1.81	-1.37	-0.70	0.70	1.37	1.81	2.23	2.76	3.17
11	-3.11	-2.72	-2.20	-1.80	-1.36	-0.70	0.70	1.36	1.80	2.20	2.72	3.11
12	-3.05	-2.68	-2.18	-1.78	-1.36	-0.70	0.70	1.36	1.78	2.18	2.68	3.05
13	-3.01	-2.65	-2.16	-1.77	-1.35	-0.69	0.69	1.35	1.77	2.16	2.65	3.01
14	-2.98	-2.62	-2.14	-1.76	-1.35	-0.69	0.69	1.35	1.76	2.14	2.62	2.98
15	-2.95	-2.60	-2.13	-1.75	-1.34	-0.69	0.69	1.34	1.75	2.13	2.60	2.95
16	-2.92	-2.58	-2.12	-1.75	-1.34	-0.69	0.69	1.34	1.75	2.12	2.58	2.92
17	-2.90	-2.57	-2.11	-1.74	-1.33	-0.69	0.69	1.33	1.74	2.11	2.57	2.90
18	-2.88	-2.55	-2.10	-1.73	-1.33	-0.69	0.69	1.33	1.73	2.10	2.55	2.88
19	-2.86	-2.54	-2.09	-1.73	-1.33	-0.69	0.69	1.33	1.73	2.09	2.54	2.86
20	-2.85	-2.53	-2.09	-1.72	-1.33	-0.69	0.69	1.33	1.72	2.09	2.53	2.85
21	-2.83	-2.52	-2.08	-1.72	-1.32	-0.69	0.69	1.32	1.72	2.08	2.52	2.83
22	-2.82	-2.51	-2.07	-1.72	-1.32	-0.69	0.69	1.32	1.72	2.07	2.51	2.82

A.2 Quantiles of the standard t distribution family (Continued)

ν	Left tail area[a]											
	0.005	0.01	0.025	0.05	0.1	0.25	0.75	0.9	0.95	0.975	0.99	0.995
23	-2.81	-2.50	-2.07	-1.71	-1.32	-0.69	0.69	1.32	1.71	2.07	2.50	2.81
24	-2.80	-2.49	-2.06	-1.71	-1.32	-0.68	0.68	1.32	1.71	2.06	2.49	2.80
25	-2.79	-2.49	-2.06	-1.71	-1.32	-0.68	0.68	1.32	1.71	2.06	2.49	2.79
26	-2.78	-2.48	-2.06	-1.71	-1.31	-0.68	0.68	1.31	1.71	2.06	2.48	2.78
27	-2.77	-2.47	-2.05	-1.70	-1.31	-0.68	0.68	1.31	1.70	2.05	2.47	2.77
28	-2.76	-2.47	-2.05	-1.70	-1.31	-0.68	0.68	1.31	1.70	2.05	2.47	2.76
29	-2.76	-2.46	-2.05	-1.70	-1.31	-0.68	0.68	1.31	1.70	2.05	2.46	2.76
30	-2.75	-2.46	-2.04	-1.70	-1.31	-0.68	0.68	1.31	1.70	2.04	2.46	2.75
31	-2.74	-2.45	-2.04	-1.70	-1.31	-0.68	0.68	1.31	1.70	2.04	2.45	2.74
32	-2.74	-2.45	-2.04	-1.69	-1.31	-0.68	0.68	1.31	1.69	2.04	2.45	2.74
33–34	-2.73	-2.44	-2.03	-1.69	-1.31	-0.68	0.68	1.31	1.69	2.03	2.44	2.73
35	-2.72	-2.44	-2.03	-1.69	-1.31	-0.68	0.68	1.31	1.69	2.03	2.44	2.72
36	-2.72	-2.43	-2.03	-1.69	-1.31	-0.68	0.68	1.31	1.69	2.03	2.43	2.72
37	-2.72	-2.43	-2.03	-1.69	-1.30	-0.68	0.68	1.30	1.69	2.03	2.43	2.72
38–39	-2.71	-2.43	-2.02	-1.69	-1.30	-0.68	0.68	1.30	1.69	2.02	2.43	2.71
40–43	-2.70	-2.42	-2.02	-1.68	-1.30	-0.68	0.68	1.30	1.68	2.02	2.42	2.70
44	-2.69	-2.41	-2.02	-1.68	-1.30	-0.68	0.68	1.30	1.68	2.02	2.41	2.69
45–46	-2.69	-2.41	-2.01	-1.68	-1.30	-0.68	0.68	1.30	1.68	2.01	2.41	2.69
47–48	-2.68	-2.41	-2.01	-1.68	-1.30	-0.68	0.68	1.30	1.68	2.01	2.41	2.68
49–51	-2.68	-2.40	-2.01	-1.68	-1.30	-0.68	0.68	1.30	1.68	2.01	2.40	2.68
52–53	-2.67	-2.40	-2.01	-1.67	-1.30	-0.68	0.68	1.30	1.67	2.01	2.40	2.67
54–55	-2.67	-2.40	-2.00	-1.67	-1.30	-0.68	0.68	1.30	1.67	2.00	2.40	2.67
100	-2.63	-2.36	-1.98	-1.66	-1.29	-0.68	0.68	1.29	1.66	1.98	2.36	2.63
Normal	-2.58	-2.33	-1.96	-1.64	-1.28	-0.67	0.67	1.28	1.64	1.96	2.33	2.58

a.Row stubs are numbers of degrees of freedom, column heads are left tail areas, and entries in the body of the table are quantiles that cut off the specified tail area. Left tail areas not listed in the table can be computed with the spreadsheet tTailArea.xls available for download on the textbook website.

B

Introduction to WinBUGS

B.1 INTRODUCTION

WinBUGS (the MS Windows operating system version of BUGS: Bayesian Analysis Using Gibbs Sampling) is a versatile package that has been designed to carry out Markov chain Monte Carlo (MCMC) computations for a wide variety of Bayesian models. The software is currently distributed electronically from the BUGS Project website. The address is

http://www.mrc-bsu.cam.ac.uk/bugs/overview/contents.shtml

(click the WinBUGS link). If this address fails, a current link is maintained on the textbook website, or try the search words: "WinBUGS Gibbs" in a search engine. The downloaded software is restricted to fairly small models, but can be made fully functional by acquiring a license, currently for no fee, from the BUGS Project website. Versions of BUGS for other operating systems can be found at the BUGS project website.

The WinBUGS installation contains an extensive user manual (Spiegelhalter et al. 2002) and many completely worked examples. The manual and examples are under the "Help" pull-down menu on the main WinBUGS screen (Fig. B.1). The user manual is a detailed and helpful programming and syntax reference;

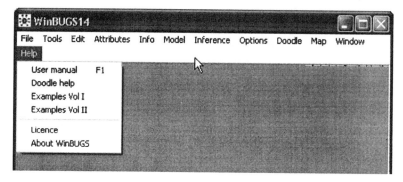

Figure B.1   The main WinBUGS screen, showing pull-down help menu.

305

however, the quickest way to become familiar with WinBUGS programming and syntax is to work through a few of the examples.

WinBUGS implements various MCMC algorithms to generate simulated observations from the posterior distribution of the unknown quantities (parameters or nodes) in the statistical model. The idea is that with sufficiently many simulated observations, it is possible to get an accurate picture of the distribution; for example, by displaying the simulated observations as a histogram as in Fig. 11.3 on page 232.

A WinBUGS analysis — model specification, data, initial values, and output — is contained in a single *compound document*. Analytic tools are available as pull-down menus and dialog boxes. Data files are entered as lists (or can be embedded as sub documents). Output is listed in a separate window but can be embedded in the compound document to help maintain a paper trail of the analysis. Any part of the compound document can be folded out of sight to make the document easier to work with. Data can be expressed in *list structures* or as rectangular tables in plain text format; however, WinBUGS cannot read data from an external file.

B.2   SPECIFYING THE MODEL — PRIOR AND LIKELIHOOD

To calculate a posterior distribution it is necessary to tell WinBUGS what prior distribution to use and what likelihood distribution to use. Distributions and likelihoods available in WinBUGS are listed in Table I of the WinBUGS user manual, and some of them are described in this section. Notice that all distribution and likelihood names begin with the letter "d" (for "distribution").

dnorm(μ, τ) is the normal distribution with parameters μ and $\tau = 1/\sigma^2$.

It is important to understand that WinBUGS specifies the normal distribution in terms of the mean μ and precision τ, rather than in terms of mean and standard deviation σ. The relationship between standard deviation and precision is $\sigma = 1/\sqrt{\tau}$. An important special case is dnorm(0, 0), which is flat over the entire number line. This distribution is *improper* in the sense that there is infinite area under the curve, and in practice a dnorm(0, ϵ) is used to represent ignorance, where ϵ is a small number such as 0.001.

dbin(p, *n*) is the binomial distribution with parameters *n* and p.

dbin is the distribution of the number of successes in *n* observations of a Bernoulli process with parameter p; for example the number of heads in 100 coin tosses has a dbin(0.5,100) distribution, and the number of black marbles in a sample of size *n* from a box in which the proportion of black marbles is p has a dbin(p, *n*) distribution.

dbeta(*a*, *b*) is the beta distribution with parameters *a* and *b*.

dbeta is a very flexible distribution family; it applies to an unknown quantity that takes values between 0 and 1 — for example, a success rate. An important special case is dbeta(1, 1), which is the uniform (flat) prior distribution over the interval (0,1). However, the dbeta(0, 0) distribution is more often used to represent complete ignorance about an unknown rate p because it implies that the log odds, $\ln(p/(1-p))$ has a uniform distribution over the entire number line. dbeta(0, 0) is an improper distribution with infinite curve area, and in practice dbeta(ε, ε) is used, with ε a small number such as 0.001.

dgamma(*a*, *s*) is the gamma distribution.

dgamma is a very flexible distribution family. It applies to unknown quantities that take values between 0 and ∞; for example, the unknown precision τ of an unknown quantity. Complete ignorance about a positive-valued unknown quantity is generally represented as a dgamma(0, 0) distribution. Since this distribution is improper, dgamma(ε, ε) is used in practice, with ε a small number such as 0.001.

B.3 INFERENCE ABOUT A SINGLE PROPORTION

The instructor in a statistics class spun a new Lincoln penny $n = 25$ times and observed "heads" $x = 11$ times. I am to obtain the posterior distribution of p, the rate at which a penny spun this way will land heads. I might profess complete ignorant about the unknown quantity p (the rate that the coin lands heads), or I might have some prior knowledge. In any case it is most convenient to represent my prior opinion as a beta distribution. For example, a flat prior is specified this way:

$$p \sim \text{dbeta}(1, 1)$$

The tilde ($\sim$) is pronounced "has a ___ distribution." Thus, my prior belief about p has a beta(1,1) distribution.

The second thing that WinBUGS needs to be told is the likelihood of the data x. Since x (the number of heads) can be modeled as the number of black marbles in a sample of size $n = 25$, the likelihood of x successes in n observations of a Bernoulli process (such as spinning a coin) is specified this way:

$$x \sim \text{dbin}(p,n)$$

Fig. B.2 is the WinBUGS program that makes use of these statements to analyze the coin spinning data ($x=11$ heads in $n=25$ spins). The word "MODEL" is not mandatory; WinBUGS treats everything between the opening and closing braces { } as a description of the statistical model, that is, a description of the prior

```
MODEL {
  p ~ dbeta(a,b)
  x ~ dbin(p,n)
}
DATA list(a=1,b=1,x=11,n=25)
```

Figure B.2   WinBUGS program to compute the posterior distribution of the success rate p based on 11 successes in n trials.

distribution of p and the likelihood of x. Observed data are entered by means of a list separated by commas. The word "list" and the parentheses are required, but the word DATA is treated as a comment. The data list contains the inputs to the analysis: the analyst's prior belief (a=1, b=1) and the observed data (x = 11 heads in n = 25 spins).

B.3.1   Setting Up the Model

Launch WinBUGS.   The icon, which resembles a spider, is in the directory where WinBUGS was installed — it is convenient to drag a shortcut to the desktop.   Read and then close the license agreement window, and open a new document window (pull down: file/new).

Specify the Model.   Type the contents of Fig. B.2 in the document window you just opened. To save the program, select the window containing the program you just typed in, pull down the File menu, select Save as, enter a file name (WinBUGS will add the extension .odc), navigate to an appropriate directory, and save the file.

Check the Syntax.   Pull down the Model menu, and select "Specification." The *specification tool* window will open (Fig. B.3) Single-click anywhere in the model (between the curly braces), and then click the "check model" button. Look in the message bar along the bottom of the WinBUGS window. You should see the phrase, "model is syntactically correct." Syntax errors produce a variety of error messages. For example, in Fig. B.4 the programmer typed "p = dbeta(1,1)" instead of "p ~ dbeta"). Note that the cursor | is positioned somewhere after the symbol that caused the error. The default cursor is hard to see but can be made more visible by checking the box for "Thick Caret" in the Edit\Preferences\ dialog box. Correct any syntax errors and repeat the "check model" and "load data" steps above until the model is free of syntax errors.

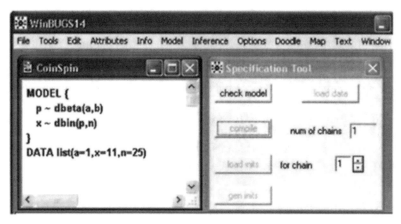

Figure B.3 The specification tool. Position the cursor in the program window, then click "check model" in the specification tool window.

Enter Data. In the DATA statement, highlight any part of the word "list" and then click "load data" in the specification tool window. If the highlighting extends beyond the word "list," there will be an error message. Correct any data errors, and repeat the "check model," "compile," and "load data" steps.

Right: DATA list(a=

Wrong: DATA list(a=·

Figure B.4 A syntax error. The programmer typed "p = "instead of "p ~ ". WinBUGS positions the cursor after the character that caused the error.

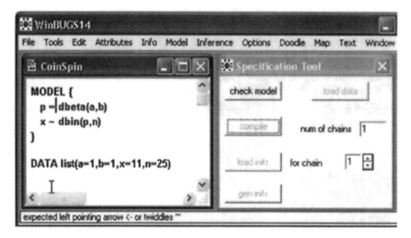

Compile the Model.   Click "compile" and look for the words "model compiled" in the message bar across the bottom of the WinBUGS window. Fig. B.5 shows a compilation error caused by not providing a value for the parameter b in the data list. Another common compilation error is misspelling a variable name — WinBUGS is case-sensitive, which means that it interprets b and B as different symbols. Inconsistent spelling of the same variable is one of the most common errors in WinBUGS programs.

Notice that WinBUGS used the word "node" in the error message in Fig. B.5. A *node* is any variable or constant that is mentioned in the model. In this case the nodes are a, b, n, p, and x. Node p is an unknown quantity; the other four are known quantities entered via the data statement.

Generate Initial Values.   Click "gen inits" in the Specification Tool. Sometimes WinBUGS will display an error message indicating that it is unable to generate initial values. In such cases it is up to the programmer to provide initial values. We'll learn how to do this later.

B.3.2   Computing the Posterior Distribution

Select the Nodes (Unknown Quantities) to be Monitored.   *Monitoring* a node means asking that WinBUGS keep a file of the simulated values of that node. In this case we must monitor node p, the unknown success rate. To do this pull down the Inference menu and select "Samples." The Sample Monitor Tool will

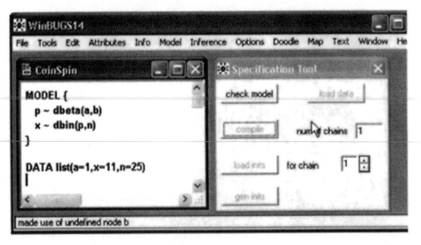

Figure B.5   A compilation error. The programmer failed to provide a value for node "b".

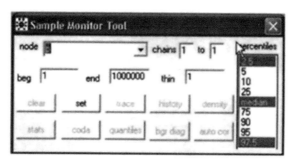

Figure B.6   The Sample Monitor Tool. Enter each node to be monitored clicking "set" after each entry

appear (Fig. B.6). In the "node" field type "p" (without quotes) and click "set." In more complicated models repeat these steps for each of the unknown quantities of interest. If the "set" button does not darken after you type a node name, check for a spelling error (perhaps you used the wrong case).

Generate Simulated Values of All Unknown Quantities.   Pull   down   the Model menu and select "Update"; the *Update Tool* will appear (Fig. B.7). In the "updates" field enter the desired number of simulations (for example 5000) as shown in Fig. B.7. In a complex model it is a good practice to start with 100 in the "updates" field and 10 in the "refresh" field to get some idea of how fast the simulation runs. Click "update" to start the simulations. The simulation can be stopped and restarted by clicking the "update" button. Several thousand to hundreds of thousands of simulations are required to get reasonably accurate posterior probabilities, moments, and quantiles. The updates field controls how often the display is refreshed – changing it has no effect on the speed of simulations; making it smaller, however, reduces the amount of time that WinBUGS is unresponsive.

Occasionally, a "Trap" display such as Fig. B.8 will appear during simulations. If this happens, try clicking the "update" button twice to restart the

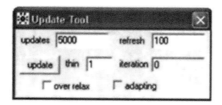

Figure B.7   The Update Tool. Enter the desired number of simulations in the "updates" field. If simulations are generated slowly, enter a smaller number in the "refresh" field.

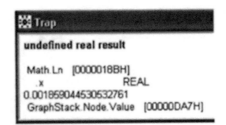

Figure B.8   An error trap may be transitory or may require tightening the prior.

simulations. If the trap continues to reappear, the model will have to be modified, typically by making prior distributions more informative.

Examine the Posterior Distribution.   Return to the Sample Monitor Tool, and enter 1001 in the "beg" field – this instructs WinBUGS to discard the first 1000 simulations to get past any initial transients. In the "node" field enter the name of the unknown quantity that you want to examine and click "density" to see a graph of its posterior density, and then click "stats" to see quantiles and moments of the posterior distribution. The default display is the posterior mean and standard deviation, along with the median and 95% credible interval; however, you can select other percentiles by clicking any number of choices in the "percentile" window of the Sample Monitor Tool. To select a percentile, position the arrow cursor over the desired percentile and ctrl-click the left mouse button. Fig. B.9 shows how to request the 5th and 95th percentiles, which are the endpoints of the 90% credible interval.

A complete WinBUGS session is displayed in Fig. B.10. The *kernel density* graph in Fig. B.10 is a smoothed histogram of the simulations and is an approximation of the posterior distribution. The *node statistics* table lists the mean

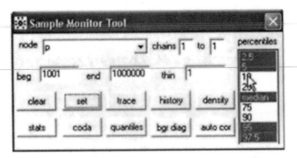

Figure B.9   The Sample Monitor Tool. The user has requested five percentiles. The arrow cursor is positioned to select the 10th percentile (ctrl-click the left mouse button).

and standard deviation of the posterior distribution of each monitored quantity as well as selected percentiles. The default display includes the median and the 2.5th and 97.5th percentiles but other percentiles can be requested as explained in the previous paragraph. The columns of the display are labeled as follows:

- **node**   The name of the unknown quantity
- **mean**   The average of the simulations, an approximation of the μ of the posterior distribution of the unknown quantity
- **sd**   The standard deviation of the simulations, an approximation of the σ of the posterior distribution
- **MC error**   The computational accuracy of the mean
- **2.5%**   the 2.5th percentile of the simulations, an approximation of the lower endpoint of the 95% credible interval
- **median**   The median or 50th percentile of the simulations, and
- **97.5%**   The 97.5th percentile of the simulations, an approximation of the upper endpoint of the 95% credible interval
- **start**   The starting simulation (after discarding the start-up)
- **sample**   The number of simulations used to approximate the posterior distribution

The MC error is purely technical, like round-off error, and can be made as small as desired by increasing the number of simulations (reported under "sample" in the node statistics table). On the other hand, the posterior standard deviation, the analog of the standard error in conventional statistical inference, represents genuine uncertainty and cannot be reduced other than by obtaining additional real data. Note also that the number 50,000 in the "samples" column in the Node statistics table is the number of simulations, *not* the sample size of the data n = 25.

The WinBUGS output displayed in Fig. B.10 indicates that the posterior distribution of p, the rate of occurrence of heads in penny spinning, is approximately normal (judging from the graph) with $\mu = 0.4441$ and $\sigma = 0.0939$. These numbers are computationally accurate to about ± 0.0004 (MC error); consequently it would be more appropriate to report $\mu = 0.444$ and $\sigma = 0.094$.

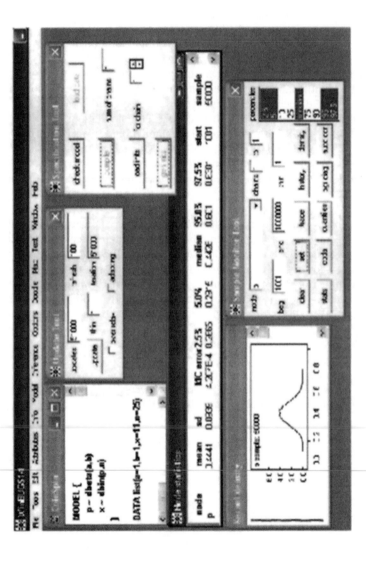

Figure B.10 A complete WinBUGS session. The posterior distribution of p, the rate of occurrence of heads in coin spinning, is approximately normal with $\mu = 0.444$ and $\sigma = 0.094$.

B.4 TWO RATES – DIFFERENCE, RELATIVE RISK, AND ODDS RATIO

In a study comparing radiation therapy vs. surgery, cancer of the larynx remained uncontrolled in 3 of 18 radiation patients and 2 of 23 surgery patients. Fig. B.11 shows the analysis. The unknowns p_{rad} and p_{srg} are the rates of failure of radiation and surgery, respectively, and x_{rad}, n_{rad}, x_{srg}, and n_{srg} are the observed data. The prior distributions of the unknown quantities have been given dbeta(.5, .5), which makes the difference ($\Delta = p_{rad} - p_{srg}$) have a more nearly uniform prior. Note that text following a pound sign, #, is interpreted as a comment.

This example illustrates the "arrow" symbol for assigning values to logical nodes; that is, unknown quantities, such as the odds ratio, that are computed from more basic unknown quantities p_{rad} and p_{srg}. Here the logical nodes are the difference of the two failure probabilities (DIFF), the relative risk (RR), the odds ratio (OR), and a Boolean variable (ppos, explained below) that counts the number of times the research question "Is radiation less effective?" is true:

$$\text{Research question: Is it true that } p_{srg} > p_{rad} ? \qquad (B.1)$$

The fundamental unknown quantities are the success rates for the two different treatment modes; however, answering the research question requires contrasting the two rates. In Chapter 7 we learned three ways of contrasting two rates: the difference (DIFF or Δ), the odds ratio (OR), and the relative risk (RR).

```
Radiation vs Surgery

MODEL RadVSurg {
  # Prior
      p.rad ~ dbeta(.5,.5)
      p.srg ~ dbeta(.5,.5)
  # Likelihood
      x.rad ~ dbin(p.rad,n.rad)
      x.srg ~ dbin(p.srg,n.srg)
  # Contrasts
      DIFF <- p.rad - p.srg
      RR <- p.rad/p.srg
      OR <- (p.rad/(1 - p.rad)) / (p.srg/(1 - p.srg))
      lnOR <- log(OR)
      ppos <- step(OR - 1)
}

DATA list(x.rad=3,n.rad=18,
          x.srg=2,n.srg=23)
```

Figure B.11   Comparing two rates – illustrating computed (logical) nodes.

The research question can be stated in terms of any one of the contrasts:

$$\text{Is } \Delta > 0?$$
$$\text{Is OR} > 1? \qquad (B.2)$$
$$\text{Is RR} > 1?$$

The contrasts D, OR, and RR are functions of the two success rates, p_{rad} and p_{srg}. In algebra, functional relationships are written with an = sign, however, in many computer languages a distinction is made between two variables occupying the same location in memory (=) and a value being computed and assigned to a variable. In WinBUGS, the identity symbol (=) is used only in data lists, whereas the assignment symbol "<–" (often pronounced "gets") is used in the model to indicated that one node gets its value from other nodes. For example, RR gets its value by dividing the surgical cure rate by the radiation cure rate: RR <– p_{srg} / p_{rad}.

The twiddle symbol indicates that a node has a particular distribution. For example, x ~ dbin(p,n) means that "x is distributed like the number of successes in n observations of a Bernoulli process." Inadvertent use of an "=" sign instead of a twiddle or an arrow is one of the most common reasons for a compilation error message. The equal sign is never used in a WinBUGS model, although it is used in data lists.

The WinBUGS program in Fig. B.11 uses the step( ) function to create a Boolean variable that counts the number of simulations in which the sentence "$p_{rad} \geq p_{srg}$" is true. Here's how it works: if V is any node, then step(V) equals 1 if $V \geq 0$ and equals 0 if $V < 0$. Consequently, step(a – U) equals 1 if a – U $\geq$ 0; that is, if a $\geq$ U. The step( ) function can be used to compute left- or right-tail areas:

$$
\begin{aligned}
P(U \leq a): &\quad \text{new\_node\_1} <– \text{step}(a – U) \\
P(U < a): &\quad \text{new\_node\_2} <– 1 – \text{step}(U – a) \qquad (B.3) \\
P(U \geq b): &\quad \text{new\_node\_3} <– \text{step}(U – b) \\
P(U > b): &\quad \text{new\_node\_4} <– 1 – \text{step}(b – U)
\end{aligned}
$$

The word "new_node" is generic; each tail request must have a unique node name such as "ppos" in Fig. B.11. The mean value of a Boolean node such as those in Equation (B.3) is a probability; for example, the mean of new_node_1 is the Monte-Carlo estimate of $P(U \leq a)$.

Table B.1 is a list of some of the other functions available in WinBUGS; a complete list is found in Table II of the WinBUGS user manual (Spiegelhalter et al. 2002). Note that the natural logarithm, ln( ), is called log( ) in WinBUGS, and the pow function is used to raise a number to a power

$$U^c: \qquad \text{new\_node} <– \text{pow}(U, c)$$

Name	Action
step(x)	1 if $x \geq 0$, otherwise 0
log(x)	ln(x)
logit(p)	ln(p/(1-p))
exp(x)	exp(x)
abs(x)	$\lvert x \rvert$
pow(x,c)	x^c
sqrt(x)	$\sqrt{x}$

To compute the posterior distributions of DIFF, RR, and OR, follow the steps in Section B.3 with this change: under the heading "Select the Nodes ... to be Monitored" on page 310 it is necessary to enter four node names, one at a time. First type DIFF in the "node" field and click "set," then do the same for OR, RR, and ppos. Be careful about capitalization, since WinBUGS is case-sensitive. After entering the node names, continue with the instructions.

A second change is required at the last step, "Examine the Posterior Distribution," page 312. To graph the posterior distributions and compute moments, quantiles, and credible intervals, enter an asterisk in the node field of the Sample Monitor Tool (Fig. B.12); then click "stats" and "density."

The raw output, displayed in Fig. B.12, has been edited and annotated for greater clarity in Fig. B.13. The edits involved clarifying the meaning of the nodes, and not reporting unreliable digits. (For example, the statistics for DIFF are unreliable beyond the 4th decimal place, because the MC error is about 0.0005). Although each of the contrasts suggests that surgery has the lower failure rate, none of them rules out equality.

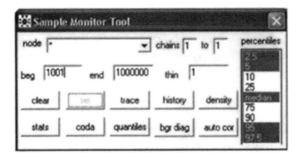

Figure B.12 How to request posterior distribution statistics for all four monitored nodes. Enter an asterisk in the node field then click stats and density.

node	mean	sd	MC error	2.5%	median	97.5%
DIFF	0.07993	0.1064	4.893E-4	-0.1232	0.07542	0.2997
OR	3.444	5.489	0.02497	0.3227	2.049	14.77
RR	2.815	4.013	0.01825	0.3726	1.849	10.93
lnOR	0.736	0.9696	0.004443	-1.131	0.7173	2.693
ppos	0.7804	0.4139	0.001837	0.0	1.0	1.0

Figure B.13 Node statistics for comparing surgical and radiological failure rates based on 50,000 simulated values.

The 95% credible intervals for the odds ratio and relative risk include 1 (meaning equal rates), and the 95% credible intervals for the difference and log odds ratio include 0 (also meaning equal rates). On the other hand, odds ratios as high as 15 and relative risks as high as 11 cannot be ruled out. The posterior probability that radiation has the higher failure rate is about 78%.

Posterior distributions are graphed in Fig. B.14. As expected the odds ratio and relative risk have heavily skewed distributions, but the difference and log odds ratio appear to have nearly normal distributions. The distribution of the node, ppos, requires some explanation. Recall that ppos was produced by the step( ) function, which means that it is a Boolean (0 or 1) variable. The value ppos = 1 identifies simulations in which surgery is better than radiation. The histogram of ppos has only two bars — at 0 and 1 — and the height of the bar at 1 is the proportion of simulations for which the ppos = 1, that is, the proportion of times the sentence "Rad > Srg" was true, which approximates the posterior probability that it is true. That proportion is also equal to the mean and therefore the mean is the only meaningful descriptive statistic for a Boolean variable. The mean (i.e. the proportion of 1's) completely describes the data, and for that reason the histogram and statistics other than the mean are confusing and have been suppressed in Table B.2.

Table B.2 Surgery vs. radiation: posterior moments and quantiles.

Failure Rate Comparisons	μ	σ	median	95% Credible Interval	
Difference (Rad - Srg)	0.0799	0.1064	0.0754	-0.1232	0.2997
Odds Ratio (Rad/Srg)	3.44	5.49	2.05	0.32	14.77
Relative Risk (Rad/Srg)	2.82	4.01	1.85	0.37	10.93
ln(Odds Ratio)	0.736	0.97	0.717	-1.131	2.693
P(Rad > Srg | Data)	0.78				

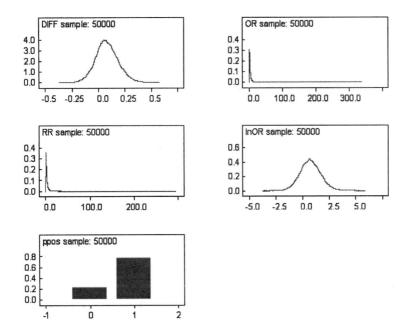

Figure B.14   Posterior distributions of five nodes. Probability nodes are entirely described by the mean.

B.5 "FOR" LOOPS

The purpose of WinBUGS model specification language is to specify the prior distributions of the unknown parameters and the likelihood function of the observed data. It is not a programming language. It does not specify a series of commands to be executed in sequence. In fact, model specification statements can be written in almost any order without changing the meaning of the model. Repetitive model components, as in a hierarchical model, can be specified using "for" loops but conditional branching structures such as "if ... then ... else" are not available and, indeed, have no meaning in model specification.

Using the "For" Structure: The Pediatric Mortality Study .   This example was described in Section 11.7. The data are numbers of patients and numbers of deaths in 12 hospitals. The WinBUGS program in Fig. B.15 uses the "for" programming structure to specify the model more compactly. The following segment illustrates how the "for" structure makes the model specification much more compact:

```
for (i in 1:k) {
    #Prior distribution of Hospital i's True Rate
    p[i] ~ dbeta(a,b)
    #Likelihood of Hospital i's Data                        (B.4)
    x[i] ~ dbin(p[i],n[i])
}
```

The variables (nodes) in this fragment are the true mortality rate in the *i*th hospital, p[i]; and the observed number of patients, n[i], and deaths, x[i], in that hospital. Without the "for" structure, it would have taken 24 lines to specify the prior and likelihood:

```
p[1] ~ dbeta(a,b)
x[1] ~ dbin(p[1],n[1])
p[2] ~ dbeta(a,b)
x[2] ~ dbin(p[2],n[2])
p[3] ~ dbeta(a,b)
x[3] ~ dbin(p[3],n[3])
... 16 lines omitted ...
p[12] ~ dbeta(a,b)
x[12] ~ dbin(p[12],n[12])
```

WinBUGS uses square brackets to denote subscripts. Thus p[i] in program fragment (B.4) is what we would ordinarily write as p_i, the unknown true long-term morality rate in the *i*th hospital, and n[i] and x[i] are what we would write as n_i and x_i. Subscript *i* is the loop *index* and the expression 1:k is its *range*. The range must include only positive integers such as 1:12 or 3:7. A range can be specified in terms of numbers or integer-valued variables.

```
MODEL Hospital {
    #Hyperprior for the Box of Rates
        a~dgamma(.001,.001)
        b~dgamma(.001,.001)
    #Prior Distribution of the True Rates
    for (i in 1:k) {
        #Prior distribution of Hospital i's True Rate
        p[i] ~ dbeta(a,b)
        #Likelihood of Hospital i's Data
        x[i] ~ dbin(p[i],n[i])
    }
}
DATA list(k=12,
    n = c(47,148,119,810,211,196,148,215,207,97,256,360),
    x = c( 0,  18,  8, 46,  8, 13,  9, 31, 14,  8, 29, 24))
INITIAL VALUES list(a=1,b=1)
```

Figure B.15 WinBUGS program for the hospital study.

B.6 DATA ENTRY

Data for a subscripted variable can be entered as a list or in a table. For example, list input is used in Figure B.15,

```
DATA list(k=12,
n = c(47,148,119,810,211,196,148,215,207,97,256,360),
x = c( 0,  18,   8,  46,   8,  13,   9,  31,  14,  8,  29,  24))
```

Individual numbers such as *k* are entered as k=12, for example. Data for the subscripted variables *n* and *x* are specified as *collectives*, indicated by the letter "c" followed by a parenthetical, comma-separated list, for example, n = c(47,...., 360). List input is convenient if there are only a few data items; however, for large data sets it can be more convenient to enter the data in the form of an *embedded table*.

B.6.1 Embedding a Data Table in WinBUGS

WinBUGS allows documents to be embedded in a *compound document*, thus providing convenient way to save the program, data, and output in a single computer file. The first step is to prepare a plain text file of data arranged (in this case) as a 12 by 2 matrix. First type or paste the data matrix in a new WinBUGS document window (pull down: file/new to create a new document window), then follow these instructions:

Step 1: Create the program and data files in two windows:

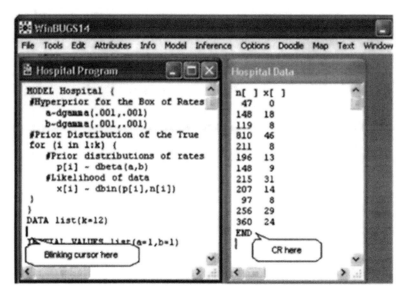

Step 2: Create a fold at the bottom of the program file:

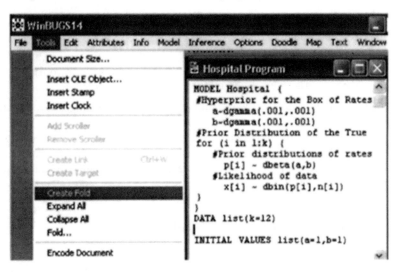

Step 3: Open the fold and enter a blank line:

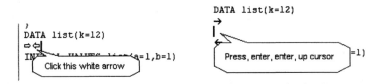

Step 4: Copy the data document:

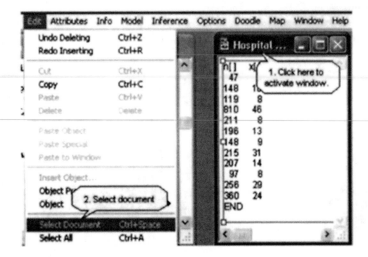

Step 5: Copy, and paste the data into the fold, resize the "hairy border" around the data table, and close and label the fold:

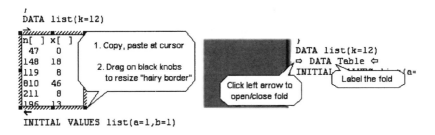

```
DATA list(k=12)
```

INITIAL VALUES list(a=1,b=1)

B.6.2   Loading Data from an Embedded Table

After checking the model, if there is a data list as well as a data table, load the list in the usual way, and then open the fold containing the data table. Click anywhere in the data table, but do not highlight any text, as that will produce an error message. The data table should be surrounded by a "hairy border." Click "load data," close the fold and proceed with the compilation. Note that the last line in the data table must be the word END on a separate line followed by a carriage return. If this is missing, WinBUGS will report that there is an incomplete data line.

B.7   PLACING OUTPUT IN A FOLD

It is a good idea to paste output tables and graphs into the compound document containing the model specification and data. This is easy to do and creates a complete record of the analysis in a single document that can be saved and, if desired, re-opened for modification or additional analyses. For example, the node statistics table is initially reported in a separate document. Click anywhere in that document, and copy it (pull down Edit/select all, then Edit/copy). Note that you must choose "select all", not "select document" to copy the node statistics table. Insert and label a fold in the main document. Open the fold, and paste in the node statistics table.

B.8   ADDITIONAL RESOURCES

Readers are urged to look at the user manual
(under the Help menu), which has a tutorial
chapter and provides much more detail on
setting up WinBUGS analyses. The examples
"Vol I" and "Vol II" (under the Help menu) are
also worth a look. A good way to learn to use
WinBUGS with your own data is to imitate an
example similar to the analysis that you want to
do. WinBUGS also has the option to set up the
model (prior and likelihood) in graphical form.
See "Doodle help" under the Help menu as well

as the excellent introduction to graphical models in Fryback et al. (2001).

B.9   WEIBULL PROPORTIONAL HAZARDS REGRESSION

This section is significantly more difficult and can be skipped without losing continuity.

Multiple myeloma survival data.   The data to be analyzed is a subset of the
multiple myeloma survival data in Table 2 of Krall et al. (1975). The response
variable is survival time (time from diagnosis to death). Some patients were alive
at the end of the observation period and their survival times are therefore
truncated (known only to be longer than the observation period). The regression
model uses ln(BUN) as a continuous explanatory variable and has separate slopes
and intercepts for men and women. The data in raw form and they must be
arranged for WinBUGS are shown in Table B.3.

Notice that the survival time variable must be split into two variables t.obs,
corresponding to subjects who died during the observation period, and t.cen,
corresponding to censored cases that were still alive at the end of the observation
period. For censored cases, t.obs is recorded as "NA," which is WinBUGS
representation of an unknown data value. The censoring time variable t.cen is
recorded as 0 for uncensored cases. Note also that the explanatory variables, BUN
and Sex, have been converted into *doubly subscripted* design variables:

$x[,1]$ 1 if female, 0 if male
$x[,2]$ ln(BUN) if female, 0 if male
$x[,3]$ 1 if Male, 0 if Female
$x[,4]$ ln(BUN) if male, 0 if female

Thus $x[,1]$ and $x[,2]$ are design variables for the female intercept and slope, and
$x[,3]$ and $x[,4]$ are design variables for the male intercept and slope. The raw data
are not entered in the WinBUGS data table.

Table B.3 Multiple myeloma survival data.[a]

Raw Data				Time (months)		Design Variables			
Time	Dead	BUN	Sex	t.obs[ ]	t.cen[ ]	x[,1]	x[,3]	x[,2]	x[,4]
3	1	35	F	3	0	1	3.56	0	0
4	0	84	F	NA	4	1	4.43	0	0
5	1	172	F	5	0	1	5.15	0	0
6	1	130	F	6	0	1	4.87	0	0
6	1	26	F	6	0	1	3.26	0	0
7	1	11	F	7	0	1	2.40	0	0
7	1	15	F	7	0	1	2.71	0	0
7	0	13	F	NA	7	1	2.56	0	0
8	0	12	F	NA	8	1	2.48	0	0
11	1	12	F	11	0	1	2.48	0	0
12	0	14	F	NA	12	1	2.64	0	0
12	0	25	F	NA	12	1	3.22	0	0
13	1	6	F	13	0	1	1.79	0	0
13	0	46	F	NA	13	1	3.83	0	0
16	1	21	F	16	0	1	3.04	0	0
18	1	28	F	18	0	1	3.33	0	0
19	1	18	F	19	0	1	2.89	0	0
19	0	21	F	NA	19	1	3.04	0	0
24	1	20	F	24	0	1	3.00	0	0
26	1	17	F	26	0	1	2.83	0	0
28	0	17	F	NA	28	1	2.83	0	0
41	1	14	F	41	0	1	2.64	0	0
41	0	57	F	NA	41	1	4.04	0	0
52	1	10	F	52	0	1	2.30	0	0
58	1	16	F	58	0	1	2.77	0	0
88	1	15	F	88	0	1	2.71	0	0
92	1	27	F	92	0	1	3.30	0	0
1.25	1	165	M	1.25	0	0	0	1	5.11
1.25	1	87	M	1.25	0	0	0	1	4.47
2	1	33	M	2	0	0	0	1	3.50
2	1	56	M	2	0	0	0	1	4.03
2	1	20	M	2	0	0	0	1	3.00
4	0	90	M	NA	4	0	0	1	4.50
5	1	48	M	5	0	0	0	1	3.87
6	1	23	M	6	0	0	0	1	3.14
6	1	13	M	6	0	0	0	1	2.56
7	1	95	M	7	0	0	0	1	4.55
7	0	34	M	NA	7	0	0	1	3.53
9	1	53	M	9	0	0	0	1	3.97
11	1	13	M	11	0	0	0	1	2.56
11	1	17	M	11	0	0	0	1	2.83
11	1	20	M	11	0	0	0	1	3.00
11	1	37	M	11	0	0	0	1	3.61
11	0	41	M	NA	11	0	0	1	3.71
14	1	25	M	14	0	0	0	1	3.22
15	1	40	M	15	0	0	0	1	3.69
16	1	22	M	16	0	0	0	1	3.09
16	0	14	M	NA	16	0	0	1	2.64
17	1	17	M	17	0	0	0	1	2.83
17	1	39	M	17	0	0	0	1	3.66
19	1	12	M	19	0	0	0	1	2.48
19	0	21	M	NA	19	0	0	1	3.04
25	1	10	M	25	0	0	0	1	2.30
32	1	21	M	32	0	0	0	1	3.04
35	1	13	M	35	0	0	0	1	2.56
37	1	40	M	37	0	0	0	1	3.69
41	1	10	M	41	0	0	0	1	2.30
51	1	37	M	51	0	0	0	1	3.61
53	0	13	M	NA	53	0	0	1	2.56
54	1	18	M	54	0	0	0	1	2.89
57	0	18	M	NA	57	0	0	1	2.89
66	1	28	M	66	0	0	0	1	3.33
67	1	21	M	67	0	0	0	1	3.04
77	0	12	M	NA	77	0	0	1	2.48
89	1	21	M	89	0	0	0	1	3.04

a. Source of raw data: Krall, et al. (1975), Table 2.

Weibull Proportional Hazards Regression.   The basic WinBUGS commands
to do any Weibull proportional hazards regression analysis are listed in Figure
B.16. The output in panel B of the figure shows that the female intercept is
greater than the male intercept and therefore that females with low BUN (blood
urea nitrogen) are at greater risk of death than males. However, the male slope is
substantially greater than the female slope; consequently, at some BUN value, the
males will "catch up" and have greater risk of death. Beyond that general
observation, it is difficult to interpret the output without estimating survival rates
for males and females at various BUN values. This requires adding some
specialized instructions to the generic instructions in Fig. B.16.

Computing Survival Rates and Female/Male Contrasts.   Instructions
directing WinBUGS to compute survival rates and make male-female contrasts
are listed in Fig. B.16 with output in Fig. B.5; however, the first step is to create an

A. Basic WinBUGS commands for Weibull regression

```
MODEL Weibull PHR {
    # Prior distribution of baseline hazard function
        shape ~ dgamma(1,.001)

    # Prior distribution of the regression coefficients
        for (i in 1:k) { beta[i] ~ dnorm(0,.001) }

    # Likelihood of the survival time data
        for (j in 1:n) {
        HRx[j] <- exp(inprod(x[j,],beta[]))
        Scale[j] <- HRx[j]
        t.obs[j] ~ dweib(shape,Scale[j])I(t.cen[j],)
        }
    # Insert any additional commands starting here.
}
```

B. Output[a]: Posterior Moments and Quantiles of Regression Coefficients.

Description	Node	Mean	sd	MC error	2.5%	Median	97.5%
Female Intercept	beta[1]	−4.84	1.44	0.05	−7.64	−4.84	−1.97
Female Slope	beta[2]	0.21	0.43	0.015	−0.69	0.24	0.98
Male Intercept	beta[3]	−8.00	1.42	0.06	−10.92	−7.96	−5.56
Male Slope	beta[4]	1.26	0.36	0.015	0.57	1.26	1.99
Weibull shape	shape	1.16	0.13	0.004	0.93	1.15	1.42

a. Based on 60,000 simulations.

Figure B.16   Weibull regression analysis of multiple myeloma survival data.

Table B.4 "Requests" file[a]

Name	Description: survival rate at 24 months for	Raw Data Month	BUN	SEX	WinBUGS formatted data t.r[]	x.r[,1]	x.r[,3]	x.r[,2]	x.r[,4]
Sr[1]	Males with BUN = 6	24	6	M	24	0	0	1	1.79
Sr[2]	Females with BUN = 6	24	6	F	24	1	1.79	0	0
Sr[3]	Males with BUN = 14	24	14	M	24	0	0	1	2.64
Sr[4]	Females with BUN = 14	24	14	F	24	1	2.64	0	0
Sr[5]	Males with BUN = 21	24	21	M	24	0	0	1	3.04
Sr[6]	Females with BUN = 21	24	21	F	24	1	3.04	0	0
Sr[7]	Males with BUN = 37	24	37	M	24	0	0	1	3.61
Sr[8]	Females with BUN = 37	24	37	F	24	1	3.61	0	0
Sr[9]	Males with BUN = 172	24	172	M	24	0	0	1	5.15
Sr[10]	Females with BUN = 172	24	172	F	24	1	5.15	0	0

a. This file, in conjunction with the commands in Fig. B.17, instructs WinBUGS to compute the indicated survival rates. Only the WinBUGS formatted data are submitted; the name, description and raw data are not included in the file.

additional data file (Table B.4) that specifies the time point and design variable values at which the survival function is to be calculated. The design variable values in this instance are the minimum, maximum, median, and quartile BUN values actually observed in the data set.

WinBUGS commands to compute the requested survival rates and male/female contrasts are shown in Fig. B.17; these commands are to be inserted in the program in Panel A of Fig. B.16 at the place indicated. Output is listed in Table B.5, symbols used in the program are explained in the output table. Thus, HRfm[1] is the female-vs.-male hazard ratio for patients at the minimum BUN value, 6. At this BUN level, females are at five time the risk of death of males; at the median, BUN=21, males and females have nearly the same risk, and at

```
# Requested survival rates for this analysis:
    for (j in 1:m) {
      HRr[j] <- exp(inprod(x.r[j,],beta[]))
        Sr[j] <- exp(-HRr[j]*pow(t.r[j],shape))
    }


# Contrasts of interest in this analysis:
    #Female/Male RR's & HR's
    for (j in 1:5) {
        RRfm[j] <- (1-Sr[2*j])/(1-Sr[2*j-1])
        HRfm[j] <- HRr[2*j]/HRr[2*j-1]
    }
```

Figure B.17 Special requests specific to this analysis: survival rates at specific BUN levels for males and females, and female/male contrasts.

Table B.5 Output of Special Requests

Explanatory Variables		Node[a]	Mean	sd	MC error[b]	2.5%	Median	97.5%
Sex	BUN		Female/Male Hazard Ratios					
	6	HRfm[1]	5.099	5.187	0.1644	0.7541	3.545	18.56
	14	HRfm[2]	1.619	0.7098	0.01883	0.6511	1.481	3.359
	21	HRfm[3]	1.025	0.3307	0.004344	0.5213	0.9776	1.807
	37	HRfm[4]	0.5829	0.2383	0.004737	0.2279	0.5461	1.146
	172	HRfm[5]	0.2007	0.2736	0.007766	0.009167	0.1135	0.918
			Female/Male 24-Month Relative Risks					
	6	RRfm[1]	3.852	3.119	0.104	0.7788	2.972	12.09
	14	RRfm[2]	1.438	0.4951	0.01366	0.7125	1.358	2.617
	21	RRfm[3]	1.007	0.2334	0.003134	0.615	0.9838	1.531
	37	RRfm[4]	0.7067	0.1815	0.003861	0.37	0.7009	1.078
	172	RRfm[5]	0.6222	0.2656	0.008312	0.1163	0.6473	0.999
			Male and Female 24-Month Survival					
F	6	Sr[1]	0.8631	0.07329	0.002712	0.6823	0.8779	0.9633
M	6	Sr[2]	0.6067	0.1669	0.005439	0.2328	0.6285	0.8665
F	14	Sr[3]	0.6805	0.07974	0.002664	0.5141	0.6848	0.8248
M	14	Sr[4]	0.5684	0.09397	0.002243	0.3799	0.5700	0.7455
F	21	Sr[5]	0.5357	0.07127	0.001667	0.3956	0.5359	0.6751
M	21	Sr[6]	0.543	0.08071	8.823E-4	0.3823	0.5433	0.6977
F	37	Sr[7]	0.2831	0.07449	0.001148	0.1525	0.2785	0.4411
M	37	Sr[8]	0.4992	0.1146	0.002497	0.2848	0.4965	0.7279
F	172	Sr[9]	0.01022	0.03385	9.349E-4	5.162E-15	1.336E-4	0.1008
M	172	Sr[10]	0.3851	0.2605	0.008167	0.008338	0.3586	0.8853

a. WinBUGS uses the word "node" to mean an unknown quantity in the model.
t. Based on 50.000 simulations.

BUN=172, the maximum level in the data set, females have about one-fifth the risk of males. Cumulative relative risk at 24 months, defined as $(1-S_{female}(24))/(1-S_{male}(24))$ shows a similar pattern. The rather large MC error is caused by "ill conditioning" of the design matrix. This could be cured by reexpressing log BUN as a deviation from its average.

Executing the WinBUGS Program. All that remains is to assemble the program and data in a WinBUGS document and follow steps 1 through 24 in panels A through H on the following pages. Panel A shows how to assemble the model specification, data list, and data tables in a compound document. The data are in one list and two tables. The data list that specifies the number of cases (n=65), the number of requests (n=10) and the number of design variables (k=4). The Data table comprises the WinBUGS columns of Table B.3, and the requests table the WinBUGS columns of Table B.4. Instructions for pasting data tables into folds are in Section B.6.1. Remember to type the word END, followed by a carriage return, at the bottom of each data table.

The procedure documented in panels A through H shows every mouse-click of the process and looks quite time-consuming; however, in practice it takes less than a minute to get to the simulation step (step 22). The simulation step itself can take several minutes depending on the number of simulations requested and the speed of the analyst's computer. Output from the program is shown in Fig. B.16 and Table B.5. For a discussion of the meaning of the output, see Section 12.4.3.

A. Check the model

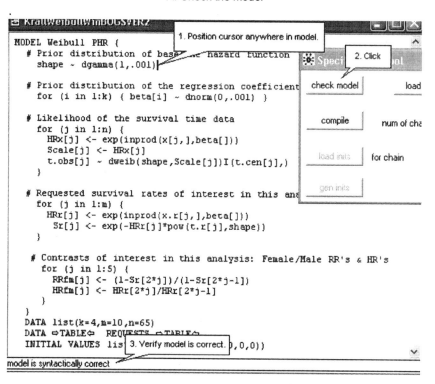

B. Load the data list and open the requests table

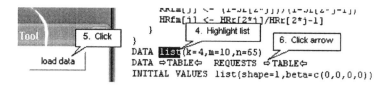

C. Load the requests table and close it

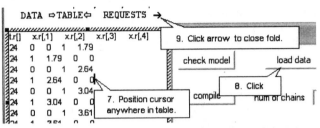

D. Load the Data Table, Close It, and Compile the Model

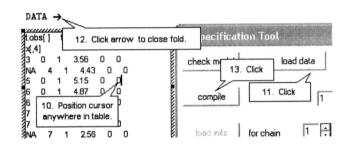

E. Verify compilation, load and generate initial values

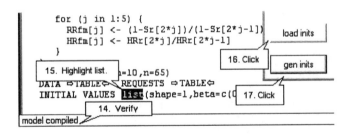

F. Verify initialization

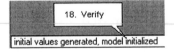

G. Enter the node names (repeat 19 and 20 for shape, Sr, HRfm, and RRfm)

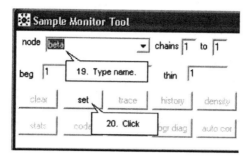

H. Generate the simulations and compute posterior moments and quantiles

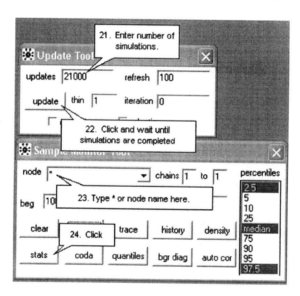

B.10 REFERENCES

Fryback D.G., Stout, N.K.,and Rosenberg, M.A., "An elementary introduction to Bayesian computing using WinBUGS," *International Journal of Technology Assessment in Health Care*, Vol 17, No. 1 (Winter 2001), pp. 98–113.

Krall, John M., Uthoff, Vincent A, and Harley, John B., "A Step-up Procedure for Selecting Variables Associated with Survival," *Biometrics*, Vol. 31, No. 1 (1975), pp. 49–57.

Spiegelhalter, D., Thomas, A., Best, N., and Lunn, D., *WinBUGS User Manual Version 1.4*, Cambridge, UK: MRC Biostatistics Unit (2002).

C

Introduction to SAS System Software

SAS®, the Statistical Analysis System, a product of the SAS Institute[9], is a general purpose statistical package that provides extensive statistical capabilities for both basic data management and both simple and highly specialized statistical analyses. Almost all analysis procedures in the SAS software package use conventional, not Bayesian, statistical methods. However, as we have seen many conventional analyses can be approximately interpreted as Bayesian analyses with flat or non-informative prior distributions. SAS software is extensively used in biomedical research. Versions of SAS software are available for several operating systems, including UNIX and Microsoft Windows ®. This chapter is an introduction to SAS version 9.x for Windows.

SAS software is launched by double clicking the SAS icon, which resembles an inverted pyramid. The opening screen displays the *log window*, the program *editor* window, and the file *explorer:*

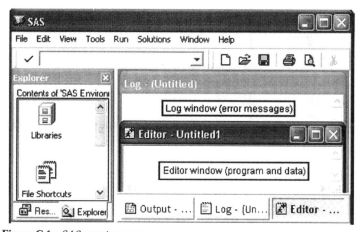

Figure C.1   SAS opening screen.

9. SAS and all other SAS Institute Inc. product or service names are registered trademarks or trademarks of SAS Institute Inc. in the USA and other countries. ® indicates USA registration.

Additional windows can be opened by clicking buttons along the bottom bar. The most important of these is the *output* window. A *results* explorer can also be opened, which functions as a table of contents for the output and graph windows.

C.1 DATA ENTRY — THE DATA STEP

SAS is a statistical analysis system and a data management system. The key component of the data management system is the *DATA step*. The purpose of the DATA step is to create a SAS data file either by reading raw data or by copying data from another SAS data file. The raw data can be typed in the editor window, or read from a text file, or imported from Microsoft Excel or a variety of data-base management programs such as Microsoft Access. Fig. C.2 shows the simplest form of data entry, reading data embedded in the program window. The raw data in this example are the genders and reaction times of 10 students in a psychology class.

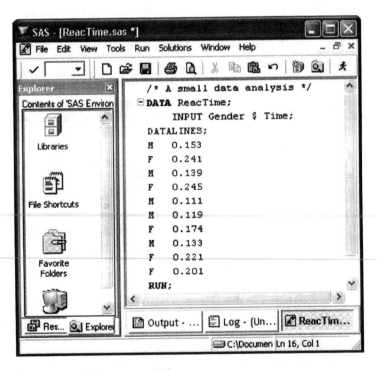

Figure C.2 Simple data entry in SAS.

The components of the data step are:

- The DATA keyword followed by the name to be assigned to the data file.
- The INPUT keyword followed by a list of variable names and (optionally) formats. The $ format indicates that gender is a non-numeric variable.
- The DATALINES keyword.
- Ten rows of gender and reaction time data.
- The RUN keyword.

With the exceptions of raw data and comments, every statement in the program must end in a semicolon. Comments are useful but optional and consist of any amount of text in the format: /* ... text... */, with no other punctuation.

The DATA step in Fig. C.2 creates a data file called ReacTime. To run the data step, highlight the entire program (click–drag from the upper left to the lower right) and click the runner icon. The file created by this program is physically located in some directory on a hard drive; however, SAS software metaphorically places the file in a drawer labeled Work in a virtual file cabinet. The official name of the file is Work.ReacTime, that is, drawer_name.file_name. If the drawer name is omitted, the SAS system assumes the file is in the Work drawer.

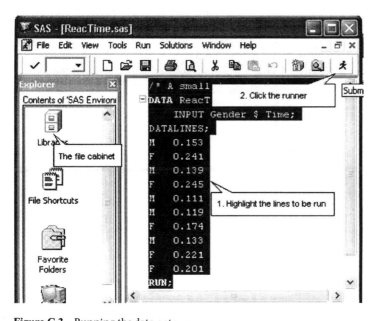

Figure C.3   Running the data entry program.

To open the file drawer, double-click on the file cabinet and see the file drawers. Double click on the Work drawer and see the contents of that drawer. In this case there is only one file, called ReacTime, which was created by the Data step in Fig. C.2. This file can be opened for examination by double-clicking on it. With small files, this is a convenient way to verify

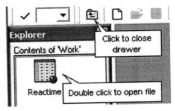

that the data were read correctly; however, it is not recommended for large data files. An open SAS data file must be closed before attempting any analysis.

Data files in the Work drawer are temporary and are discarded at the end of the current SAS session; however, it is possible to create a permanent file drawer that is not emptied at the end of the session and can be reopened in a later session.

Creating a Permanent SAS System Data File. To create a permanent data file it is necessary to associate a virtual file drawer with a real directory on a disk. To do this, the program in Fig. C.2 on page 334 requires two changes. First there must be a LIBNAME, which names a new file drawer and specifies where it is to be physically saved. For example, if an analyst wanted to call the file drawer MyDATA and wanted to save the file drawer on a floppy disk in the computer's

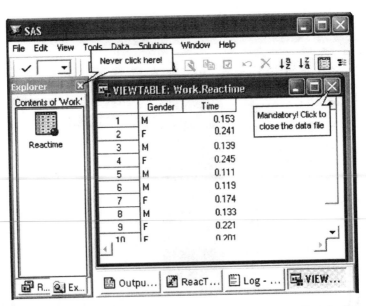

Figure C.4 Looking at the data file to verify correct data entry. Close the file before attempting any statistical analyses.

A drive, the command would be

LIBNAME MyData "A:\";   (C.1)

The semicolon is mandatory at the end of the LIBNAME statement. The second modification is to direct the data file into the new file drawer. This is done by adding a prefix to the file name in the DATA statement,

DATA MyData.ReacTime;   (C.2)

The modified program is

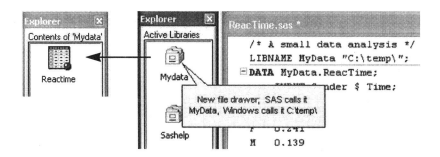

Figure C.5   Use LIBNAME to save a permanent SAS data file in C:\temp.

In this example, the analyst wanted to store the SAS system data file in the directory C:\temp\ rather than on the A drive.

Reopening a permanent SAS data file.   A saved SAS system data file can be re-used in a later session without having to go through the trouble of re-entering the raw data. Simply include the LIBNAME statement at the top of the later session and always add the drawer name prefix to any reference to the data file.

C.2   STATISTICAL ANALYSES — THE PROC STEP

SAS has a large repertory of basic and specialized statistical analyses. Each has its own syntax and somewhat idiosyncratic placement of punctuation. The procedures used in this book are MEANS, UNIVARIATE, FREQ, LOGISTIC, LIFETEST, and PHREG. The first three are in the SAS/BASIC®

module, and the others are in the SAS/STAT® module. Details of each procedure are found in SAS Help and Documentation under the Help menu (sidebar).

We will begin with the MEANS procedure, which computes means and other descriptive statistics. To find help on this procedure, open SAS Help and Documentation and double-click on SAS Products. In SAS Products double click on Base SAS, and in Base SAS double-click on SAS Procedures. At this point you will see a list of procedures including MEANS, UNIVARIATE, and FREQ. Double click on The MEANS Procedure to open the help page (Fig. C.6). Instructions on how to set up the procedure are found under Syntax: Means Procedure. However, the first thing to do is read the overview and then scroll about halfway down and look at the examples.

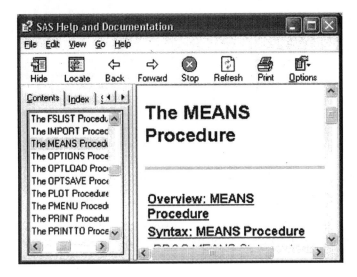

Figure C.6 Help page for the MEANS procedure. The path was: Help/SAS Help and Documentation/SAS Products/Base SAS/SAS Procedures/The MEANS Procedure.

The MEANS Procedure.

The basic syntax of the MEANS procedure is

```
PROC MEANS DATA = libname.data_file_name;
        CLASS categorical_vars;
        VAR measurement_vars;
RUN;
```
(C.3)

The CLASS statement requests a breakdown into categories defined by the categorical variable(s). For example, "CLASS gender" requests a breakdown by gender, and "CLASS gender race" would request a breakdown by gender and race (of course race would have to be a variable in the data file). If a breakdown is not wanted, leave out the CLASS statement. The CLASS statement must not be used with a continuous variable.

The VAR statement specifies the variable or variables to be analyzed; the analysis variable(s) must be quantitative, not categorical. For example, to compute statistics on height and weight (assuming they are in the data file), the VAR statement would be

```
VAR height weight;
```

Here is how to set up the MEANS procedure to calculate reaction time statistics broken down by gender.

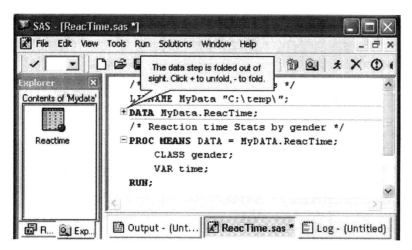

Figure C.7   The MEANS procedure set up to compute reaction time statistics broken down by gender.

Fig. C.8 shows how to run the procedure and navigate to the output. The output window (not shown) will open on top of the program window and the file explorer window will switch to the output explorer. Here is the output edited to fit the page:

Analysis Variable: Time

Gender	Obs	N	Mean	Std Dev	Minimum	Maximum
F	5	5	0.2164	0.02949	0.174	0.245
M	5	5	0.1310	0.01655	0.111	0.153

The mean reaction time was 216 ms for women and 131 ms for men.

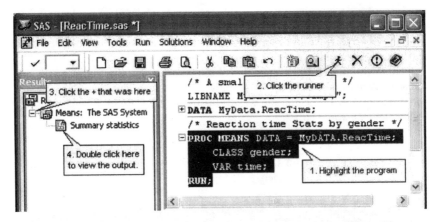

Figure C.8   How to run the MEANS procedure. Use the results explorer in the left panel to open a view of the output.

C.3   IMPORTING FILES INTO THE SAS SYSTEM

SAS software is able to read files in the following file formats and others:

Format	Usual extension	Method of opening
SAS system binary data file	.sas7bdat	DATA / SET
Tab-delimited text	.txt	PROC IMPORT
MS Excel table	.xls	PROC IMPORT

C.3.1   Permanent SAS System Binary Files

A SAS system binary file created in a data step can be directed to a permanent file drawer created with the LIBNAME statement (see Fig. C.5). Such files can be identified by the extension .sas7bdat and can be transferred, downloaded, sent as e-mail attachments, etc.; however, do not attempt to open a .sas7bdat in a word processor, as the result will be unintelligible gibberish. SAS binary files are ready for analysis and do not need to be read. All that is required is to notify SAS where the file is located. To do this, first place the .sas7bdat file in a convenient directory, then include the following as the first line of your SAS program:

LIBNAME library_name "Directory_containing_file_to_be_read";

For example, suppose the file ReacTime.sas7bdat is located in the directory C:\temp\biostat\.

In that case the LIBNAME statement would be

LIBNAME biostat "C:\temp\biostat\";

Fig. C.9 shows how to access a permanent SAS system binary file in a PROC step — in this case for a calculation of means and standard deviations broken down by gender.

```
LIBNAME MyData "C:\temp\";
/* Reaction time Stats by gender */
PROC MEANS DATA = MyData.ReacTime;
    CLASS gender;
    VAR time;
    RUN;
```

Figure C.9   Reopening a saved SAS binary file. A DATA step is not required.

The file ReacTime.sas7bdat used in this analysis was downloaded from the textbook website and saved in the directory C:\temp\; however, any convenient directory could be used. The LIBNAME statement assigns the name "MyData" to this directory, which will appear to be a file drawer named "MyData" in the SAS system file cabinet.

If the analyst does not know the names of the variable in the file, she should first run only the LIBNAME statement and then navigate in the SAS system Explorer window to the file drawer MyData, left-click on the file ReacTime, and right-click on "View Columns" to see the list of variable names, as shown here:

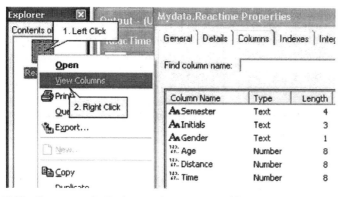

Figure C.10   How to use the Explorer to determine variable names in a SAS system binary file.

C.3.2 Importing Data Base Files and Tab-Delimited Files

SAS can import MS Excel tables as well as tab-delimited text files, provided that the first row contains the variable names and the remaining rows are data. SAS expects a header row containing variable names, followed by rows containing numeric or alphabetic data values. If there are extraneous rows above the data and variable names, add the phrase

DATAROW=number

where the number is that of the first row containing data. Insert this phrase in the place indicated in the program below (the angle brackets < > mean "optional" and should not be typed in the program).

The syntax for PROC IMPORT is

```
PROC IMPORT
    DATAFILE="filename"
    OUT=sas-data-set
    < DATAROW=number >
    DBMS=identifier REPLACE;
GETNAMES=YES;
```

filename is the path and name of the file to be read and should
be in quotes, for example "C:\Temp\ReacTime.xls".
sas-data-set is the name of the SAS data set to be created
identifier tells SAS what kind of file is being read: EXCEL or TAB

Fig. C.11 shows an example. The file Survey.xls contains the results of a classroom survey described on the second page of the file Survey.xls, which is available for download from the textbook website. The analyst downloaded the file to a floppy disk in the "A" drive and ran the following program in Fig. C.11 to compute the percents of men and women who reported having tattoos or piercings. The output, a *crosstabulation* of 196 students by gender and piercing, is shown in the figure. One or the other of the two variables was missing for three students. The crosstabulation reports absolute frequencies and relative frequencies expressed as percents of the row totals, the column totals, and the grand total. It shows that a greater percentage of women (57.35%) than men (28.33%) reported having a piercing.

```
PROC IMPORT OUT= WORK.SURVEY
    DATAFILE= "A:\Survey.xls"
    DBMS=EXCEL REPLACE;
    GETNAMES=YES;  /* read variable names from the first row */
RUN;

PROC FREQ DATA=WORK.SURVEY;
    TABLES GENDER_*PIERCED_ GENDER_*TATTOO_ ;
RUN;
```

```
        Gender(Gender)        Pierced(Pierced)

        Frequency
        Percent
        Row Pct
        Col Pct    N         Y            Total

        F
                      58        78          136
                   29.59     39.80       69.39
                   42.65     57.35
                   57.43     82.11

        M
                      43        17           60
                   21.94      8.67        30.61
                   71.67     28.33
                   42.57     17.89

        Total        101        95          196
                   51.53     48.47      100.00

        Frequency Missing = 3
```

Figure C.11   Program and output from PROC FREQ. Each cell contains the absolute frequency as well as relative frequencies expressed as percents of the grand total, row totals, and column totals. 57% of women and 28% of men reported having a piercing.

Index

345

WILEY SERIES IN PROBABILITY AND STATISTICS
ESTABLISHED BY WALTER A. SHEWHART AND SAMUEL S. WILKS

Editors: *David J. Balding, Noel A. C. Cressie, Nicholas I. Fisher,*
Iain M. Johnstone, J. B. Kadane, Geert Molenberghs. Louise M. Ryan,
David W. Scott, Adrian F. M. Smith, Jozef L. Teugels
Editors Emeriti: *Vic Barnett, J. Stuart Hunter, David G. Kendall*

The *Wiley Series in Probability and Statistics* is well established and authoritative. It covers many topics of current research interest in both pure and applied statistics and probability theory. Written by leading statisticians and institutions, the titles span both state-of-the-art developments in the field and classical methods.

Reflecting the wide range of current research in statistics, the series encompasses applied, methodological and theoretical statistics, ranging from applications and new techniques made possible by advances in computerized practice to rigorous treatment of theoretical approaches.

This series provides essential and invaluable reading for all statisticians, whether in academia, industry, government, or research.

*Now available in a lower priced paperback edition in the Wiley Classics Library.

* BELSLEY, KUH, and WELSCH · Regression Diagnostics: Identifying Influential Data and Sources of Collinearity

BENDAT and PIERSOL · Random Data: Analysis and Measurement Procedures, *Third Edition*

BERRY, CHALONER, and GEWEKE · Bayesian Analysis in Statistics and Econometrics: Essays in Honor of Arnold Zellner

BERNARDO and SMITH · Bayesian Theory

BHAT and MILLER · Elements of Applied Stochastic Processes, *Third Edition*

BHATTACHARYA and WAYMIRE · Stochastic Processes with Applications

BILLINGSLEY · Convergence of Probability Measures, *Second Edition*

BILLINGSLEY · Probability and Measure, *Third Edition*

BIRKES and DODGE · Alternative Methods of Regression

BLISCHKE AND MURTHY (editors) · Case Studies in Reliability and Maintenance

BLISCHKE AND MURTHY · Reliability: Modeling, Prediction, and Optimization

BLOOMFIELD · Fourier Analysis of Time Series: An Introduction, *Second Edition*

BOLLEN · Structural Equations with Latent Variables

BOROVKOV · Ergodicity and Stability of Stochastic Processes

BOULEAU · Numerical Methods for Stochastic Processes

BOX · Bayesian Inference in Statistical Analysis

BOX · R. A. Fisher, the Life of a Scientist

BOX and DRAPER · Empirical Model-Building and Response Surfaces

*BOX and DRAPER · Evolutionary Operation: A Statistical Method for Process Improvement

BOX, HUNTER, and HUNTER · Statistics for Experimenters: An Introduction to Design, Data Analysis, and Model Building

BOX and LUCEÑO · Statistical Control by Monitoring and Feedback Adjustment

BRANDIMARTE · Numerical Methods in Finance: A MATLAB-Based Introduction

BROWN and HOLLANDER · Statistics: A Biomedical Introduction

BRUNNER, DOMHOF, and LANGER · Nonparametric Analysis of Longitudinal Data in Factorial Experiments

BUCKLEW · Large Deviation Techniques in Decision, Simulation, and Estimation

CAIROLI and DALANG · Sequential Stochastic Optimization

CASTILLO, HADI, BALAKRISHNAN, and SARABIA · Extreme Value and Related Models with Applications in Engineering and Science

CHAN · Time Series: Applications to Finance

CHATTERJEE and HADI · Sensitivity Analysis in Linear Regression

CHATTERJEE and PRICE · Regression Analysis by Example, *Third Edition*

CHERNICK · Bootstrap Methods: A Practitioner's Guide

CHERNICK and FRIIS · Introductory Biostatistics for the Health Sciences

CHILÈS and DELFINER · Geostatistics: Modeling Spatial Uncertainty

CHOW and LIU · Design and Analysis of Clinical Trials: Concepts and Methodologies, *Second Edition*

CLARKE and DISNEY · Probability and Random Processes: A First Course with Applications, *Second Edition*

*COCHRAN and COX · Experimental Designs, *Second Edition*

CONGDON · Applied Bayesian Modelling

CONGDON · Bayesian Statistical Modelling

CONOVER · Practical Nonparametric Statistics, *Third Edition*

COOK · Regression Graphics

COOK and WEISBERG · Applied Regression Including Computing and Graphics

COOK and WEISBERG · An Introduction to Regression Graphics

CORNELL · Experiments with Mixtures, Designs, Models, and the Analysis of Mixture Data, *Third Edition*

*Now available in a lower priced paperback edition in the Wiley Classics Library.

*Now available in a lower priced paperback edition in the Wiley Classics Library.

GOLDSTEIN and LEWIS · Assessment: Problems, Development, and Statistical Issues
GREENWOOD and NIKULIN · A Guide to Chi-Squared Testing
GROSS and HARRIS · Fundamentals of Queueing Theory, *Third Edition*
*HAHN and SHAPIRO · Statistical Models in Engineering
HAHN and MEEKER · Statistical Intervals: A Guide for Practitioners
HALD · A History of Probability and Statistics and their Applications Before 1750
HALD · A History of Mathematical Statistics from 1750 to 1930
HAMPEL · Robust Statistics: The Approach Based on Influence Functions
HANNAN and DEISTLER · The Statistical Theory of Linear Systems
HEIBERGER · Computation for the Analysis of Designed Experiments
HEDAYAT and SINHA · Design and Inference in Finite Population Sampling
HELLER · MACSYMA for Statisticians
HINKELMAN and KEMPTHORNE: · Design and Analysis of Experiments, Volume 1:
    Introduction to Experimental Design
HOAGLIN, MOSTELLER, and TUKEY · Exploratory Approach to Analysis
    of Variance
HOAGLIN, MOSTELLER, and TUKEY · Exploring Data Tables, Trends and Shapes
*HOAGLIN, MOSTELLER, and TUKEY · Understanding Robust and Exploratory
    Data Analysis
HOCHBERG and TAMHANE · Multiple Comparison Procedures
HOCKING · Methods and Applications of Linear Models: Regression and the Analysis
    of Variance, *Second Edition*
HOEL · Introduction to Mathematical Statistics, *Fifth Edition*
HOGG and KLUGMAN · Loss Distributions
HOLLANDER and WOLFE · Nonparametric Statistical Methods, *Second Edition*
HOSMER and LEMESHOW · Applied Logistic Regression, *Second Edition*
HOSMER and LEMESHOW · Applied Survival Analysis: Regression Modeling of
    Time to Event Data
HUBER · Robust Statistics
HUBERTY · Applied Discriminant Analysis
HUNT and KENNEDY · Financial Derivatives in Theory and Practice
HUSKOVA, BERAN, and DUPAC · Collected Works of Jaroslav Hajek—
    with Commentary
HUZURBAZAR · Flowgraph Models for Multistate Time-to-Event Data
IMAN and CONOVER · A Modern Approach to Statistics
JACKSON · A User's Guide to Principle Components
JOHN · Statistical Methods in Engineering and Quality Assurance
JOHNSON · Multivariate Statistical Simulation
JOHNSON and BALAKRISHNAN · Advances in the Theory and Practice of Statistics: A
    Volume in Honor of Samuel Kotz
JOHNSON and BHATTACHARYYA · Statistics: Principles and Methods, *Fifth Edition*
JOHNSON and KOTZ · Distributions in Statistics
JOHNSON and KOTZ (editors) · Leading Personalities in Statistical Sciences: From the
    Seventeenth Century to the Present
JOHNSON, KOTZ, and BALAKRISHNAN · Continuous Univariate Distributions,
    Volume 1, *Second Edition*
JOHNSON, KOTZ, and BALAKRISHNAN · Continuous Univariate Distributions,
    Volume 2, *Second Edition*
JOHNSON, KOTZ, and BALAKRISHNAN · Discrete Multivariate Distributions
JOHNSON, KOTZ, and KEMP · Univariate Discrete Distributions, *Second Edition*
JUDGE, GRIFFITHS, HILL, LÜTKEPOHL, and LEE · The Theory and Practice of
    Econometrics, *Second Edition*
JUREČKOVÁ and SEN · Robust Statistical Procedures: Aymptotics and Interrelations

*Now available in a lower priced paperback edition in the Wiley Classics Library.

JUREK and MASON · Operator-Limit Distributions in Probability Theory

KADANE · Bayesian Methods and Ethics in a Clinical Trial Design

KADANE AND SCHUM · A Probabilistic Analysis of the Sacco and Vanzetti Evidence

KALBFLEISCH and PRENTICE · The Statistical Analysis of Failure Time Data, *Second Edition*

KASS and VOS · Geometrical Foundations of Asymptotic Inference

KAUFMAN and ROUSSEEUW · Finding Groups in Data: An Introduction to Cluster Analysis

KEDEM and FOKIANOS · Regression Models for Time Series Analysis

KENDALL, BARDEN, CARNE, and LE · Shape and Shape Theory

KHURI · Advanced Calculus with Applications in Statistics, *Second Edition*

KHURI, MATHEW, and SINHA · Statistical Tests for Mixed Linear Models

*KISH · Statistical Design for Research

KLEIBER and KOTZ · Statistical Size Distributions in Economics and Actuarial Sciences

KLUGMAN, PANJER, and WILLMOT · Loss Models: From Data to Decisions, *Second Edition*

KLUGMAN, PANJER, and WILLMOT · Solutions Manual to Accompany Loss Models: From Data to Decisions, *Second Edition*

KOTZ, BALAKRISHNAN, and JOHNSON · Continuous Multivariate Distributions, Volume 1, *Second Edition*

KOTZ and JOHNSON (editors) · Encyclopedia of Statistical Sciences: Volumes 1 to 9 with Index

KOTZ and JOHNSON (editors) · Encyclopedia of Statistical Sciences: Supplement Volume

KOTZ, READ, and BANKS (editors) · Encyclopedia of Statistical Sciences: Update Volume 1

KOTZ, READ, and BANKS (editors) · Encyclopedia of Statistical Sciences: Update Volume 2

KOVALENKO, KUZNETZOV, and PEGG · Mathematical Theory of Reliability of Time-Dependent Systems with Practical Applications

LACHIN · Biostatistical Methods: The Assessment of Relative Risks

LAD · Operational Subjective Statistical Methods: A Mathematical, Philosophical, and Historical Introduction

LAMPERTI · Probability: A Survey of the Mathematical Theory, *Second Edition*

LANGE, RYAN, BILLARD, BRILLINGER, CONQUEST, and GREENHOUSE · Case Studies in Biometry

LARSON · Introduction to Probability Theory and Statistical Inference, *Third Edition*

LAWLESS · Statistical Models and Methods for Lifetime Data, *Second Edition*

LAWSON · Statistical Methods in Spatial Epidemiology

LE · Applied Categorical Data Analysis

LE · Applied Survival Analysis

LEE and WANG · Statistical Methods for Survival Data Analysis, *Third Edition*

LePAGE and BILLARD · Exploring the Limits of Bootstrap

LEYLAND and GOLDSTEIN (editors) · Multilevel Modelling of Health Statistics

LIAO · Statistical Group Comparison

LINDVALL · Lectures on the Coupling Method

LINHART and ZUCCHINI · Model Selection

LITTLE and RUBIN · Statistical Analysis with Missing Data, *Second Edition*

LLOYD · The Statistical Analysis of Categorical Data

MAGNUS and NEUDECKER · Matrix Differential Calculus with Applications in Statistics and Econometrics, *Revised Edition*

MALLER and ZHOU · Survival Analysis with Long Term Survivors

MALLOWS · Design, Data, and Analysis by Some Friends of Cuthbert Daniel

MANN, SCHAFER, and SINGPURWALLA · Methods for Statistical Analysis of Reliability and Life Data

*Now available in a lower priced paperback edition in the Wiley Classics Library.

*Now available in a lower priced paperback edition in the Wiley Classics Library.

RAUSAND and HØYLAND · System Reliability Theory: Models, Statistical Methods, and Applications, *Second Edition*
RENCHER · Linear Models in Statistics
RENCHER · Methods of Multivariate Analysis, *Second Edition*
RENCHER · Multivariate Statistical Inference with Applications
* RIPLEY · Spatial Statistics
RIPLEY · Stochastic Simulation
ROBINSON · Practical Strategies for Experimenting
ROHATGI and SALEH · An Introduction to Probability and Statistics, *Second Edition*
ROLSKI, SCHMIDLI, SCHMIDT, and TEUGELS · Stochastic Processes for Insurance and Finance
ROSENBERGER and LACHIN · Randomization in Clinical Trials: Theory and Practice
ROSS · Introduction to Probability and Statistics for Engineers and Scientists
ROUSSEEUW and LEROY · Robust Regression and Outlier Detection
RUBIN · Multiple Imputation for Nonresponse in Surveys
RUBINSTEIN · Simulation and the Monte Carlo Method
RUBINSTEIN and MELAMED · Modern Simulation and Modeling
RYAN · Modern Regression Methods
RYAN · Statistical Methods for Quality Improvement, *Second Edition*
SALTELLI, CHAN, and SCOTT (editors) · Sensitivity Analysis
*SCHEFFE · The Analysis of Variance
SCHIMEK · Smoothing and Regression: Approaches, Computation, and Application
SCHOTT · Matrix Analysis for Statistics
SCHOUTENS · Levy Processes in Finance: Pricing Financial Derivatives
SCHUSS · Theory and Applications of Stochastic Differential Equations
SCOTT · Multivariate Density Estimation: Theory, Practice, and Visualization
*SEARLE · Linear Models
SEARLE · Linear Models for Unbalanced Data
SEARLE · Matrix Algebra Useful for Statistics
SEARLE, CASELLA, and McCULLOCH · Variance Components
SEARLE and WILLETT · Matrix Algebra for Applied Economics
SEBER and LEE · Linear Regression Analysis, *Second Edition*
*SEBER · Multivariate Observations
SEBER and WILD · Nonlinear Regression
SENNOTT · Stochastic Dynamic Programming and the Control of Queueing Systems
*SERFLING · Approximation Theorems of Mathematical Statistics
SHAFER and VOVK · Probability and Finance: It's Only a Game!
SILVAPULLE and SEN · Constrained Statistical Inference: Order, Inequality and Shape Constraints
SMALL and McLEISH · Hilbert Space Methods in Probability and Statistical Inference
SRIVASTAVA · Methods of Multivariate Statistics
STAPLETON · Linear Statistical Models
STAUDTE and SHEATHER · Robust Estimation and Testing
STOYAN, KENDALL, and MECKE · Stochastic Geometry and Its Applications, *Second Edition*
STOYAN and STOYAN · Fractals, Random Shapes and Point Fields: Methods of Geometrical Statistics
STYAN · The Collected Papers of T. W. Anderson: 1943–1985
SUTTON, ABRAMS, JONES, SHELDON, and SONG · Methods for Meta-Analysis in Medical Research
TANAKA · Time Series Analysis: Nonstationary and Noninvertible Distribution Theory
THOMPSON · Empirical Model Building
THOMPSON · Sampling, *Second Edition*
THOMPSON · Simulation: A Modeler's Approach

*Now available in a lower priced paperback edition in the Wiley Classics Library.

*Now available in a lower priced paperback edition in the Wiley Classics Library.